SISSA Springer Series

Volume 2

The **SISSA Springer Series** publishes research monographs, contributed volumes, conference proceedings and lectures notes in English language resulting from workshops, conferences, courses, schools, seminars, and research activities carried out by SISSA: https://www.sissa.it/.

The books in the series will discuss recent results and analyze new trends focusing on the following areas: geometry, mathematical analysis, mathematical modelling, mathematical physics, numerical analysis and scientific computing, showing a fruitful collaboration of scientists with researchers from other fields.

The series is aimed at providing useful reference material to students, academic and researchers at an international level.

More information about this series at http://www.springer.com/series/16109

Nicola Gigli • Enrico Pasqualetto

Lectures on Nonsmooth Differential Geometry

Springer

Nicola Gigli
Department of Mathematics
International School for Advanced
Studies (SISSA)
Trieste, Italy

Enrico Pasqualetto
Department of Mathematics and Statistics
University of Jyväskylä
Jyväskylä, Finland

ISSN 2524-857X ISSN 2524-8588 (electronic)
SISSA Springer Series
ISBN 978-3-030-38615-3 ISBN 978-3-030-38613-9 (eBook)
https://doi.org/10.1007/978-3-030-38613-9

This Springer imprint is published by the registered company Springer Nature Switzerland AG.
The registered company address is: Gewerbestrasse 11, 6330 Cham, Switzerland

Preface

These are the lecture notes of the Ph.D. level course 'Nonsmooth Differential Geometry' given by the first author at SISSA (Trieste, Italy) from October 2017 to March 2018. The material discussed in the classroom has been collected and reorganised by the second author.

The course was intended for students with no prior exposure to non-smooth calculus and aimed at giving a rather complete picture of first-order Sobolev calculus on general metric measure spaces and a glimpse at second order calculus on RCD spaces. For this reason the first lectures covered basic material like the concept of absolutely continuous curve or Bochner integration. This material is collected in Chap. 1.

A great deal of time has been spent at introducing the by-now classical concept of real valued Sobolev function on a metric measure space. Out of the several equivalent definitions, the approach chosen in the course has been the one based on the concept of 'test plan' introduced in [4] as it better fits what comes next. The original approach by relaxation due to Cheeger [13] and the one by Shanmugalingam [28] based on the concept of 'modulus of a family of curves' are presented, but for time constraint the equivalence of these notions with the one related to test plans has not been proved. These topics are covered in Chap. 2.

The definition of Sobolev map on a metric measure space does not come with a notion of differential, as it happens in the Euclidean setting, but rather with an object, called minimal weak upper gradient, which plays the role of 'modulus of the distributional differential'. One of the recent achievements of the theory, obtained in [17], has been to show that actually a well-defined notion of differential exists also in this setting: its introduction is based on the concept of L^∞/L^0-normed module. Chapter 3 investigates these structures from a rather abstract perspective without insisting on their use in non-smooth analysis.

The core of the course is then covered in Chap. 4, where first-order calculus is studied in great detail and the key notions of tangent/cotangent modules are introduced. Beside the notion of differential of a Sobolev map, other topics discussed are the dual concept of divergence of a vector field and how these behave under transformation of the metric measure structures. For simplicity, some of the constructions, like the one of speed of a test plan, are presented only in the technically convenient case of infinitesimally Hilbertian metric measure spaces, i.e. those for which the corresponding Sobolev space $W^{1,2}$ is Hilbert.

A basic need in most branches of mathematical analysis is that of a regularisation procedure. In working on a non-smooth environment this is true more than ever and classical tools like covering arguments are typically unavailable if one does not assume at least a doubling property at the metric level. Instead the key, and often only, tool one has at disposal is that of regularisation via the heat flow (which behaves particularly well under a lower Ricci curvature bound, a situation which the theory presented here aims to cover). Such flow can be introduced in a purely variational way as gradient flow of the 'Dirichlet energy' (in this setting called Cheeger energy) in the Hilbert space L^2, and thus can be defined in general metric measure spaces. In Chap. 5 we present a quick overview of the general theory of gradient flows in Hilbert spaces and then we discuss its application to the study of the heat flow in the 'linear' case of infinitesimally Hilbertian spaces.

Finally, the last lessons aimed at a quick guided tour in the world of RCD spaces and second order calculus on them. This material is collected in Chap. 6, where:

- We define $\mathsf{RCD}(K, \infty)$ spaces.
- Prove some better estimates for the heat flow on them.
- Introduce the algebra of 'test functions' on RCD spaces, which is the 'largest algebra of smooth functions' that we have at disposal in this environment, in a sense.
- Quickly develop the second-order differential calculus on RCD spaces, by building on top of the first-order one. Meaningful and 'operative' definitions (among others) of Hessian, covariant derivative, exterior derivative and Hodge Laplacian are discussed.

These lecture notes are mostly self-contained and should be accessible to any Ph.D. student with a standard background in analysis and geometry: having basic notions of measure theory, functional analysis and Riemannian geometry suffices to navigate this text. Hopefully, this should provide a hands-on guide to recent mathematical theories accessible to the widest possible audience.

The most recent research-level material contained here comes, to a big extent, from the paper [17], see also the survey [19]. With respect to these presentations, the current text offers a gentler introduction to all the topics, paying little in terms of generality: as such it is the most suitable source for the young researcher who is willing to learn about this fast growing research direction. The presentation is also complemented by a collection of exercises scattered through the text; since these are at times essential for the results presented, their solutions are reported (or, sometimes, just sketched) in Appendix B.

We wish to thank Emanuele Caputo, Francesco Nobili, Francesco Sapio and Ivan Yuri Violo for their careful reading of a preliminary version of this manuscript.

Trieste, Italy
Jyväskylä, Finland
January 2019

Nicola Gigli
Enrico Pasqualetto

Contents

About the Authors

Nicola Gigli studied Mathematics at the Scuola Normale Superiore of Pisa and is Professor of Mathematical Analysis at SISSA, Trieste. He is interested in calculus of variations, optimal transport, and geometric and nonsmooth analysis, with particular focus on properties of spaces with curvature bounded from below.

Enrico Pasqualetto earned his Ph.D. degree in Mathematics at SISSA (Trieste) in 2018 and is currently a postdoctoral researcher at the University of Jyväskylä. His fields of interest consist in functional analysis and geometric measure theory on non-smooth metric structures, mainly in the presence of synthetic curvature bounds. Within this framework, his research topics include rectifiability properties, Sobolev spaces, and sets of finite perimeter.

Chapter 1
Preliminaries

In this chapter we introduce several classic notions that will be needed in the sequel. Namely, in Sect. 1.1 we review the basics of measure theory, with a particular accent on the space $L^0(\mathfrak{m})$ of Borel functions considered up to $\mathfrak{m}$-almost everywhere equality (see Sect. 1.1.2); in Sect. 1.2 we discuss about continuous, absolutely continuous and geodesic curves on metric spaces; in Sect. 1.3 we collect the most important results about Bochner integration. Some functional analytic tools will be treated in Appendix A.

1.1 General Measure Theory

1.1.1 Borel Probability Measures

Given a complete and separable metric space (X, d), let us denote

$$\begin{aligned} \mathscr{P}(\mathrm{X}) &:= \big\{\text{Borel probability measures on } (\mathrm{X}, \mathsf{d})\big\}, \\ C_b(\mathrm{X}) &:= \big\{\text{bounded continuous functions } f : \mathrm{X} \to \mathbb{R}\big\}. \end{aligned} \tag{1.1}$$

We can define a topology on $\mathscr{P}(\mathrm{X})$, called *weak topology*, as follows:

Definition 1.1.1 (Weak Topology) The *weak topology* on $\mathscr{P}(\mathrm{X})$ is defined as the coarsest topology on $\mathscr{P}(\mathrm{X})$ such that:

$$\text{the function } \mathscr{P}(\mathrm{X}) \ni \mu \longmapsto \int f \, \mathrm{d}\mu \text{ is continuous,} \quad \text{for every } f \in C_b(\mathrm{X}). \tag{1.2}$$

N. Gigli, E. Pasqualetto, *Lectures on Nonsmooth Differential Geometry*,
SISSA Springer Series 2, https://doi.org/10.1007/978-3-030-38613-9_1

Remark 1.1.2 If a sequence of measures $(\mu_n)_n$ weakly converges to a limit measure μ, then

$$\mu(\Omega) \leq \varliminf_{n\to\infty} \mu_n(\Omega) \qquad \text{for every } \Omega \subseteq \mathrm{X} \text{ open.} \tag{1.3}$$

Indeed, let $f_k := k\,\mathsf{d}(\cdot, \mathrm{X} \setminus \Omega) \wedge 1 \in C_b(\mathrm{X})$ for $k \in \mathbb{N}$. Hence $f_k(x) \nearrow \chi_\Omega(x)$ for all $x \in \mathrm{X}$, so that $\mu(\Omega) = \sup_k \int f_k \, \mathrm{d}\mu$ by monotone convergence theorem. Since $\nu \mapsto \int f_k \, \mathrm{d}\nu$ is continuous for any k, we deduce that the function $\nu \mapsto \nu(\Omega)$ is lower semicontinuous as supremum of continuous functions, thus yielding (1.3).

In particular, if a sequence $(\mu_n)_n \subseteq \mathscr{P}(\mathrm{X})$ weakly converges to some $\mu \in \mathscr{P}(\mathrm{X})$, then

$$\mu(C) \geq \varlimsup_{n\to\infty} \mu_n(C) \qquad \text{for every } C \subseteq \mathrm{X} \text{ closed.} \tag{1.4}$$

To prove it, just apply (1.3) to $\Omega := \mathrm{X} \setminus C$. ■

Remark 1.1.3 We claim that if $\int f \, \mathrm{d}\mu = \int f \, \mathrm{d}\nu$ for every $f \in C_b(\mathrm{X})$, then $\mu = \nu$. Indeed, $\mu(C) = \nu(C)$ for any $C \subseteq \mathrm{X}$ closed as a consequence of (1.4), whence $\mu = \nu$ by the monotone class theorem. ■

Remark 1.1.4 Given any Banach space V, we denote by V' its dual Banach space. Then

$$\mathscr{P}(\mathrm{X}) \text{ is continuously embedded into } C_b(X)'. \tag{1.5}$$

Such embedding is given by the operator sending $\mu \in \mathscr{P}(\mathrm{X})$ to the map $C_b(\mathrm{X}) \ni f \mapsto \int f \, \mathrm{d}\mu$, which is injective by Remark 1.1.3 and linear by definition. Finally, continuity stems from the inequality $\left|\int f \, \mathrm{d}\mu\right| \leq \|f\|_{C_b(\mathrm{X})}$, which holds for any $f \in C_b(\mathrm{X})$. ■

Fix a countable dense subset $(x_n)_n$ of X. Let us define

$$\begin{aligned} \mathcal{A} &:= \Big\{ \big(a - b\,\mathsf{d}(\cdot, x_n)\big) \vee c \;:\; a, b, c \in \mathbb{Q},\ n \in \mathbb{N} \Big\}, \\ \widetilde{\mathcal{A}} &:= \big\{ f_1 \vee \ldots \vee f_n \;:\; n \in \mathbb{N},\ f_1, \ldots, f_n \in \mathcal{A} \big\}. \end{aligned} \tag{1.6}$$

Observe that $\mathcal{A}$ and $\widetilde{\mathcal{A}}$ are countable subsets of $C_b(\mathrm{X})$. We claim that:

$$f(x) = \sup\big\{ g(x) \;:\; g \in \mathcal{A},\ g \leq f \big\} \qquad \text{for every } f \in C_b(\mathrm{X}) \text{ and } x \in \mathrm{X}. \tag{1.7}$$

Indeed, the inequality $\geq$ is trivial, while to prove $\leq$ fix $x \in \mathrm{X}$ and $\varepsilon > 0$. The function f being continuous, there is a neighbourhood U of x such that $f(y) \geq f(x) - \varepsilon$ for all $y \in U$. Then we can easily build a function $g \in \mathcal{A}$ such that $g \leq f$ and $g(x) \geq f(x) - 2\,\varepsilon$. By arbitrariness of $x \in \mathrm{X}$ and $\varepsilon > 0$, we thus proved the validity of (1.7).

Exercise 1.1.5 Suppose that X is compact. Prove that if a sequence $(f_n)_n \subseteq C(\mathrm{X})$ satisfies $f_n(x) \searrow 0$ for every $x \in \mathrm{X}$, then $f_n \to 0$ uniformly on X. ■

Corollary 1.1.6 *Suppose that* X *is compact. Then* $\widetilde{\mathcal{A}}$ *is dense in* $C(\mathrm{X}) = C_b(\mathrm{X})$. *In particular, the space* $C(\mathrm{X})$ *is separable.*

Proof Fix $f \in C(\mathrm{X})$. Enumerate $\{g \in \mathcal{A} : g \leq f\}$ as $(g_n)_n$. Call $h_n := g_1 \vee \ldots \vee g_n \in \widetilde{\mathcal{A}}$ for each $n \in \mathbb{N}$, thus $h_n(x) \nearrow f(x)$ for all $x \in \mathrm{X}$ by (1.7). Hence $(f - h_n)(x) \searrow 0$ for all $x \in \mathrm{X}$ and accordingly $f - h_n \to 0$ in $C(\mathrm{X})$ by Exercise 1.1.5, proving the statement. □

The converse implication holds true as well:

Exercise 1.1.7 Let (X, d) be a complete and separable metric space. Prove that if $C_b(\mathrm{X})$ is separable, then the space X is compact. ■

Corollary 1.1.8 *It holds that*

$$\int f\, \mathrm{d}\mu = \sup\left\{\int g\, \mathrm{d}\mu \,\middle|\, g \in \widetilde{\mathcal{A}},\ g \leq f\right\} \quad \textit{for every } \mu \in \mathscr{P}(\mathrm{X}) \textit{ and } f \in C_b(\mathrm{X}). \tag{1.8}$$

Proof Call $(g_n)_n = \{g \in \mathcal{A} : g \leq f\}$ and put $h_n := g_1 \vee \ldots \vee g_n \in \widetilde{\mathcal{A}}$, thus $h_n(x) \nearrow f(x)$ for all $x \in \mathrm{X}$ and accordingly $\int f\, \mathrm{d}\mu = \lim_n \int h_n\, \mathrm{d}\mu$, proving (1.8). □

We endow $\mathscr{P}(\mathrm{X})$ with a distance δ. Enumerate $\{g \in \widetilde{\mathcal{A}} \cup (-\widetilde{\mathcal{A}}) : \|g\|_{C_b(\mathrm{X})} \leq 1\}$ as $(f_i)_i$. Then for any $\mu, \nu \in \mathscr{P}(\mathrm{X})$ we define

$$\delta(\mu, \nu) := \sum_{i=0}^{\infty} \frac{1}{2^i} \left| \int f_i\, \mathrm{d}(\mu - \nu) \right|. \tag{1.9}$$

Proposition 1.1.9 *The weak topology on* $\mathscr{P}(\mathrm{X})$ *is induced by the distance* δ.

Proof To prove one implication, we want to show that for any $f \in C_b(\mathrm{X})$ the map $\mu \mapsto \int f\, \mathrm{d}\mu$ is δ-continuous. Fix $\mu, \nu \in \mathscr{P}(\mathrm{X})$. Given any $\varepsilon > 0$, there exists a map $g \in \widetilde{\mathcal{A}}$ such that $g \leq f$ and $\int g\, \mathrm{d}\mu \geq \int f\, \mathrm{d}\mu - \varepsilon$, by Corollary 1.1.8. Let $i \in \mathbb{N}$ be such that $f_i = g/\|g\|_{C_b(\mathrm{X})}$. Then

$$\int f\, \mathrm{d}\nu - \int f\, \mathrm{d}\mu \geq \|g\|_{C_b(\mathrm{X})} \int f_i\, \mathrm{d}(\nu - \mu) - \varepsilon \geq -\|g\|_{C_b(\mathrm{X})}\, 2^i\, \delta(\nu, \mu) - \varepsilon,$$

whence $\underline{\lim}_{\delta(\nu,\mu)\to 0} \int f\, \mathrm{d}(\nu - \mu) \geq 0$ by arbitrariness of ε, i.e. the map $\mu \mapsto \int f\, \mathrm{d}\mu$ is δ-lower semicontinuous. Its δ-upper semicontinuity can be proved in an analogous way.

Conversely, fix $\mu \in \mathscr{P}(\mathrm{X})$ and $\varepsilon > 0$. Choose $N \in \mathbb{N}$ such that $2^{-N} < \varepsilon/2$. Then there is a weak neighbourhood W of μ such that $\left|\int f_i \,\mathrm{d}(\mu - \nu)\right| < \varepsilon/4$ for all $i = 0, \ldots, N$ and $\nu \in W$. Therefore

$$\delta(\mu, \nu) \leq \sum_{i=0}^{N} \frac{1}{2^i}\left|\int f_i \,\mathrm{d}(\mu - \nu)\right| + \sum_{i=N+1}^{\infty} \frac{1}{2^i} \leq \frac{\varepsilon}{2} + \frac{1}{2^N} < \varepsilon \quad \text{for every } \nu \in W,$$

proving that W is contained in the open δ-ball of radius ε centered at μ. □

Remark 1.1.10 Suppose that X is compact. Then $C(\mathrm{X}) = C_b(\mathrm{X})$, thus accordingly $\mathscr{P}(\mathrm{X})$ is weakly compact by (1.5) and Banach-Alaoglu theorem. Conversely, for X non-compact this is in general no longer true. For instance, take $\mathrm{X} := \mathbb{R}$ and $\mu_n := \delta_n$. Suppose by contradiction that a subsequence $(\mu_{n_m})_m$ weakly converges to some limit $\mu \in \mathscr{P}(\mathbb{R})$. For any $k \in \mathbb{N}$ we have that $\mu\big((-k, k)\big) \leq \underline{\lim}_m \, \delta_{n_m}\big((-k, k)\big) = 0$, so that $\mu(\mathbb{R}) = \lim_{k\to\infty} \mu\big((-k, k)\big) = 0$, which leads to a contradiction. This proves that $\mathscr{P}(\mathbb{R})$ is not weakly compact. ■

Definition 1.1.11 (Tightness) A set $\mathcal{K} \subseteq \mathscr{P}(\mathrm{X})$ is said to be *tight* provided for every $\varepsilon > 0$ there exists a compact set $K_\varepsilon \subseteq \mathrm{X}$ such that $\mu(K_\varepsilon) \geq 1 - \varepsilon$ for every $\mu \in \mathcal{K}$.

Theorem 1.1.12 (Prokhorov) *Let $\mathcal{K} \subseteq \mathscr{P}(\mathrm{X})$ be fixed. Then $\mathcal{K}$ is weakly relatively compact if and only if $\mathcal{K}$ is tight.*

Proof In light of Proposition 1.1.9, compactness and sequential compactness are equivalent. We separately prove the two implications:
SUFFICIENCY. Fix $\mathcal{K} \subseteq \mathscr{P}(\mathrm{X})$ tight. Without loss of generality, suppose that $\mathcal{K} = (\mu_i)_{i\in\mathbb{N}}$. For any $n \in \mathbb{N}$, choose a compact set $K_n \subseteq \mathrm{X}$ such that $\mu_i(K_n) \geq 1 - 1/n$ for all i. By a diagonalization argument we see that, up to a not relabeled subsequence, $\mu_i|_{K_n}$ converges to some measure ν_n in duality with $C_b(K_n)$ for all $n \in \mathbb{N}$, as a consequence of Remark 1.1.10. We now claim that:

$$\begin{aligned} &\nu_n \to \nu \ \text{ in total variation norm,} \quad \text{for some measure } \nu, \\ &\mu_i \rightharpoonup \nu \ \text{ in duality with } C_b(\mathrm{X}). \end{aligned} \tag{1.10}$$

To prove the former, recall (cf. Remark 1.1.15 below) that for any $m \geq n \geq 1$ one has

$$\|\nu_n - \nu_m\|_{\mathsf{TV}} = \sup\left\{\int f \,\mathrm{d}(\nu_n - \nu_m) \;\middle|\; f \in C_b(\mathrm{X}),\ \|f\|_{C_b(\mathrm{X})} \leq 1\right\}.$$

Then fix $f \in C_b(\mathrm{X})$ with $\|f\|_{C_b(\mathrm{X})} \leq 1$. We can assume without loss of generality that $(K_n)_n$ is increasing. We deduce from (1.3) that $\nu_m(K_m \setminus K_n) \leq \underline{\lim}_i \, \mu_i|_{K_m}(\mathrm{X} \setminus K_n) \leq 1/n$. Therefore

$$\int f \,\mathrm{d}(\nu_n - \nu_m) \leq \lim_{i\to\infty}\left(\int f \,\mathrm{d}\mu_i - \int f \,\mathrm{d}\mu_i\right) + \frac{1}{n} + \frac{1}{m} = \frac{1}{n} + \frac{1}{m},$$

proving that $(\nu_n)_n$ is Cauchy with respect to $\|\cdot\|_{\mathrm{TV}}$ and accordingly the first in (1.10). For the latter, notice that for any $f \in C_b(\mathrm{X})$ it holds that

$$\begin{aligned}\left|\int f \,\mathrm{d}(\mu_i - \nu)\right| &= \left|\int_{K_n} f \,\mathrm{d}(\mu_i - \nu_n) - \int_{K_n} f \,\mathrm{d}(\nu - \nu_n)\right. \\ &\qquad \left. + \int_{\mathrm{X}\setminus K_n} f \,\mathrm{d}\mu_i - \int_{\mathrm{X}\setminus K_n} f \,\mathrm{d}\nu\right| \\ &\leq \left|\int_{K_n} f \,\mathrm{d}(\mu_i - \nu_n)\right| + \|f\|_{C_b(\mathrm{X})} \|\nu - \nu_n\|_{\mathrm{TV}} + \frac{2\,\|f\|_{C_b(\mathrm{X})}}{n}.\end{aligned}$$

By first letting $i \to \infty$ and then $n \to \infty$, we obtain that $\lim_i \left|\int f \,\mathrm{d}(\mu_i - \nu)\right| = 0$, showing the second in (1.10). Hence sufficiency is proved.
NECESSITY. Fix $\mathcal{K} \subseteq \mathscr{P}(\mathrm{X})$ weakly relatively sequentially compact. Choose $\varepsilon > 0$ and a sequence $(x_n)_n$ that is dense in X. Arguing by contradiction, we aim to prove that

$$\forall i \in \mathbb{N} \quad \exists N_i \in \mathbb{N}: \quad \mu\left(\bigcup_{j=1}^{N_i} \bar{B}_{1/i}(x_j)\right) \geq 1 - \frac{\varepsilon}{2^i} \quad \forall \mu \in \mathcal{K}. \tag{1.11}$$

If not, there exist $i_0 \in \mathbb{N}$ and $(\mu_m)_m \subseteq \mathcal{K}$ such that $\mu_m\big(\bigcup_{j=1}^m \bar{B}_{1/i_0}(x_j)\big) < 1 - \varepsilon/2^{i_0}$ holds for every $m \in \mathbb{N}$. Up to a not relabeled subsequence $\mu_m \rightharpoonup \mu \in \mathscr{P}(\mathrm{X})$ and accordingly

$$\mu\left(\bigcup_{j=1}^{n} B_{1/i_0}(x_j)\right) \overset{(1.3)}{\leq} \varliminf_{m\to\infty} \mu_m\left(\bigcup_{j=1}^{m} \bar{B}_{1/i_0}(x_j)\right) \leq 1 - \varepsilon/2^{i_0} \quad \text{for any } n \in \mathbb{N},$$

which contradicts the fact that $\lim_{n\to\infty} \mu\big(\bigcup_{j=1}^n B_{1/i_0}(x_j)\big) = \mu(\mathrm{X}) = 1$. This proves (1.11).

Now define $K := \bigcap_{i\in\mathbb{N}} \bigcup_{j=1}^{N_i} \bar{B}_{1/i}(x_j)$. Such set is compact, as it is closed and totally bounded by construction. Moreover, for any $\mu \in \mathcal{K}$ one has that

$$\mu(\mathrm{X} \setminus K) \leq \sum_i \mu\left(\bigcap_{j=1}^{N_i} \mathrm{X} \setminus \bar{B}_{1/i}(x_j)\right) \overset{(1.11)}{\leq} \varepsilon \sum_i \frac{1}{2^i} = \varepsilon,$$

thus proving also necessity. □

Remark 1.1.13 We have that a set $\mathcal{K} \subseteq \mathscr{P}(\mathrm{X})$ is tight if and only if

$$\exists\, \Psi : \mathrm{X} \to [0, +\infty], \text{ with compact sublevels, such that } s := \sup_{\mu\in\mathcal{K}} \int \Psi \,\mathrm{d}\mu < +\infty. \tag{1.12}$$

To prove sufficiency, first notice that Ψ is Borel as its sublevels are closed sets. Now fix $\varepsilon > 0$ and choose $C > 0$ such that $s/C < \varepsilon$. Moreover, by applying Čebyšëv's inequality we obtain that $C\,\mu\{\Psi > C\} \leq \int \Psi \,\mathrm{d}\mu \leq s$ for all $\mu \in \mathcal{K}$, whence $\mu\big(\{\Psi \leq C\}\big) \geq 1 - s/C > 1 - \varepsilon$.

To prove necessity, suppose $\mathcal{K}$ tight and choose a sequence $(K_n)_n$ of compact sets such that $\mu(\mathrm{X} \setminus K_n) \leq 1/n^3$ for all $n \in \mathbb{N}$ and $\mu \in \mathcal{K}$. Define $\Psi(x) := \inf\big\{n \in \mathbb{N} : x \in K_n\big\}$ for every $x \in \mathrm{X}$. Clearly Ψ has compact sublevels by construction. Moreover, it holds that

$$\sup_{\mu\in\mathcal{K}} \int \Psi \,\mathrm{d}\mu = \sup_{\mu\in\mathcal{K}} \sum_n \int_{K_{n+1}\setminus K_n} \Psi \,\mathrm{d}\mu \leq \sum_n \frac{n+1}{n^3} < +\infty,$$

as required. ■

Remark 1.1.14 Let $\mu \geq 0$ be a finite non-negative Borel measure on X. Then for any Borel set $E \subseteq \mathrm{X}$ one has

$$\mu(E) = \sup\big\{\mu(C) \,:\, C \subseteq E \text{ closed}\big\} = \inf\big\{\mu(\Omega) \,:\, \Omega \supseteq E \text{ open}\big\}. \tag{1.13}$$

To prove it, it suffices to show that the family of all Borel sets E satisfying (1.13), which we shall denote by $\mathcal{E}$, forms a σ-algebra containing all open subsets of X. Then fix $\Omega \subseteq \mathrm{X}$ open. Call $C_n := \big\{x \in \Omega \,:\, \mathsf{d}(x, \mathrm{X} \setminus \Omega) \geq 1/n\big\}$ for all $n \in \mathbb{N}$, whence $(C_n)_n$ is an increasing sequence of closed sets and $\mu(\Omega) = \lim_n \mu(C_n)$ by continuity from below of μ. This grants that $\Omega \in \mathcal{E}$.

It only remains to show that $\mathcal{E}$ is a σ-algebra. It is obvious that $\emptyset \in \mathcal{E}$ and that $\mathcal{E}$ is stable by complements. Now fix $(E_n)_n \subseteq \mathcal{E}$ and $\varepsilon > 0$. There exist $(C_n)_n$ closed and $(\Omega_n)_n$ open such that $C_n \subseteq E_n \subseteq \Omega_n$ and $\mu(\Omega_n) - \varepsilon\,2^{-n} \leq \mu(E_n) \leq \mu(C_n) + \varepsilon\,2^{-n}$ for every $n \in \mathbb{N}$. Let us denote $\Omega := \bigcup_n \Omega_n$. Moreover, continuity from above of μ yields the existence of $N \in \mathbb{N}$ such that $\mu\big(\bigcup_{n\in\mathbb{N}} C_n \setminus C\big) \leq \varepsilon$, where we put $C := \bigcup_{n=1}^N C_n$. Notice that Ω is open, C is closed and $C \subseteq \bigcup_n E_n \subseteq \Omega$. Finally, it holds that

$$\mu\bigg(\bigcup_{n=1}^{\infty} E_n \setminus C\bigg) \leq \sum_{n=1}^{\infty} \mu(E_n \setminus C_n) + \varepsilon \leq \sum_{n=1}^{\infty} \frac{\varepsilon}{2^n} + \varepsilon = 2\,\varepsilon,$$

$$\mu\bigg(\Omega \setminus \bigcup_{n=1}^{\infty} E_n\bigg) \leq \sum_{n=1}^{\infty} \mu(\Omega_n \setminus E_n) \leq \sum_{n=1}^{\infty} \frac{\varepsilon}{2^n} = \varepsilon.$$

This grants that $\bigcup_n E_n \in \mathcal{E}$, concluding the proof. ■

Remark 1.1.15 (Total Variation Norm) During the proof of Theorem 1.1.12, we needed the following two properties of the *total variation norm*:

$$\begin{aligned} &\|\mu\|_{\mathsf{TV}} = \sup\left\{\int f \,\mathrm{d}\mu \;\middle|\; f \in C_b(\mathrm{X}),\ \|f\|_{C_b(\mathrm{X})} \leq 1\right\} && \text{for any signed Borel measure } \mu \text{ on X},\\ &\big(\mathscr{P}(\mathrm{X}), \|\cdot\|_{\mathsf{TV}}\big) \quad \text{is complete.} \end{aligned} \tag{1.14}$$

In order to prove them, we proceed as follows. Given a signed measure μ, let us consider its *Hahn-Jordan decomposition* $\mu = \mu^+ - \mu^-$, where $\mu^\pm$ are non-negative measures with $\mu^+ \perp \mu^-$, which satisfy $\mu(P) = \mu^+(\mathrm{X})$ and $\mu(P^c) = -\mu^-(\mathrm{X})$ for a suitable Borel set $P \subseteq \mathrm{X}$. Hence by definition the total variation norm is defined as

$$\|\mu\|_{\mathsf{TV}} := \mu^+(\mathrm{X}) + \mu^-(\mathrm{X}). \tag{1.15}$$

Such definition is well-posed, since the Hahn-Jordan decomposition (μ^+, μ^-) of μ is unique.

To prove the first in (1.14), we start by noticing that $\int f \,\mathrm{d}\mu \leq \int |f| \,\mathrm{d}(\mu^+ + \mu^-) \leq \|\mu\|_{\mathsf{TV}}$ holds for any $f \in C_b(\mathrm{X})$ with $\|f\|_{C_b(\mathrm{X})} \leq 1$, proving one inequality. To show the converse one, let $\varepsilon > 0$ be fixed. By Remark 1.1.14, we can choose two closed sets $C \subseteq P$ and $C' \subseteq P^c$ such that $\mu^+(P \setminus C), \mu^-(P^c \setminus C') < \varepsilon$. Call $f_n := \big(1 - n\,\mathsf{d}(\cdot, C)\big)^+$ and $g_n := \big(1 - n\,\mathsf{d}(\cdot, C')\big)^+$, so that $f_n \searrow \chi_C$ and $g_n \searrow \chi_{C'}$ as $n \to \infty$. Now define $h_n := f_n - g_n$. Since $|h_n| \leq 1$, we have that $(h_n)_n \subseteq C_b(\mathrm{X})$ and $\|h_n\|_{C_b(\mathrm{X})} \leq 1$ for every $n \in \mathbb{N}$. Moreover, it holds that

$$\begin{aligned} \varliminf_{n\to\infty} \int h_n \,\mathrm{d}\mu &= \varliminf_{n\to\infty} \left[\int f_n \,\mathrm{d}\mu^+ - \int f_n \,\mathrm{d}\mu^- - \int g_n \,\mathrm{d}\mu^+ + \int g_n \,\mathrm{d}\mu^- \right] \\ &= \mu^+(C) + \mu^-(C') \geq \mu^+(P) + \mu^-(P^c) - 2\,\varepsilon = \|\mu\|_{\mathsf{TV}} - 2\,\varepsilon. \end{aligned}$$

By arbitrariness of $\varepsilon > 0$, we conclude that $\varliminf_n \int h_n \,\mathrm{d}\mu \geq \|\mu\|_{\mathsf{TV}}$, proving the first in (1.14).

To show the second, fix a sequence $(\mu_n)_n \subseteq \mathscr{P}(\mathrm{X})$ that is $\|\cdot\|_{\mathsf{TV}}$-Cauchy. Notice that

$$\big|\mu(E)\big| \leq \|\mu\|_{\mathsf{TV}} \quad \text{for every signed measure } \mu \text{ and Borel set } E \subseteq \mathrm{X}.$$

Indeed, $\big|\mu(E)\big| \leq \mu^+(E) + \mu^-(E) \leq \mu^+(\mathrm{X}) + \mu^-(\mathrm{X}) = \|\mu\|_{\mathsf{TV}}$. Therefore

$$\big|\mu_n(E) - \mu_m(E)\big| \leq \|\mu_n - \mu_m\|_{\mathsf{TV}} \quad \text{for every } n, m \in \mathbb{N} \text{ and } E \subseteq \mathrm{X} \text{ Borel}. \tag{1.16}$$

In particular, $\big(\mu_n(E)\big)_n$ is Cauchy for any $E \subseteq \mathrm{X}$ Borel, so that $\lim_n \mu_n(E) = \mathsf{L}(E)$ for some limit $\mathsf{L}(E) \in [0, 1]$. We thus deduce from (1.16) that

$$\forall \varepsilon > 0 \quad \exists \bar{n}_\varepsilon \in \mathbb{N}: \quad \big|\mathsf{L}(E) - \mu_n(E)\big| \leq \varepsilon \quad \forall n \geq \bar{n}_\varepsilon \quad \forall E \subseteq \mathrm{X} \text{ Borel}. \tag{1.17}$$

We claim that L is a probability measure. Clearly, $\mathsf{L}(\emptyset) = 0$ and $\mathsf{L}(\mathrm{X}) = 1$. For any E, F Borel with $E \cap F = \emptyset$, we have $\mathsf{L}(E \cup F) = \lim_n \mu_n(E \cup F) = \lim_n \mu_n(E) + \lim_n \mu_n(F) = \mathsf{L}(E) + \mathsf{L}(F)$, which grants that L is finitely additive. To show that it is also σ-additive, fix a sequence $(E_i)_i$ of pairwise disjoint Borel sets. Let us call

$U_N := \bigcup_{i=1}^N E_i$ for all $N \in \mathbb{N}$ and $U := \bigcup_{i=1}^\infty E_i$. Given any $\varepsilon > 0$, we infer from (1.17) that for any $n \geq \bar{n}_\varepsilon$ one has

$$\begin{aligned}\overline{\lim_{N\to\infty}} \big|\mathsf{L}(U) - \mathsf{L}(U_N)\big| &\leq \big|\mathsf{L}(U) - \mu_n(U)\big| \\ &\quad + \overline{\lim_{N\to\infty}} \big|\mu_n(U) - \mu_n(U_N)\big| + \overline{\lim_{N\to\infty}} \big|\mu_n(U_N) - \mathsf{L}(U_N)\big| \\ &\leq 2\,\varepsilon + \overline{\lim_{N\to\infty}} \big|\mu_n(U) - \mu_n(U_N)\big| = 2\,\varepsilon,\end{aligned}$$

where the last equality follows from the continuity from below of μ_n. By letting $\varepsilon \to 0$ in the previous formula, we thus obtain that $\mathsf{L}(U) = \lim_N \mathsf{L}(U_N) = \sum_{i=1}^\infty \mathsf{L}(E_i)$, so that $\mathsf{L} \in \mathscr{P}(\mathrm{X})$. Finally, we aim to prove that $\lim_n \|\mathsf{L} - \mu_n\|_{\mathsf{TV}} = 0$. For any $n \in \mathbb{N}$, choose a Borel set $P_n \subseteq \mathrm{X}$ satisfying $(\mathsf{L} - \mu_n)(P_n) = (\mathsf{L} - \mu_n)^+(\mathrm{X})$ and $(\mathsf{L} - \mu_n)(P_n^c) = -(\mathsf{L} - \mu_n)^-(\mathrm{X})$. Now fix $\varepsilon > 0$. Hence (1.17) guarantees that for every $n \geq \bar{n}_\varepsilon$ it holds that

$$\|\mathsf{L} - \mu_n\|_{\mathsf{TV}} = (\mathsf{L} - \mu_n)(P_n) - (\mathsf{L} - \mu_n)(P_n^c) = \big|(\mathsf{L} - \mu_n)(P_n)\big| + \big|(\mathsf{L} - \mu_n)(P_n^c)\big| \leq 2\,\varepsilon.$$

Therefore μ_n converges to L in the $\|\cdot\|_{\mathsf{TV}}$-norm. Since $\mathsf{L} \geq 0$ by construction, the proof of (1.14) is achieved. ■

We now present some consequences of Theorem 1.1.12:

Corollary 1.1.16 (Ulam's Theorem) *Any $\mu \in \mathscr{P}(\mathrm{X})$ is concentrated on a σ-compact set.*

Proof Clearly the singleton $\{\mu\}$ is weakly relatively compact, so it is tight by Theorem 1.1.12. Thus for any $n \in \mathbb{N}$ we can choose a compact set $K_n \subseteq \mathrm{X}$ such that $\mu(\mathrm{X} \setminus K_n) < 1/n$. In particular, μ is concentrated on $\bigcup_n K_n$, yielding the statement. □

Corollary 1.1.17 *Let $\mu \in \mathscr{P}(\mathrm{X})$ be given. Then μ is* inner regular, *i.e.*

$$\mu(E) = \sup\big\{\mu(K) \,:\, K \subseteq E \text{ compact}\big\} \quad \text{for every } E \subseteq \mathrm{X} \text{ Borel.} \tag{1.18}$$

In particular, μ is a Radon measure.

Proof By Corollary 1.1.16, there exists an increasing sequence $(K_n)_n$ of compact sets such that $\lim_n \mu\big(\mathrm{X} \setminus K_n\big) = 0$. Any closed subset C of X that is contained in some K_n is clearly compact, whence

$$\begin{aligned}\mu(E) = \lim_{n\to\infty} \mu(E \cap K_n) &= \lim_{n\to\infty} \sup\big\{\mu(C) \,:\, C \subseteq E \cap K_n \text{ closed}\big\} \\ &\leq \sup\big\{\mu(K) \,:\, K \subseteq E \text{ compact}\big\} \quad \text{for every } E \subseteq \mathrm{X} \text{ Borel,}\end{aligned}$$

proving (1.18), as required. □

Given any function $f : \mathrm{X} \to \mathbb{R}$, let us define

$$\mathrm{Lip}(f) := \sup_{\substack{x,y\in \mathrm{X}\\ x\neq y}} \frac{|f(x)-f(y)|}{\mathsf{d}(x,y)} \in [0,+\infty]. \tag{1.19}$$

We say that f is *Lipschitz* provided $\mathrm{Lip}(f) < +\infty$ and we define

$$\begin{aligned} \mathrm{LIP}(\mathrm{X}) &:= \big\{ f : \mathrm{X} \to \mathbb{R} \ :\ \mathrm{Lip}(f) < +\infty \big\}, \\ \mathrm{LIP}_{bs}(\mathrm{X}) &:= \big\{ f \in \mathrm{LIP}(\mathrm{X}) \ :\ \mathrm{spt}(f) \text{ is bounded} \big\} \subseteq C_b(\mathrm{X}). \end{aligned} \tag{1.20}$$

We point out that continuous maps having bounded support are not necessarily bounded.

Proposition 1.1.18 (Separability of $L^p(\mu)$ for $p < \infty$) *Let $\mu \in \mathscr{P}(\mathrm{X})$ and $p \in [1,\infty)$. Then the space $\mathrm{LIP}_{bs}(\mathrm{X})$ is dense in $L^p(\mu)$. In particular, the space $L^p(\mu)$ is separable.*

Proof First, notice that $\mathrm{LIP}_{bs}(\mathrm{X}) \subseteq L^\infty(\mu) \subseteq L^p(\mu)$. Call $\mathscr{C}$ the $L^p(\mu)$-closure of $\mathrm{LIP}_{bs}(\mathrm{X})$.

STEP 1. We claim that $\big\{\chi_C \ :\ C \subseteq \mathrm{X} \text{ closed bounded}\big\}$ is contained in the set $\mathscr{C}$. Indeed, called $f_n := \big(1 - n\,\mathsf{d}(\cdot, C)\big)^+ \in \mathrm{LIP}_{bs}(\mathrm{X})$ for any $n \in \mathbb{N}$, one has $f_n \to \chi_C$ in $L^p(\mu)$ by dominated convergence theorem.

STEP 2. We also have that $\big\{\chi_E \ :\ E \subseteq \mathrm{X} \text{ Borel}\big\} \subseteq \mathscr{C}$. Indeed, we can pick an increasing sequence $(C_n)_n$ of closed subsets of E such that $\mu(E) = \lim_n \mu(C_n)$, as seen in (1.13). Then one has that $\|\chi_E - \chi_{C_n}\|_{L^p(\mu)} = \mu(E \setminus C_n)^{1/p} \to 0$, whence $\chi_E \in \mathscr{C}$ by STEP 1.

STEP 3. To prove that $L^p(\mu) \subseteq \mathscr{C}$, fix $f \in L^p(\mu)$, without loss of generality say $f \geq 0$. Given any $n, i \in \mathbb{N}$, let us define $E_{ni} := f^{-1}\big([i/2^n, (i+1)/2^n[\big)$. Observe that $(E_{ni})_i$ is a Borel partition of X, thus it makes sense to define $f_n := \sum_{i\in\mathbb{N}} i\,2^{-n}\chi_{E_{ni}} \in L^p(\mu)$. Given that we have $f_n(x) \nearrow f(x)$ for μ-a.e. $x \in \mathrm{X}$, it holds $f_n \to f$ in $L^p(\mu)$ by dominated convergence theorem. We aim to prove that $(f_n)_n \subseteq \mathscr{C}$, which would immediately imply that $f \in \mathscr{C}$. Then fix $n \in \mathbb{N}$. Notice that f_n is the $L^p(\mu)$-limit of $f_n^N := \sum_{i=1}^N i\,2^{-n}\chi_{E_{ni}}$ as $N \to \infty$, again by dominated convergence theorem. Given that each $f_n^N \in \mathscr{C}$ by STEP 2, we get that f_n is in $\mathscr{C}$ as well. Hence $\mathrm{LIP}_{bs}(\mathrm{X})$ is dense in $L^p(\mu)$.

STEP 4. Finally, we prove separability of $L^p(\mu)$. We can take an increasing sequence $(K_n)_n$ of compact subsets of X such that the measure μ is concentrated on $\bigcup_n K_n$, by Corollary 1.1.16. Since $\chi_{K_n} f \to f$ in $L^p(\mu)$ for any $f \in L^p(\mu)$, we see that

$$\bigcup_{n\in\mathbb{N}} \underbrace{\big\{f \in L^p(\mu) \ :\ f = 0 \ \ \mu\text{-a.e. in } \mathrm{X}\setminus K_n\big\}}_{=:S_n} \quad \text{is dense in } L^p(\mu).$$

To conclude, it is sufficient to show that each S_n is separable. Observe that $C(K_n)$ is separable by Corollary 1.1.6, thus accordingly its subset $\mathrm{LIP}_{bs}(K_n)$ is separable with respect to $\|\cdot\|_{C_b(K_n)}$. In particular, $\mathrm{LIP}_{bs}(K_n)$ is separable with respect to $\|\cdot\|_{L^p(\mu)}$. Moreover, $\mathrm{LIP}_{bs}(K_n)$ is dense in $L^p(\mu_{|K_n}) \cong S_n$ by the first part of the statement, therefore each S_n is separable. □

1.1.2 *The Space* $L^0(\mathfrak{m})$

By *metric measure space* we mean a triple $(\mathrm{X}, \mathsf{d}, \mathfrak{m})$, where

$$\begin{aligned} &(\mathrm{X}, \mathsf{d}) && \text{is a complete and separable metric space,} \\ &\mathfrak{m} \neq 0 && \text{is a non-negative Borel measure on } (\mathrm{X}, \mathsf{d}), \text{ which is finite on balls.} \end{aligned} \tag{1.21}$$

Let us denote by $L^0(\mathfrak{m})$ the vector space of all Borel functions $f : \mathrm{X} \to \mathbb{R}$, which are considered modulo $\mathfrak{m}$-a.e. equality. Then $L^0(\mathfrak{m})$ becomes a topological vector space when endowed with the following distance: choose any Borel probability measure $\mathfrak{m}' \in \mathscr{P}(\mathrm{X})$ such that $\mathfrak{m} \ll \mathfrak{m}' \ll \mathfrak{m}$ (for instance, pick any Borel partition $(E_n)_n$ made of sets having finite positive $\mathfrak{m}$-measure and set $\mathfrak{m}' := \sum_n \frac{\chi_{E_n}\mathfrak{m}}{2^n\,\mathfrak{m}(E_n)}$) and define

$$\mathsf{d}_{L^0}(f, g) := \int |f - g| \wedge 1 \, \mathrm{d}\mathfrak{m}' \quad \text{for every } f, g \in L^0(\mathfrak{m}). \tag{1.22}$$

Such distance may depend on the choice of $\mathfrak{m}'$, but its induced topology does not, as we are going to show in the next result:

Proposition 1.1.19 *A sequence* $(f_n)_n \subseteq L^0(\mathfrak{m})$ *is* d_{L^0}*-Cauchy if and only if*

$$\varlimsup_{n,m\to\infty} \mathfrak{m}\Big(E \cap \big\{|f_n - f_m| > \varepsilon\big\}\Big) = 0 \qquad \begin{array}{l} \textit{for every } \varepsilon > 0 \textit{ and } E \subseteq \mathrm{X} \\ \textit{Borel with } \mathfrak{m}(E) < +\infty. \end{array} \tag{1.23}$$

Proof We separately prove the two implications:
NECESSITY. Suppose that (1.23) holds. Fix $\varepsilon > 0$. Choose any point $\bar{x} \in \mathrm{X}$, then there exists $R > 0$ such that $\mathfrak{m}'\big(B_R(\bar{x})\big) \geq 1 - \varepsilon$. Recall that $\mathfrak{m}$ is finite on bounded sets by hypothesis, so that $\mathfrak{m}\big(B_R(\bar{x})\big) < +\infty$. Moreover, since $\mathfrak{m}'$ is a finite measure, we clearly have that $\chi_{B_R(\bar{x})} \frac{\mathrm{d}\mathfrak{m}'}{\mathrm{d}\mathfrak{m}} \in L^1(\mathfrak{m})$. Now let us call $A_{nm}(\varepsilon)$ the set $B_R(\bar{x}) \cap \big\{|f_n - f_m| > \varepsilon\big\}$. Then property (1.23) grants that $\chi_{A_{nm}(\varepsilon)} \to 0$ in $L^1(\mathfrak{m})$ as $n, m \to \infty$, whence an application of the dominated convergence theorem yields

$$\varlimsup_{n,m\to\infty} \mathfrak{m}'\big(A_{nm}(\varepsilon)\big) = \varlimsup_{n,m\to\infty} \int \chi_{A_{nm}(\varepsilon)}\, \chi_{B_R(\bar{x})} \frac{\mathrm{d}\mathfrak{m}'}{\mathrm{d}\mathfrak{m}} \, \mathrm{d}\mathfrak{m} = 0. \tag{1.24}$$

Therefore we deduce that

$$\begin{aligned}\int |f_n - f_m| \wedge 1 \,\mathrm{d}\mathfrak{m}' &= \int_{\mathrm{X}\setminus B_R(\bar{x})} |f_n - f_m| \wedge 1 \,\mathrm{d}\mathfrak{m}' + \int_{B_R(\bar{x})} |f_n - f_m| \wedge 1 \,\mathrm{d}\mathfrak{m}' \\ &\leq \varepsilon + \int_{B_R(\bar{x})\cap\{|f_n - f_m| \leq \varepsilon\}} |f_n - f_m| \wedge 1 \,\mathrm{d}\mathfrak{m}' \\ &\quad + \int_{A_{nm}(\varepsilon)} |f_n - f_m| \wedge 1 \,\mathrm{d}\mathfrak{m}' \\ &\leq 2\,\varepsilon + \mathfrak{m}'\big(A_{nm}(\varepsilon)\big),\end{aligned}$$

from which we see that $\overline{\lim}_{n,m} \mathsf{d}_{L^0}(f_n, f_m) \leq 2\,\varepsilon$ by (1.24). By arbitrariness of $\varepsilon > 0$, we conclude that $\lim_{n,m} \mathsf{d}_{L^0}(f_n, f_m) = 0$, which shows that the sequence $(f_n)_n$ is d_{L^0}-Cauchy.

SUFFICIENCY. Suppose that $(f_n)_n$ is d_{L^0}-Cauchy. Fix any $\varepsilon \in (0, 1)$ and a Borel set $E \subseteq \mathrm{X}$ with $\mathfrak{m}(E) < +\infty$. Hence the Čebyšëv inequality yields

$$\mathfrak{m}'\big(\{|f_n - f_m| > \varepsilon\}\big) = \mathfrak{m}'\big(\{|f_n - f_m| \wedge 1 > \varepsilon\}\big) \leq \frac{1}{\varepsilon}\int |f_n - f_m| \wedge 1 \,\mathrm{d}\mathfrak{m}' = \frac{\mathsf{d}_{L^0}(f_n, f_m)}{\varepsilon},$$

so that $\overline{\lim}_{n,m} \mathfrak{m}'\big(\{|f_n - f_m| > \varepsilon\}\big) = 0$. Finally, observe that $\chi_E \frac{\mathrm{d}\mathfrak{m}}{\mathrm{d}\mathfrak{m}'} \in L^1(\mathfrak{m}')$, whence

$$\mathfrak{m}\Big(E \cap \big\{|f_n - f_m| > \varepsilon\big\}\Big) = \int \chi_E \frac{\mathrm{d}\mathfrak{m}}{\mathrm{d}\mathfrak{m}'} \chi_{\{|f_n - f_m| > \varepsilon\}} \,\mathrm{d}\mathfrak{m}' \overset{n,m}{\longrightarrow} 0$$

by dominated convergence theorem. Therefore (1.23) is proved. □

Remark 1.1.20 Recall that two metrizable spaces with the same Cauchy sequences have the same topology, while the converse implication does not hold in general. For instance, consider the real line $\mathbb{R}$ endowed with the following two distances:

$$\begin{aligned}\mathsf{d}_1(x, y) &:= |x - y|, \\ \mathsf{d}_2(x, y) &:= \big|\arctan(x) - \arctan(y)\big|,\end{aligned} \qquad \text{for every } x, y \in \mathbb{R}.$$

Then d_1 and d_2 induce the same topology on $\mathbb{R}$, but the d_2-Cauchy sequence $(x_n)_n \subseteq \mathbb{R}$ defined by $x_n := n$ is not d_1-Cauchy. ■

We now show that the distance d_{L^0} metrizes the 'local convergence in measure':

Proposition 1.1.21 *Let $f \in L^0(\mathfrak{m})$ and $(f_n)_n \subseteq L^0(\mathfrak{m})$. Then the following are equivalent:*

i) *It holds that $\mathsf{d}_{L^0}(f_n, f) \to 0$ as $n \to \infty$.*
ii) *Given any subsequence $(n_m)_m$, there exists a further subsequence $(n_{m_k})_k$ such that the limit $\lim_k f_{n_{m_k}}(x) = f(x)$ is verified for $\mathfrak{m}$-a.e. $x \in \mathrm{X}$.*

iii) *We have that* $\overline{\lim}_n \mathfrak{m}\big(E \cap \{|f_n - f| > \varepsilon\}\big) = 0$ *is satisfied for every* $\varepsilon > 0$ *and* $E \subseteq \mathrm{X}$ *Borel with* $\mathfrak{m}(E) < +\infty$.
iv) *We have that* $\overline{\lim}_n \mathfrak{m}'\big(\{|f_n - f| > \varepsilon\}\big) = 0$ *for every* $\varepsilon > 0$.

Proof The proof goes as follows:

i) $\Longrightarrow$ ii) Since $|f_{n_m} - f| \wedge 1 \to 0$ in $L^1(\mathfrak{m}')$, there is $(n_{m_k})_k$ such that $|f_{n_{m_k}} - f|(x) \wedge 1 \to 0$ for $\mathfrak{m}'$-a.e. $x \in \mathrm{X}$, or equivalently $f_{n_{m_k}}(x) \to f(x)$ for $\mathfrak{m}$-a.e. $x \in \mathrm{X}$.

ii) $\Longrightarrow$ iii) Fix $(n_m)_m$, $\varepsilon > 0$ and $E \subseteq \mathrm{X}$ Borel with $\mathfrak{m}(E) < +\infty$. Since $\chi_{\{|f_{n_{m_k}} - f| > \varepsilon\}} \to 0$ pointwise $\mathfrak{m}$-a.e. for some $(m_k)_k$ and $\chi_E \in L^1(\mathfrak{m})$, we apply the dominated convergence theorem to deduce that $\lim_k \int \chi_E\, \chi_{\{|f_{n_{m_k}} - f| > \varepsilon\}}\, \mathrm{d}\mathfrak{m} = 0$, i.e. $\lim_n \mathfrak{m}\big(E \cap \{|f_n - f| > \varepsilon\}\big) = 0$.

iii) $\Longrightarrow$ iv) Fix $\delta > 0$ and $\bar{x} \in \mathrm{X}$, then there is $R > 0$ such that $\mathfrak{m}'\big(\mathrm{X} \setminus B_R(\bar{x})\big) < \delta$. Exactly as we did in (1.24), we can prove that $\lim_n \mathfrak{m}\big(B_R(\bar{x}) \cap \{|f_n - f| > \varepsilon\}\big) = 0$ implies that the limit $\lim_n \mathfrak{m}'\big(B_R(\bar{x}) \cap \{|f_n - f| > \varepsilon\}\big) = 0$ holds as well. Therefore

$$\overline{\lim_{n\to\infty}}\ \mathfrak{m}'\big(\{|f_n - f| > \varepsilon\}\big) \leq \delta + \overline{\lim_{n\to\infty}}\ \mathfrak{m}'\big(B_R(\bar{x}) \cap \{|f_n - f| > \varepsilon\}\big) = \delta.$$

By letting $\delta \searrow 0$, we thus conclude that $\overline{\lim}_n \mathfrak{m}'\big(\{|f_n - f| > \varepsilon\}\big) = 0$, as required.

iv) $\Longrightarrow$ i) Take any $\varepsilon \in (0, 1)$. Notice that

$$\begin{aligned} \mathsf{d}_{L^0}(f_n, f) &= \int |f_n - f| \wedge 1\, \mathrm{d}\mathfrak{m}' = \int_{\{|f_n - f| \leq \varepsilon\}} |f_n - f| \wedge 1\, \mathrm{d}\mathfrak{m}' \\ &\quad + \int_{\{|f_n - f| > \varepsilon\}} |f_n - f| \wedge 1\, \mathrm{d}\mathfrak{m}' \\ &\leq \varepsilon + \mathfrak{m}'\big(\{|f_n - f| > \varepsilon\}\big), \end{aligned}$$

whence $\overline{\lim}_n \mathsf{d}_{L^0}(f_n, f) \leq \varepsilon$, thus accordingly $\lim_n \mathsf{d}_{L^0}(f_n, f) = 0$ by arbitrariness of ε.

□

In particular, Proposition 1.1.21 grants that the completeness of $L^0(\mathfrak{m})$ does not depend on the particular choice of the measure $\mathfrak{m}'$.

Remark 1.1.22 The inclusion map $L^p(\mathfrak{m}) \hookrightarrow L^0(\mathfrak{m})$ is continuous for every $p \in [1, \infty]$.

Indeed, choose any $\mathfrak{m}' \in \mathscr{P}(\mathrm{X})$ with $\mathfrak{m} \ll \mathfrak{m}' \leq \mathfrak{m}$ and define d_{L^0} as in (1.22). Now take any sequence $(f_n)_n$ in $L^p(\mathfrak{m})$ that $L^p(\mathfrak{m})$-converges to some limit $f \in$

$L^p(\mathfrak{m})$. In particular, we have that $f_n \to f$ in $L^p(\mathfrak{m}')$, so that

$$\mathsf{d}_{L^0}(f_n, f) = \int |f_n - f| \wedge 1 \, \mathrm{d}\mathfrak{m}' \leq \int |f_n - f| \, \mathrm{d}\mathfrak{m}' \leq \|f_n - f\|_{L^p(\mathfrak{m}')} \xrightarrow{n} 0,$$

which proves the claim. ■

Exercise 1.1.23 Prove that $L^p(\mathfrak{m})$ is dense in $L^0(\mathfrak{m})$ for every $p \in [1, \infty]$. ■

Proposition 1.1.24 *The space* $\big(L^0(\mathfrak{m}), \mathsf{d}_{L^0}\big)$ *is complete and separable.*

Proof The proof goes as follows:
COMPLETENESS. Fix a d_{L^0}-Cauchy sequence $(f_n)_n \subseteq L^0(\mathfrak{m})$ and some $\varepsilon > 0$. Then there exists a subsequence $(n_k)_k$ such that $\mathfrak{m}'\big(\{|f_{n_{k+1}} - f_{n_k}| > 1/2^k\}\big) < \varepsilon/2^k$ holds for all k. Let us call $A_k := \big\{|f_{n_{k+1}} - f_{n_k}| > 1/2^k\big\}$ and $A := \bigcup_k A_k$, so that $\mathfrak{m}'(A) \leq \varepsilon$. Given $x \in \mathrm{X} \setminus A$, it holds that $\big|f_{n_{k+1}}(x) - f_{n_k}(x)\big| \leq 1/2^k$ for all k, in other words $\big(f_{n_k}(x)\big)_k \subseteq \mathbb{R}$ is a Cauchy (thus also converging) sequence, say $f_{n_k}(x) \to f(x)$ for some $f(x) \in \mathbb{R}$. Up to performing a diagonalisation argument, we have that $f_{n_k} \to f$ pointwise $\mathfrak{m}'$-a.e. for some $f \in L^0(\mathfrak{m})$. Therefore Proposition 1.1.21 grants that $\mathsf{d}_{L^0}(f_n, f) \to 0$, as required.
SEPARABILITY. Fix $f \in L^0(\mathfrak{m})$. Take any increasing sequence $(E_n)_n$ of Borel subsets of X having finite $\mathfrak{m}$-measure and such that $\mathrm{X} = \bigcup_n E_n$. Denote $f_n := \big((\chi_{E_n} f) \wedge n\big) \vee (-n)$ for every $n \in \mathbb{N}$. By dominated convergence theorem, we have that $f_n \to f$ in $L^0(\mathfrak{m})$. Moreover, it holds that $(f_n)_n \subseteq L^1(\mathfrak{m})$. Hence we get the statement by recalling Remark 1.1.22 and the fact that $L^1(\mathfrak{m})$ is separable. □

Remark 1.1.25 Notice that $\mathsf{d}_{L^0}(f, g) = \mathsf{d}_{L^0}(f + h, g + h)$ for every $f, g, h \in L^0(\mathfrak{m})$. However, the distance d_{L^0} is not induced by any norm, as shown by the fact that $\mathsf{d}_{L^0}(\lambda f, 0)$ differs from $|\lambda| \, \mathsf{d}_{L^0}(f, 0)$ for some $\lambda \in \mathbb{R}$ and $f \in L^0(\mathfrak{m})$. ■

Exercise 1.1.26 Suppose that the measure $\mathfrak{m}$ has no atoms. Let $L : L^0(\mathfrak{m}) \to \mathbb{R}$ be linear and continuous. Then $L = 0$. ■

Exercise 1.1.27 Let $(\mathrm{X}, \mathsf{d}, \mathfrak{m})$ be any metric measure space. Then the topology of $L^0(\mathfrak{m})$ comes from a norm if and only if $\mathfrak{m}$ has finite support. ■

1.1.3 Pushforward of Measures

Consider two complete separable metric spaces $(\mathrm{X}, \mathsf{d}_\mathrm{X})$, $(\mathrm{Y}, \mathsf{d}_\mathrm{Y})$ and a Borel map $T : \mathrm{X} \to \mathrm{Y}$. Given a Borel measure $\mu \geq 0$ on X, we define the *pushforward measure* $T_*\mu$ as

$$T_*\mu(E) := \mu\big(T^{-1}(E)\big) \quad \text{for every } E \subseteq \mathrm{X} \text{ Borel.} \tag{1.25}$$

It can be readily checked that $T_*\mu$ is a Borel measure on Y.

Remark 1.1.28 In general, if μ is a Radon measure then $T_*\mu$ is not necessarily Radon. However, if μ is a finite Radon measure then $T_*\mu$ is Radon by Corollary 1.1.17. ■

Example 1.1.29 Let us consider the projection map $\mathbb{R}^2 \ni (x, y) \mapsto \pi^1(x, y) := x \in \mathbb{R}$. Given any Borel subset E of $\mathbb{R}$, it clearly holds that $\pi^1_*\mathcal{L}^2(E) = 0$ if $\mathcal{L}^1(E) = 0$ and $\pi^1_*\mathcal{L}^2(E) = +\infty$ if $\mathcal{L}^1(E) > 0$. ■

Proposition 1.1.30 *Let $\nu \geq 0$ be a Borel measure on* Y. *Then $\nu = T_*\mu$ if and only if*

$$\int f \, d\nu = \int f \circ T \, d\mu \quad \textit{for every } f : \mathrm{X} \to [0, +\infty] \textit{ Borel.} \tag{1.26}$$

We shall call (1.26) *the* change-of-variable formula.

Proof Given $E \subseteq \mathrm{Y}$ Borel and supposing the validity of (1.26), we have that

$$\nu(E) = \int \chi_E \, d\nu = \int \chi_E \circ T \, d\mu = \int \chi_{T^{-1}(E)} \, d\mu = \mu\big(T^{-1}(E)\big) = T_*\mu(E),$$

proving sufficiency. On the other hand, by Cavalieri's principle we see that

$$\int f \, dT_*\mu = \int_0^{+\infty} T_*\mu\big(\{f \geq t\}\big) \, dt = \int_0^{+\infty} \mu\big(\{f \circ T \geq t\}\big) \, dt = \int f \circ T \, d\mu$$

is satisfied for any Borel map $f : \mathrm{X} \to [0, +\infty]$, granting also necessity. □

Remark 1.1.31 Observe that

$$\begin{aligned} T = \tilde{T} \quad \mu\text{-a.e.} \quad &\Longrightarrow \quad T_*\mu = \tilde{T}_*\mu, \\ f = \tilde{f} \quad (T_*\mu)\text{-a.e.} \quad &\Longrightarrow \quad f \circ T = \tilde{f} \circ T \quad \mu\text{-a.e.} \end{aligned} \tag{1.27}$$

Moreover, if $\nu \geq 0$ is a Borel measure on Y satisfying $T_*\mu \leq C\nu$ for some $C > 0$ and $p \in [1, \infty]$, then the operator $L^p(\nu) \ni f \mapsto f \circ T \in L^p(\mu)$ is well-defined, linear and continuous. Indeed, we have for any $f \in L^p(\nu)$ that

$$\int |f \circ T|^p \, d\mu = \int |f|^p \circ T \, d\mu \overset{(1.26)}{=} \int |f|^p \, dT_*\mu \leq C \int |f|^p \, d\nu.$$

In particular, the operator $L^p(T_*\mu) \ni f \mapsto f \circ T \in L^p(\mu)$ is an isometry. ■

1.2 Spaces of Curves

We equip the space $C([0, 1], \mathrm{X})$ of all continuous curves in X with the *sup distance*:

$$\underline{\mathsf{d}}(\gamma, \tilde{\gamma}) := \max_{t \in [0,1]} \mathsf{d}(\gamma_t, \tilde{\gamma}_t) \quad \text{for every } \gamma, \tilde{\gamma} \in C([0, 1], \mathrm{X}). \tag{1.28}$$

Proposition 1.2.1 *Let* (X, d) *be a complete (resp. separable) metric space. Then the metric space* $\big(C([0,1],\mathrm{X}), \underline{\mathsf{d}}\big)$ *is complete (resp. separable).*

Proof The proof goes as follows:
COMPLETENESS. Take a $\underline{\mathsf{d}}$-Cauchy sequence $(\gamma^n)_n \subseteq C([0,1],\mathrm{X})$. Hence for any $\varepsilon > 0$ there exists $n_\varepsilon \in \mathbb{N}$ such that $\underline{\mathsf{d}}(\gamma^n, \gamma^m) < \varepsilon$ for all $n, m \geq n_\varepsilon$. In particular, $(\gamma_t^n)_n$ is d-Cauchy for each $t \in [0,1]$, so that $\lim_n \gamma_t^n = \gamma_t$ with respect to d for a suitable $\gamma_t \in \mathrm{X}$, by completeness of (X, d). Given any $\varepsilon > 0$ and $n \geq n_\varepsilon$, we have $\sup_t \mathsf{d}(\gamma_t^n, \gamma_t) \leq \sup_t \lim_m \mathsf{d}(\gamma_t^n, \gamma_t^m) \leq \varepsilon$ and

$$\begin{aligned}\varlimsup_{s\to t} \mathsf{d}(\gamma_s, \gamma_t) &\leq \varlimsup_{s\to t}\big[\mathsf{d}(\gamma_s, \gamma_s^n) + \mathsf{d}(\gamma_s^n, \gamma_t^n) + \mathsf{d}(\gamma_t^n, \gamma_t)\big]\\ &\leq 2\,\varepsilon + \lim_{s\to t} \mathsf{d}(\gamma_s^n, \gamma_t^n) = 2\,\varepsilon \quad \forall t \in [0,1],\end{aligned}$$

proving that γ is continuous and $\lim_n \underline{\mathsf{d}}(\gamma^n, \gamma) = 0$. Then $\big(C([0,1],\mathrm{X}), \underline{\mathsf{d}}\big)$ is complete.
SEPARABILITY. Fix $(x_n)_n \subseteq \mathrm{X}$ dense. Given $k, n \in \mathbb{N}$ and $f : \{0, \ldots, n-1\} \to \mathbb{N}$, we let

$$\begin{aligned}A_{k,n,f} := \Big\{\gamma \in C([0,1],\mathrm{X}) \;\Big|\; &\mathsf{d}(\gamma_t, x_{f(i)}) < 1/2^k \quad \forall i = 0, \ldots, n-1,\\ &\times t \in \big[i/n, (i+1)/n\big]\Big\}.\end{aligned}$$

We then claim that

$$\begin{aligned}\bigcup_{n,f} A_{k,n,f} &= C([0,1],\mathrm{X}) \qquad \text{for every } k \in \mathbb{N},\\ \underline{\mathsf{d}}(\gamma, \tilde{\gamma}) &\leq \frac{1}{2^{k-1}} \qquad \text{for every } \gamma, \tilde{\gamma} \in A_{k,n,f}.\end{aligned} \tag{1.29}$$

To prove the first in (1.29), fix $k \in \mathbb{N}$ and $\gamma \in C([0,1],\mathrm{X})$. Since γ is uniformly continuous, there exists $\delta > 0$ such that $\mathsf{d}(\gamma_t, \gamma_s) < 1/2^{k+1}$ provided $t, s \in [0,1]$ satisfy $|t-s| < \delta$. Choose any $n \in \mathbb{N}$ such that $1/n < \delta$. Since $(x_n)_n$ is dense in X, for every $i = 0, \ldots, n-1$ we can choose $f(i) \in \mathbb{N}$ such that $\mathsf{d}(x_{f(i)}, \gamma_{i/n}) < 1/2^{k+1}$. Hence for any $i = 0, \ldots, n-1$ it holds that

$$\mathsf{d}(\gamma_t, x_{f(i)}) \leq \mathsf{d}(\gamma_t, \gamma_{i/n}) + \mathsf{d}(\gamma_{i/n}, x_{f(i)}) < \frac{1}{2^k} \quad \text{for every } t \in \left[\frac{i}{n}, \frac{i+1}{n}\right],$$

proving that $\gamma \in A_{k,n,f}$ and accordingly the first in (1.29). To prove the second, simply notice that $\mathsf{d}(\gamma_t, \tilde{\gamma}_t) \leq \mathsf{d}(\gamma_t, x_{f(i)}) + \mathsf{d}(x_{f(i)}, \tilde{\gamma}_t) < 1/2^{k-1}$ for all $i = 1, \ldots, n-1$ and $t \in \big[i/n, (i+1)/n\big]$.

In order to conclude, pick any $\gamma^{k,n,f} \in A_{k,n,f}$ for every k, n, f. The family $(\gamma^{k,n,f})_{k,n,f}$, which is clearly countable, is $\underline{\mathsf{d}}$-dense in $C([0,1], \mathrm{X})$ by (1.29), giving the statement. □

We say that $C([0,1], \mathrm{X})$ is a *Polish space*, i.e. a topological space whose topology comes from a complete and separable distance.

Exercise 1.2.2 Any open subset of a Polish space is a Polish space. ■

Definition 1.2.3 (Absolutely Continuous Curves) We say that a curve $\gamma : [0,1] \to \mathrm{X}$ is *absolutely continuous*, briefly *AC*, provided there exists a map $f \in L^1(0,1)$ such that

$$\mathsf{d}(\gamma_t, \gamma_s) \leq \int_s^t f(r)\,\mathrm{d}r \quad \text{for every } t, s \in [0,1] \text{ with } s < t. \tag{1.30}$$

Clearly, all absolutely continuous curves are continuous.

Remark 1.2.4 If $\mathrm{X} = \mathbb{R}$ then this notion of AC curve coincides with the classical one. ■

Theorem 1.2.5 (Metric Speed) *Let γ be an absolutely continuous curve in* X. *Then*

$$\exists\, |\dot{\gamma}_t| := \lim_{h \to 0} \frac{\mathsf{d}(\gamma_{t+h}, \gamma_t)}{|h|} \quad \textit{for a.e. } t \in [0,1]. \tag{1.31}$$

Moreover, the function $|\dot{\gamma}|$, which is called metric speed *of γ, belongs to $L^1(0,1)$ and is the minimal function (in the a.e. sense) that can be chosen as f in* (1.30).

Proof Fix $(x_n)_n \subseteq \mathrm{X}$ dense. We define $g_n(t) := \mathsf{d}(\gamma_t, x_n)$ for all $t \in [0,1]$. Then

$$\big|g_n(t) - g_n(s)\big| \leq \mathsf{d}(\gamma_t, \gamma_s) \leq \int_s^t f(r)\,\mathrm{d}r \quad \text{for every } t, s \in [0,1] \text{ with } s < t, \tag{1.32}$$

showing that each $g_n : [0,1] \to \mathbb{R}$ is AC. Hence g_n is differentiable a.e. and by applying the Lebesgue differentiation theorem to (1.32) we get that $\big|g_n'(t)\big| \leq f(t)$ for a.e. $t \in [0,1]$. Let us call $g := \sup_n g_n'$, so that $g \in L^1(0,1)$ with $|g| \leq f$ a.e. Moreover, one has that

$$\mathsf{d}(\gamma_t, \gamma_s) = \sup_{n \in \mathbb{N}} \big[g_n(t) - g_n(s)\big] \quad \text{for every } t, s \in [0,1]. \tag{1.33}$$

Indeed, $\mathsf{d}(\gamma_t, \gamma_s) \geq \big[g_n(t) - g_n(s)\big]$ for all n by triangle inequality. On the other hand, given any $\varepsilon > 0$ we can choose $n \in \mathbb{N}$ such that $\mathsf{d}(x_n, \gamma_s) < \varepsilon$, whence $g_n(t) - g_n(s) \geq \mathsf{d}(\gamma_t, \gamma_s) - 2\varepsilon$.

We thus deduce from (1.33) that g can substitute the function f in (1.30), because

$$\mathsf{d}(\gamma_t, \gamma_s) = \sup_{n\in\mathbb{N}} \int_s^t g_n'(r)\,\mathrm{d}r \leq \int_s^t g(r)\,\mathrm{d}r \quad \text{for every } t, s \in [0,1] \text{ with } s < t. \tag{1.34}$$

In order to conclude, it only remains to prove that g is actually the metric speed. By applying Lebesgue differentiation theorem to (1.34), we see that $\overline{\lim}_{s\to t} \mathsf{d}(\gamma_t, \gamma_s)/|t-s| \leq g(t)$ holds for almost every $t \in [0,1]$. Conversely, $\mathsf{d}(\gamma_t, \gamma_s) \geq g_n(t) - g_n(s) = \int_s^t g_n'(r)\,\mathrm{d}r$ is satisfied for every $s < t$ and $n \in \mathbb{N}$ by triangle inequality, so $\underline{\lim}_{s\to t} \mathsf{d}(\gamma_t, \gamma_s)/|t-s| \geq g_n'(t)$ is satisfied for a.e. $t \in [0,1]$ and for every $n \in \mathbb{N}$ by Lebesgue differentiation theorem. This implies that

$$g(t) \geq \overline{\lim_{s\to t}} \frac{\mathsf{d}(\gamma_t, \gamma_s)}{|t-s|} \geq \underline{\lim_{s\to t}} \frac{\mathsf{d}(\gamma_t, \gamma_s)}{|t-s|} \geq \sup_{n\in\mathbb{N}} g_n'(t) = g(t) \quad \text{for a.e. } t \in [0,1],$$

thus concluding the proof. □

Remark 1.2.6 Let us define the function $\mathsf{ms} : C([0,1], \mathrm{X}) \times [0,1] \longrightarrow [0,+\infty]$ as

$$\mathsf{ms}(\gamma, t) := \begin{cases} |\dot{\gamma}_t| = \lim_{h\to 0} \mathsf{d}(\gamma_{t+h}, \gamma_t)/|h| & \text{if such limit exists finite,} \\ +\infty & \text{otherwise.} \end{cases}$$

We claim that ms is Borel. To prove it, consider an enumeration $(r_n)_n$ of $\mathbb{Q} \cap (0,+\infty)$. Given any $\varepsilon, h > 0$ and $n \in \mathbb{N}$, we define the Borel sets $A(\varepsilon, n, h)$ and $B(\varepsilon, n)$ as follows:

$$A(\varepsilon, n, h) := \left\{ (\gamma, t) \;:\; \left| \frac{\mathsf{d}(\gamma_{t+h}, \gamma_t)}{|h|} - r_n \right| < \varepsilon \right\}, \qquad B(\varepsilon, n) := \bigcup_{0<\delta\in\mathbb{Q}} \bigcap_{h\in(0,\delta)\cap\mathbb{Q}} A(\varepsilon, n, h).$$

Hence $\lim_{h\to 0} \mathsf{d}(\gamma_{t+h}, \gamma_t)/|h|$ exists finite if and only if $(\gamma, t) \in \bigcap_{j\in\mathbb{N}} \bigcup_{n\in\mathbb{N}} B(2^{-j}, n)$. Now let us call $C(j,n) := B(2^{-j}, n) \setminus \bigcup_{i<n} B(2^{-j}, i)$ for every $j, n \in \mathbb{N}$. Then the map f_j, defined as

$$f_j(\gamma, t) := \begin{cases} r_n & \text{if } (\gamma, t) \in C(j,n) \text{ for some } n \in \mathbb{N}, \\ +\infty & \text{if } (\gamma, t) \notin \bigcup_n C(j,n), \end{cases}$$

is Borel by construction. Given that $f_j(\gamma, t) \xrightarrow{j} \mathsf{ms}(\gamma, t)$ for every (γ, t), we finally conclude that the function ms is Borel. ■

We define the *kinetic energy functional* KE : $C([0,1],\mathrm{X}) \to [0,+\infty]$ as follows:

$$\mathrm{KE}(\gamma) := \begin{cases} \int_0^1 |\dot\gamma_t|^2\,\mathrm{d}t & \text{if } \gamma \text{ is AC,} \\ +\infty & \text{if } \gamma \text{ is not AC.} \end{cases} \tag{1.35}$$

Proposition 1.2.7 *The functional* KE *is* d*-lower semicontinuous.*

Proof Fix a sequence $(\gamma^n)_n \subseteq C([0,1],\mathrm{X})$ that d-converges to some $\gamma \in C([0,1],\mathrm{X})$. We can take a subsequence $(\gamma^{n_k})_k$ satisfying $\lim_k \mathrm{KE}(\gamma^{n_k}) = \underline{\lim}_n \mathrm{KE}(\gamma^n)$. Our aim is to prove the inequality $\mathrm{KE}(\gamma) \leq \lim_k \mathrm{KE}(\gamma^{n_k})$. The case in which $\lim_k \mathrm{KE}(\gamma^{n_k}) = +\infty$ is trivial, so suppose that such limit is finite. In particular, up to discarding finitely many γ^{n_k}'s, we have that all curves γ^{n_k} are absolutely continuous with $\big(|\dot\gamma^{n_k}|\big)_k \subseteq L^2(0,1)$ bounded. Therefore, up to a not relabeled subsequence, $|\dot\gamma^{n_k}|$ converges to some limit function $G \in L^2(0,1) \subseteq L^1(0,1)$ weakly in $L^2(0,1)$. Given any $t, s \in [0,1]$ with $s < t$, we thus have that

$$\mathsf{d}(\gamma_t,\gamma_s) = \lim_{k\to\infty} \mathsf{d}(\gamma_t^{n_k},\gamma_s^{n_k}) \leq \varliminf_{k\to\infty} \int_s^t |\dot\gamma_r^{n_k}|\,\mathrm{d}r = \lim_{k\to\infty} \big\langle |\dot\gamma^{n_k}|, \chi_{[s,t]}\big\rangle_{L^2(0,1)} = \int_s^t G(r)\,\mathrm{d}r,$$

which grants that γ is absolutely continuous with $|\dot\gamma| \leq G$ a.e. by Theorem 1.2.5. Hence

$$\mathrm{KE}(\gamma) = \int_0^1 |\dot\gamma_t|^2\,\mathrm{d}t \leq \|G\|^2_{L^2(0,1)} \leq \varliminf_{k\to\infty} \int_0^1 |\dot\gamma_t^{n_k}|^2\,\mathrm{d}t = \lim_{k\to\infty} \mathrm{KE}(\gamma^{n_k}),$$

proving the statement. □

Exercise 1.2.8 Prove that

$$\mathrm{KE}(\gamma) = \sup_{0=t_0<\ldots<t_n=1} \sum_{i=0}^{n-1} \frac{\mathsf{d}(\gamma_{t_{i+1}},\gamma_{t_i})^2}{t_{i+1}-t_i} \quad \text{holds for every } \gamma \in C([0,1],\mathrm{X}). \tag{1.36}$$

■

Definition 1.2.9 (Geodesic Curve) A curve $\gamma : [0,1] \to \mathrm{X}$ is said to be a *geodesic* provided

$$\mathsf{d}(\gamma_t,\gamma_s) \leq |t-s|\,\mathsf{d}(\gamma_0,\gamma_1) \quad \text{holds for every } t,s \in [0,1]. \tag{1.37}$$

Clearly, any geodesic curve is continuous.

Proposition 1.2.10 *Let* $\gamma \in C([0,1],\mathrm{X})$ *be fixed. Then the following are equivalent:*

i) *The curve* γ *is a geodesic.*
ii) *It holds that* $\mathsf{d}(\gamma_t,\gamma_s) = |t-s|\,\mathsf{d}(\gamma_0,\gamma_1)$ *for every* $t,s \in [0,1]$.

iii) *The curve* γ *is AC, its metric speed* $|\dot{\gamma}|$ *is a.e. constant and* $\mathsf{d}(\gamma_0, \gamma_1) = \int_0^1 |\dot{\gamma}_t| \, \mathrm{d}t$.

iv) *It holds that* $\mathrm{KE}(\gamma) = \mathsf{d}(\gamma_0, \gamma_1)^2$.

Proof The proof goes as follows:

i) $\Longrightarrow$ ii) Suppose that $\mathsf{d}(\gamma_t, \gamma_s) < (t-s)\,\mathsf{d}(\gamma_0, \gamma_1)$ for some $0 \leq s < t \leq 1$, then

$$\begin{aligned}\mathsf{d}(\gamma_0, \gamma_1) &\leq \mathsf{d}(\gamma_0, \gamma_s) + \mathsf{d}(\gamma_s, \gamma_t) + \mathsf{d}(\gamma_t, \gamma_1)\\ &< \big[t + (t-s) + (1-s)\big]\,\mathsf{d}(\gamma_0, \gamma_1) = \mathsf{d}(\gamma_0, \gamma_1),\end{aligned}$$

which leads to a contradiction. Hence $\mathsf{d}(\gamma_t, \gamma_s) = |t-s|\,\mathsf{d}(\gamma_0, \gamma_1)$ for every $t, s \in [0, 1]$.

ii) $\Longrightarrow$ iii) Observe that $\mathsf{d}(\gamma_t, \gamma_s) = (t-s)\,\mathsf{d}(\gamma_0, \gamma_1) = \int_s^t \mathsf{d}(\gamma_0, \gamma_1)\,\mathrm{d}t$ holds for every $t, s \in [0, 1]$ with $s < t$, whence the curve γ is AC. Moreover, $|\dot{\gamma}_t| = \lim_{h \to 0} \mathsf{d}(\gamma_{t+h}, \gamma_t)/|h| = \mathsf{d}(\gamma_0, \gamma_1)$ holds for a.e. $t \in [0, 1]$, thus accordingly $\int_0^1 |\dot{\gamma}_t|\,\mathrm{d}t = \mathsf{d}(\gamma_0, \gamma_1)$.

iii) $\Longrightarrow$ iv) Clearly $|\dot{\gamma}_t| = \mathsf{d}(\gamma_0, \gamma_1)$ for a.e. $t \in [0, 1]$, hence $\mathrm{KE}(\gamma) = \int_0^1 |\dot{\gamma}_t|^2\,\mathrm{d}t = \mathsf{d}(\gamma_0, \gamma_1)^2$.

iv) $\Longrightarrow$ i) Notice that the function $(0, +\infty)^2 \ni (a, b) \mapsto a^2/b$ is convex and 1-homogeneous, therefore subadditive. Also, γ is AC since $\mathrm{KE}(\gamma) < \infty$. Then for all $t, s \in (0, 1)$ with $s < t$ one has

$$\begin{aligned}\mathsf{d}(\gamma_0, \gamma_1)^2 &= \int_0^s |\dot{\gamma}_r|^2\,\mathrm{d}r + \int_s^t |\dot{\gamma}_r|^2\,\mathrm{d}r + \int_t^1 |\dot{\gamma}_r|^2\,\mathrm{d}r\\ &\geq \frac{1}{s}\left(\int_0^s |\dot{\gamma}_r|\,\mathrm{d}r\right)^2 + \frac{1}{t-s}\left(\int_s^t |\dot{\gamma}_r|\,\mathrm{d}r\right)^2\\ &\quad + \frac{1}{1-t}\left(\int_t^1 |\dot{\gamma}_r|\,\mathrm{d}r\right)^2\\ &\geq \frac{\mathsf{d}(\gamma_0, \gamma_s)^2}{s} + \frac{\mathsf{d}(\gamma_s, \gamma_t)^2}{t-s} + \frac{\mathsf{d}(\gamma_t, \gamma_1)^2}{1-t}\\ &\geq \frac{\big[\mathsf{d}(\gamma_0, \gamma_s) + \mathsf{d}(\gamma_s, \gamma_t) + \mathsf{d}(\gamma_t, \gamma_1)\big]^2}{s + (t-s) + (1-t)} \geq \mathsf{d}(\gamma_0, \gamma_1)^2,\end{aligned}$$

where the last line follows from the subadditivity of the function $(0,+\infty)^2 \ni (a,b) \mapsto a^2/b$. Hence all inequalities are actually equalities, which forces $\mathsf{d}(\gamma_t,\gamma_s) = (t-s)\,\mathsf{d}(\gamma_0,\gamma_1)$.

□

Let us define

$$\mathrm{Geo}(\mathrm{X}) := \big\{\gamma \in C([0,1],\mathrm{X}) \,:\, \gamma \text{ is a geodesic}\big\}. \tag{1.38}$$

Since uniform limits of geodesic curves are geodesic, we have that Geo(X) is $\underline{\mathsf{d}}$-closed.

Definition 1.2.11 (Geodesic Space) We say (X,d) is a *geodesic space* provided for any pair of points $x, y \in \mathrm{X}$ there exists a curve $\gamma \in \mathrm{Geo}(\mathrm{X})$ such that $\gamma_0 = x$ and $\gamma_1 = y$.

Proposition 1.2.12 (Kuratowski Embedding) *Let* (X,d) *be a complete and separable metric space. Then there exists a complete, separable and geodesic metric space* $(\tilde{\mathrm{X}},\tilde{\mathsf{d}})$ *such that* X *is isometrically embedded into* $\tilde{\mathrm{X}}$.

Proof Fix $(x_n)_n \subseteq \mathrm{X}$ dense. Let us define the map $\iota : \mathrm{X} \to \ell^\infty$ as follows:

$$\iota(x) := \big(\mathsf{d}(x,x_n) - \mathsf{d}(x_0,x_n)\big)_n \quad \text{for every } x \in \mathrm{X}.$$

Since $\big|\mathsf{d}(x,x_n) - \mathsf{d}(x_0,x_n)\big| \leq \mathsf{d}(x,x_0)$ for any $n \in \mathbb{N}$, we see that $\iota(x)$ actually belongs to the space ℓ^∞ for every $x \in \mathrm{X}$. By arguing as in the proof of Theorem 1.2.5, precisely when we showed (1.33), we deduce from the density of $(x_n)_n$ in X that

$$\big\|\iota(x) - \iota(y)\big\|_{\ell^\infty} = \sup_{n\in\mathbb{N}} \big|\mathsf{d}(x,x_n) - \mathsf{d}(y,x_n)\big| = \mathsf{d}(x,y) \quad \text{holds for every } x, y \in \mathrm{X},$$

which proves that ι is an isometry. The Banach space ℓ^∞ is clearly geodesic, but it is not separable, so that we cannot just take $\tilde{\mathrm{X}} = \ell^\infty$. We thus proceed as follows: call $\mathrm{X}_0 := \iota(\mathrm{X})$ and recursively define $\mathrm{X}_{n+1} := \big\{\lambda\, x + (1-\lambda)\, y \,:\, \lambda \in [0,1],\ x, y \in \mathrm{X}_n\big\}$ for every $n \in \mathbb{N}$. Finally, let us denote $\tilde{\mathrm{X}} := \mathrm{cl}_{\ell^\infty} \bigcup_n \mathrm{X}_n$, which is the closed convex hull of X_0. Note that X is separable, so that X_0 and accordingly $\tilde{\mathrm{X}}$ are separable, and that $\iota : \mathrm{X} \to \tilde{\mathrm{X}}$ is an isometry. Since $\tilde{\mathrm{X}}$ is also complete and geodesic, we get the statement. □

1.3 Bochner Integral

Fix a Banach space $\mathbb{B}$ and a metric measure space $(\mathrm{X},\mathsf{d},\mu)$ with $\mu \in \mathscr{P}(\mathrm{X})$.

A map $f : \mathrm{X} \to \mathbb{B}$ is said to be *simple* provided it can be written as $f = \sum_{i=1}^n \chi_{E_i}\, v_i$, for some $v_1,\ldots,v_n \in \mathbb{B}$ and some Borel partition $E_1,\ldots,E_n$ of X.

Definition 1.3.1 (Strongly Borel) A map $f : \mathrm{X} \to \mathbb{B}$ is said to be *strongly Borel* (resp. *strongly μ-measurable*) provided it is Borel (resp. μ-measurable) and there exists a separable subset V of $\mathbb{B}$ such that $f(x) \in V$ for μ-a.e. $x \in \mathrm{X}$. This last condition can be briefly expressed by saying that f is *essentially separably valued.*

Lemma 1.3.2 *Let $f : \mathrm{X} \to \mathbb{B}$ be any given map. Then f is strongly Borel if and only if it is Borel and there exists a sequence $(f_n)_n$ of simple maps such that $\lim_n \|f_n(x) - f(x)\|_{\mathbb{B}} = 0$ is satisfied for μ-a.e. $x \in \mathrm{X}$.*

Proof We separately prove the two implications:
SUFFICIENCY. Choose any $V_n \subseteq \mathbb{B}$ separable such that $f_n(x) \in V_n$ for μ-a.e. $x \in \mathrm{X}$. Then the set $V := \bigcup_n V_n$ is separable and $f(x) \in V$ for μ-a.e. $x \in \mathrm{X}$, whence f is strongly Borel.
NECESSITY. We can assume without loss of generality that $f(x) \in V$ for every $x \in \mathrm{X}$. Choose a dense countable subset $(v_n)_n$ of V and notice that $V \subseteq \bigcup_n B_\varepsilon(v_n)$ for every $\varepsilon > 0$. We define $P_\varepsilon : V \to (v_n)_n$ as follows:

$$P_\varepsilon := \sum_{n\in\mathbb{N}} \chi_{C(\varepsilon,n)}\, v_n, \quad \text{where } C(\varepsilon, n) := \big(V \cap B_\varepsilon(v_n)\big) \setminus \bigcup_{i<n} B_\varepsilon(v_i). \tag{1.39}$$

Let us call $f_\varepsilon := P_\varepsilon \circ f$. Since $\|P_\varepsilon(v) - v\|_{\mathbb{B}} \leq \varepsilon$ for all $v \in V$, we have that $\|f_\varepsilon(x) - f(x)\|_{\mathbb{B}} \leq \varepsilon$ for all $x \in \mathrm{X}$, so that f can be pointwise approximated by maps taking countably many values. With a cut-off argument, we can then approximate f by simple maps, as required. □

Given a simple map $f : \mathrm{X} \to \mathbb{B}$ and a Borel set $E \subseteq \mathrm{X}$, we define

$$\int_E f \, \mathrm{d}\mu := \sum_{i=1}^n \mu(E_i \cap E)\, v_i \in \mathbb{B} \quad \text{if } f = \sum_{i=1}^n \chi_{E_i}\, v_i. \tag{1.40}$$

Exercise 1.3.3 Show that the integral in (1.40) is well-posed, i.e. it does not depend on the particular way of writing f, and that it is linear. ■

Definition 1.3.4 (Bochner Integral) A map $f : \mathrm{X} \to \mathbb{B}$ is said to be *Bochner integrable* provided there exists a sequence $(f_n)_n$ of simple maps such that each $x \mapsto \|f_n(x) - f(x)\|_{\mathbb{B}}$ is a μ-measurable function and $\lim_n \int \|f_n - f\|_{\mathbb{B}} \, \mathrm{d}\mu = 0$. In this case, we define

$$\int_E f \, \mathrm{d}\mu := \lim_{n\to\infty} \int_E f_n \, \mathrm{d}\mu \quad \text{for every } E \subseteq \mathrm{X} \text{ Borel.} \tag{1.41}$$

Remark 1.3.5 It follows from the very definition that the inequality

$$\left\| \int_E f \, \mathrm{d}\mu \right\|_{\mathbb{B}} \leq \int_E \|f\|_{\mathbb{B}} \, \mathrm{d}\mu \tag{1.42}$$

holds for every f simple. Now fix a Bochner integrable map f and a sequence $(f_n)_n$ of simple maps that converge to f as in Definition 1.3.4. Hence we have that

$$\left\| \int_E (f_n - f_m)\,\mathrm{d}\mu \right\|_{\mathbb{B}} \overset{(1.42)}{\leq} \int_E \|f_n - f\|_{\mathbb{B}}\,\mathrm{d}\mu + \int_E \|f - f_m\|_{\mathbb{B}}\,\mathrm{d}\mu \overset{n,m}{\longrightarrow} 0,$$

proving that $\left(\int_E f_n\,\mathrm{d}\mu\right)_n$ is Cauchy in $\mathbb{B}$ and accordingly the limit in (1.41) exists. Further, take another sequence $(g_n)_n$ of simple maps converging to f in the sense of Definition 1.3.4. Therefore one has that

$$\left\| \int_E (f_n - g_n)\,\mathrm{d}\mu \right\|_{\mathbb{B}} \overset{(1.42)}{\leq} \int_E \|f_n - f\|_{\mathbb{B}}\,\mathrm{d}\mu + \int_E \|f - g_n\|_{\mathbb{B}}\,\mathrm{d}\mu \overset{n}{\longrightarrow} 0,$$

which implies $\lim_n \int_E f_n\,\mathrm{d}\mu = \lim_n \int_E g_n\,\mathrm{d}\mu$. This grants that $\int_E f\,\mathrm{d}\mu$ is well-defined. ■

Proposition 1.3.6 *Let $f : \mathrm{X} \to \mathbb{B}$ be a given map. Then f is Bochner integrable if and only if it is strongly μ-measurable and $\int \|f\|_{\mathbb{B}}\,\mathrm{d}\mu < +\infty$.*

Proof Necessity is trivial. To prove sufficiency, consider the maps P_ε defined in (1.39) and call $f_\varepsilon := P_\varepsilon \circ f$. Hence we have $\int \|f_\varepsilon - f\|_{\mathbb{B}}\,\mathrm{d}\mu \leq \varepsilon$ for all $\varepsilon > 0$. Recall that the projection maps P_ε are written in the form $\sum_{n\in\mathbb{N}} \chi_{C(\varepsilon,n)}\, v_n$, so that $f_\varepsilon = \sum_{n\in\mathbb{N}} \chi_{f^{-1}(C(\varepsilon,n))}\, v_n$. Now let us define $g_\varepsilon^k := \sum_{n\leq k} \chi_{f^{-1}(C(\varepsilon,n))}\, v_n$ for all $k \in \mathbb{N}$. Given that $\sum_{n\in\mathbb{N}} \mu\big(f^{-1}(C(\varepsilon,n))\big)\, \|v_n\|_{\mathbb{B}}$ is equal to $\int \|f_\varepsilon\|_{\mathbb{B}}\,\mathrm{d}\mu$, which is smaller than $\int \|f\|_{\mathbb{B}}\,\mathrm{d}\mu + \varepsilon$ and accordingly finite, we see that

$$\int \|g_\varepsilon^k - f_\varepsilon\|_{\mathbb{B}}\,\mathrm{d}\mu = \sum_{n=k+1}^{\infty} \mu\big(f^{-1}(C(\varepsilon,n))\big)\, \|v_n\|_{\mathbb{B}} \overset{k}{\longrightarrow} 0.$$

Since the maps g_ε^k are simple, we can thus conclude by a diagonalisation argument. □

Example 1.3.7 Denote by $\mathcal{M}([0,1])$ the Banach space of all signed Radon measures on $[0,1]$, endowed with the total variation norm. Then the map $[0,1] \to \mathcal{M}([0,1])$, which sends $t \in [0,1]$ to $\delta_t \in \mathscr{P}([0,1])$, is not strongly Borel (thus also not Borel).

Indeed, notice that $\|\delta_t - \delta_s\|_{\mathsf{TV}} = 2$ for every $t, s \in [0,1]$ with $t \neq s$. Now suppose that there exists a Borel set $N \subseteq [0,1]$ with $\mathcal{L}^1(N) = 0$ such that $\{\delta_t : t \in [0,1] \setminus N\}$ is separable. Take a countable dense subset $(\mu_n)_n$ of such set. Hence for every $t \in [0,1] \setminus N$ we can choose an index $\underline{n}(t) \in \mathbb{N}$ such that $\|\delta_t - \mu_{\underline{n}(t)}\|_{\mathsf{TV}} < 1$. Clearly the function $\underline{n} : [0,1] \setminus N \to \mathbb{N}$ must be injective, which contradicts the fact that $[0,1] \setminus N$ is not countable. ■

Let us define the space $L^1(\mu; \mathbb{B})$ as follows:

$$L^1(\mu; \mathbb{B}) := \{f : \mathrm{X} \to \mathbb{B} \text{ Bochner integrable}\}/(\mu\text{-a.e. equality}). \tag{1.43}$$

Then $L^1(\mu; \mathbb{B})$ is a Banach space if endowed with the norm $\|f\|_{L^1(\mu;\mathbb{B})} := \int \|f(x)\|_{\mathbb{B}} \, \mathrm{d}\mu(x)$.

Remark 1.3.8 Given two metric spaces X, Y and a continuous map $f : \mathrm{X} \to \mathrm{Y}$, we have that the image $f(\mathrm{X})$ is separable whenever X is separable.

Indeed, if $(x_n)_n$ is dense in X, then $\big(f(x_n)\big)_n$ is dense in $f(\mathrm{X})$ by continuity of f. ■

Proposition 1.3.9 *Let $E \subseteq \mathrm{X}$ be Borel. Let $\mathbb{V}$ be another Banach space. Then:*

i) *For every $f \in L^1(\mu; \mathbb{B})$, it holds that*

$$\left\| \int_E f \, \mathrm{d}\mu \right\|_{\mathbb{B}} \leq \int_E \|f\|_{\mathbb{B}} \, \mathrm{d}\mu. \tag{1.44}$$

In particular, the map $L^1(\mu; \mathbb{B}) \to \mathbb{B}$ sending f to $\int f \, \mathrm{d}\mu$ is linear and continuous.

ii) *The space $C_b(\mathrm{X}, \mathbb{B})$ is (contained and) dense in $L^1(\mu; \mathbb{B})$.*

iii) *If $\ell : \mathbb{B} \to \mathbb{V}$ is linear continuous and $f \in L^1(\mu; \mathbb{B})$, one has that $\ell \circ f \in L^1(\mu; \mathbb{V})$ and*

$$\ell\left(\int_E f \, \mathrm{d}\mu\right) = \int_E \ell \circ f \, \mathrm{d}\mu. \tag{1.45}$$

Proof The proof goes as follows:

i) As already mentioned in (1.42), we have that the inequality (1.44) is satisfied whenever the map f is simple, because if $f = \sum_{i=1}^n \chi_{E_i} v_i$ then

$$\left\| \int_E f \, \mathrm{d}\mu \right\|_{\mathbb{B}} \leq \sum_{i=1}^n \left\| \int \chi_{E_i \cap E} \, v_i \, \mathrm{d}\mu \right\|_{\mathbb{B}} = \sum_{i=1}^n \mu(E_i \cap E) \, \|v_i\|_{\mathbb{B}} = \int_E \|f\|_{\mathbb{B}} \, \mathrm{d}\mu.$$

For f generic, choose a sequence $(f_n)_n$ of simple maps that converge to f in $L^1(\mu; \mathbb{B})$. Then

$$\left\| \int_E f \, \mathrm{d}\mu \right\|_{\mathbb{B}} = \lim_{n\to\infty} \left\| \int_E f_n \, \mathrm{d}\mu \right\|_{\mathbb{B}} \leq \lim_{n\to\infty} \int_E \|f_n\|_{\mathbb{B}} \, \mathrm{d}\mu = \int_E \|f\|_{\mathbb{B}} \, \mathrm{d}\mu,$$

thus proving the validity of (1.44).

ii) The elements of $C(\mathrm{X}, \mathbb{B})$, which are clearly Borel, are (essentially) separably valued by Remark 1.3.8, in other words they are strongly Borel. This grants that $C_b(\mathrm{X}, \mathbb{B}) \subseteq L^1(\mu; \mathbb{B})$. To prove its density, it suffices to approximate just the maps of the form $\chi_E \, v$. First choose any sequence $(C_n)_n$ of closed subsets of E with $\mu(E \setminus C_n) \searrow 0$, so that $\chi_{C_n} v \to \chi_E \, v$ with respect to the $L^1(\mu; \mathbb{B})$-norm, then for each $n \in \mathbb{N}$ notice that the maps $\big(1 - k \, \mathsf{d}(\cdot, C_n)\big)^+ v$ belong to $C_b(\mathrm{X}, \mathbb{B})$ and $L^1(\mu; \mathbb{B})$-converge to $\chi_{C_n} v$ as $k \to \infty$. So $C_b(\mathrm{X}, \mathbb{B})$ is dense in $L^1(\mu; \mathbb{B})$.

iii) In the case in which f is simple, say $f = \sum_{i=1}^{n} \chi_{E_i} v_i$, one has that

$$\ell\bigg(\int_E f \, \mathrm{d}\mu\bigg) = \sum_{i=1}^{n} \mu(E_i \cap E)\,\ell(v_i) = \int_E \ell \circ f \, \mathrm{d}\mu.$$

For a general f, choose a sequence $(f_n)_n$ of simple maps that $L^1(\mu; \mathbb{B})$-converge to f. We note that the inequality $\int \big\|\ell(f - f_n)\big\|_{\mathbb{V}}(x) \, \mathrm{d}\mu(x) \leq \|\ell\| \int \|f - f_n\|_{\mathbb{B}} \, \mathrm{d}\mu$ is satisfied, where $\|\ell\|$ stands for the operator norm of ℓ. In particular $\int_E \ell \circ f_n \, \mathrm{d}\mu \to \int_E \ell \circ f \, \mathrm{d}\mu$. Therefore

$$\ell\bigg(\int_E f \, \mathrm{d}\mu\bigg) = \lim_{n\to\infty} \ell\bigg(\int_E f_n \, \mathrm{d}\mu\bigg) = \lim_{n\to\infty} \int_E \ell \circ f_n \, \mathrm{d}\mu = \int_E \ell \circ f \, \mathrm{d}\mu,$$

proving (1.45) as required. □

Definition 1.3.10 (Closed Operator) A *closed operator* $T : \mathbb{B} \to \mathbb{V}$ is a couple $\big(D(T), T\big)$, where $D(T)$ is a linear subspace of $\mathbb{B}$ and $T : D(T) \to \mathbb{V}$ is a linear map whose graph, defined as $\mathrm{Graph}(T) := \big\{(v, Tv) \ : \ v \in D(T)\big\}$, is a closed subspace of the product space $\mathbb{B} \times \mathbb{V}$.

Closedness of Graph(T) can be equivalently stated as follows: if a sequence $(v_n)_n \subseteq D(T)$ satisfy $\lim_n \|v_n - v\|_{\mathbb{B}} = 0$ and $\lim_n \|Tv_n - w\|_{\mathbb{V}} = 0$ for some vectors $v \in B$ and $w \in \mathbb{V}$, then necessarily $v \in D(T)$ and $w = Tv$.

Example 1.3.11 (of Closed Operators) We provide three examples of closed operators:

i) Let $\mathbb{B} = \mathbb{V} = C([0,1])$. Then take $D(T_1) = C^1([0,1])$ and $T_1(f) = f'$.
ii) Let $\mathbb{B} = \mathbb{V} = L^2(0,1)$. Then take $D(T_2) = W^{1,2}(0,1)$ and $T_2(f) = f'$.
iii) Let $\mathbb{B} = L^2(\mathbb{R}^n)$ and $\mathbb{V} = \big[L^2(\mathbb{R}^n)\big]^n$. Then take $D(T_3) = W^{1,2}(\mathbb{R}^n)$ and $T_3(f)$ equal to the n-tuple $(\partial_{x_1} f, \ldots, \partial_{x_n} f)$. ■

Example 1.3.12 (of Non-closed Operator) Consider $\mathbb{B} = \mathbb{V} = L^2(\mathbb{R}^n)$, with $n > 1$. We define $D(T_4) = W^{1,2}(\mathbb{R}^n)$ and $T_4(f) = \partial_{x_1} f$. Then $\big(D(T_4), T_4\big)$ is not a closed operator. ■

Exercise 1.3.13 Prove Examples 1.3.11 and 1.3.12. ■

Remark 1.3.14 Let $f \in L^1(\mu; \mathbb{B})$ be given. Suppose there exists a closed subspace V of $\mathbb{B}$ such that $f(x) \in V$ holds for μ-a.e. $x \in \mathrm{X}$. Then $\int_E f \, \mathrm{d}\mu \in V$ for every $E \subseteq \mathrm{X}$ Borel.

We argue by contradiction: suppose $\int_E f \, \mathrm{d}\mu \notin V$, then we can choose $\ell \in \mathbb{B}'$ with $\ell = 0$ on V and $\ell\big(\int_E f \, \mathrm{d}\mu\big) = 1$ by Hahn-Banach theorem. But the fact that $(\ell \circ f)(x) = 0$ holds for μ-a.e. $x \in \mathrm{X}$ implies $\ell\big(\int_E f \, \mathrm{d}\mu\big) = \int_E \ell \circ f \, \mathrm{d}\mu = 0$ by (1.45), giving a contradiction. ■

Theorem 1.3.15 (Hille) *Let $T : \mathbb{B} \to \mathbb{V}$ be a closed operator. Consider a map $f \in L^1(\mu; \mathbb{B})$ that satisfies $f(x) \in D(T)$ for μ-a.e. $x \in \mathrm{X}$ and $T \circ f \in L^1(\mu; \mathbb{V})$. Then for every $E \subseteq \mathrm{X}$ Borel it holds that $\int_E f \, \mathrm{d}\mu \in D(T)$ and that*

$$T\left(\int_E f \, \mathrm{d}\mu\right) = \int_E T \circ f \, \mathrm{d}\mu. \tag{1.46}$$

Proof Define the map $\Phi : \mathrm{X} \to \mathbb{B} \times \mathbb{V}$ as $\Phi(x) := \big(f(x), (T \circ f)(x)\big)$ for μ-a.e. $x \in \mathrm{X}$. One can readily check that $\Phi \in L^1(\mu; \mathbb{B} \times \mathbb{V})$. Moreover, $\Phi(x) \in \mathrm{Graph}(T)$ for μ-a.e. $x \in \mathrm{X}$, whence

$$\left(\int_E f \, \mathrm{d}\mu, \int_E T \circ f \, \mathrm{d}\mu\right) = \int_E \Phi(x) \, \mathrm{d}\mu(x) \in \mathrm{Graph}(T)$$

by Remark 1.3.14. This means that $\int_E f \, \mathrm{d}\mu \in D(T)$ and that $T\big(\int_E f \, \mathrm{d}\mu\big) = \int_E T \circ f \, \mathrm{d}\mu$. □

Let us now concentrate our attention on the case in which $\mathrm{X} = [0, 1]$ and $\mu = \mathcal{L}^1|_{[0,1]}$.

Proposition 1.3.16 *Let $v : [0, 1] \to \mathbb{B}$ be an absolutely continuous curve. Suppose that*

$$v'_t := \lim_{h \to 0} \frac{v_{t+h} - v_t}{h} \in \mathbb{B} \quad \textit{exists for a.e. } t \in [0, 1]. \tag{1.47}$$

Then the map $v' : [0, 1] \to \mathbb{B}$ is Bochner integrable and satisfies

$$v_t - v_s = \int_s^t v'_r \, \mathrm{d}r \quad \textit{for every } t, s \in [0, 1] \textit{ with } s < t. \tag{1.48}$$

Proof First of all, by arguing as in Remark 1.2.6, we see that v' is Borel. Moreover, if V is a closed separable subspace of $\mathbb{B}$ such that $v_t \in V$ for a.e. $t \in [0, 1]$, then $v'_t \in V$ for a.e. $t \in [0, 1]$ as well, i.e. v' is essentially separably valued. Hence v' is a strongly Borel map. Since the function $\|v'\|_{\mathbb{B}}$ coincides a.e. with the metric speed $|\dot{v}|$, which belongs to $L^1(0, 1)$, we conclude that v' is Bochner integrable by Proposition 1.3.6. Finally, to prove (1.48) it is enough to show that $v_t = v_0 + \int_0^t v'_s \, \mathrm{d}s$ for any $t \in [0, 1]$. For every $\ell \in \mathbb{B}'$ it holds that $t \mapsto \ell(v_t) \in \mathbb{R}$ is absolutely continuous, with $\frac{\mathrm{d}}{\mathrm{d}t}\ell(v_t) = \ell(v'_t)$ for a.e. $t \in [0, 1]$. Therefore

$$\ell(v_t) = \ell(v_0) + \int_0^t \left(\frac{\mathrm{d}}{\mathrm{d}s}\ell(v_s)\right) \mathrm{d}s = \ell(v_0) + \int_0^t \ell(v'_s) \, \mathrm{d}s \overset{(1.45)}{=} \ell\left(v_0 + \int_0^t v'_s \, \mathrm{d}s\right),$$

which implies that $v_t = v_0 + \int_0^t v'_s \, \mathrm{d}s$ by arbitrariness of $\ell \in \mathbb{B}'$. Thus (1.48) is proved. □

Example 1.3.17 Let us define the map $v : [0,1] \to L^1(0,1)$ as $v_t := \chi_{[0,t]}$ for every $t \in [0,1]$. Then v is 1-Lipschitz (so also absolutely continuous), because $\|v_t - v_s\|_{L^1(0,1)} = t - s$ holds for every $t, s \in [0,1]$ with $s < t$, but v is not differentiable at any $t \in [0,1]$: the incremental ratios $h^{-1}(v_{t+h} - v_t) = h^{-1}\chi_{(t,t+h]}$ pointwise converge to 0 as $h \searrow 0$ and have $L^1(0,1)$-norm equal to 1. Notice that the probability measures $h^{-1}\chi_{(t,t+h]}\mathcal{L}^1$ weakly converges to δ_t. ■

Proposition 1.3.18 (Lebesgue Points) *Let $v : [0,1] \to \mathbb{B}$ be Bochner integrable. Then*

$$\lim_{h\searrow 0} \fint_{t-h}^{t+h} \|v_s - v_t\|_{\mathbb{B}}\, \mathrm{d}s = 0 \quad \textit{for a.e. } t \in [0,1]. \tag{1.49}$$

Proof Choose a separable set $V \subseteq \mathbb{B}$ such that $v_t \in V$ for a.e. $t \in [0,1]$ and a sequence $(w_n)_n$ that is dense in V. For any $n \in \mathbb{N}$, the map $t \mapsto \|v_t - w_n\|_{\mathbb{B}} \in \mathbb{R}$ belongs to $L^1(0,1)$, hence there exists a Borel set $N_n \subseteq [0,1]$, with $\mathcal{L}^1(N_n) = 0$, such that

$$\|v_t - w_n\|_{\mathbb{B}} = \lim_{h\searrow 0} \fint_{t-h}^{t+h} \|v_s - w_n\|_{\mathbb{B}}\, \mathrm{d}s \quad \text{holds for every } t \in [0,1] \setminus N_n,$$

by Lebesgue differentiation theorem. Call $N := \bigcup_n N_n$, which is an $\mathcal{L}^1$-negligible Borel subset of $[0,1]$. Therefore for every $t \in [0,1] \setminus N$ one has that

$$\begin{aligned} \overline{\lim_{h\searrow 0}} \fint_{t-h}^{t+h} \|v_s - v_t\|_{\mathbb{B}}\, \mathrm{d}s &\leq \inf_{n\in\mathbb{N}} \overline{\lim_{h\searrow 0}} \left[\fint_{t-h}^{t+h} \|v_s - w_n\|_{\mathbb{B}}\, \mathrm{d}s + \|v_t - w_n\|_{\mathbb{B}} \right] \\ &= \inf_{n\in\mathbb{N}} 2\,\|v_t - w_n\|_{\mathbb{B}} = 0 \end{aligned}$$

by density of $(w_n)_n$ in V. Hence (1.49) is proved, getting the statement. □

Fix two metric measure spaces $(\mathrm{X}, \mathsf{d}_{\mathrm{X}}.\mu)$, $(\mathrm{Y}, \mathsf{d}_{\mathrm{Y}}, \nu)$, with μ and ν finite measures. In the following three results we will distinguish real-valued functions from their equivalence classes up to a.e. equality: namely, we will denote by $f : \mathrm{Y} \to \mathbb{R}$ the ν-measurable maps and by $[f]$ the elements of $L^1(\nu)$.

Proposition 1.3.19 *Let $\mathrm{X} \ni x \mapsto [f_x] \in L^1(\nu)$ be any μ-measurable map. Then there exists a choice $(x,y) \mapsto \tilde{f}(x,y)$ of representatives, i.e. $\big[\tilde{f}(x,\cdot)\big] = [f_x]$ holds for μ-a.e. $x \in \mathrm{X}$, which is Borel measurable. Moreover, any two such choices agree $(\mu \times \nu)$-a.e. in $\mathrm{X} \times \mathrm{Y}$.*

Proof The statement is clearly verified when $x \mapsto [f_x]$ is a simple map. For $x \mapsto [f_x]$ generic, define $[f_x^k] := \chi_{A_k}(x)\,[f_x]$ for μ-a.e. $x \in \mathrm{X}$, where we set $A_k := \big\{x \in \mathrm{X} : \big\|[f_x]\big\|_{L^1(\nu)} \leq k\big\}$. Now let $k \in \mathbb{N}$ be fixed. Given that $[f^k]$ belongs to $L^1\big(\mu; L^1(\nu)\big)$, we can choose a sequence of simple maps $[g^n] : \mathrm{X} \to L^1(\nu)$ such that $\big\|[g^n] - [f^k]\big\|_{L^1(\mu; L^1(\nu))} \leq 2^{-2n}$ for every $n \in \mathbb{N}$. As observed in the first part of the proof, we can choose a Borel representative $\tilde{g}^n : \mathrm{X} \times \mathrm{Y} \to \mathbb{R}$ of $[g^n]$ for every $n \in \mathbb{N}$. By using Čebyšëv's inequality, we obtain that

$$\mu\Big(\big\{x \in \mathrm{X} : \big\|[g_x^n] - [f_x^k]\big\|_{L^1(\nu)} > 2^{-n}\big\}\Big) \leq \frac{1}{2^n} \quad \text{holds for every } n \in \mathbb{N}.$$

Therefore we have that

$$\mu\Big(\bigcup_{n_0 \in \mathbb{N}} \big\{x \in \mathrm{X} : \big\|[g_x^n] - [f_x^k]\big\|_{L^1(\nu)} \leq 2^{-n} \text{ for all } n \geq n_0\big\}\Big) = \mu(\mathrm{X}). \tag{1.50}$$

Then the functions $\tilde{g}^n$ converge $(\mu \times \nu)$-a.e. to some limit function $\tilde{f}^k : \mathrm{X} \times \mathrm{Y} \to \mathbb{R}$, which is accordingly a Borel representative of $[f^k]$. To conclude, let us define

$$\tilde{f}(x, y) := \sum_{k \in \mathbb{N}} \chi_{A_k \setminus \bigcup_{i<k} A_i}(x)\, \tilde{f}^k(x, y) \quad \text{for every } (x, y) \in \mathrm{X} \times \mathrm{Y}.$$

Therefore $\tilde{f}$ is the desired representative of $x \mapsto [f_x]$, whence the statement is proved. □

Proposition 1.3.20 *Consider the operator* $\Phi : L^1(\mu; L^1(\nu)) \to L^1(\mu \times \nu)$ *sending* $x \mapsto [f_x]$ *to (the equivalence class of) one of its Borel representatives* $\tilde{f}$ *found in Proposition 1.3.19. Then the map* Φ *is (well-defined and) an isometric isomorphism.*

Proof Well-posedness of Φ follows from Proposition 1.3.19 and from the fact that

$$\begin{aligned}&\big\|[f_\cdot]\big\|_{L^1(\mu; L^1(\nu))} \\ &= \iint \big|[f_x]\big|(y)\,\mathrm{d}\nu(y)\,\mathrm{d}\mu(x) = \iint |\tilde{f}|(x, y)\,\mathrm{d}\nu(y)\,\mu(x) = \int |\tilde{f}|\,\mathrm{d}(\mu \times \nu)\end{aligned}$$

where the last equality is a consequence of Fubini theorem. The same equalities also guarantee that Φ is an isometry. Moreover, the map Φ is linear, continuous and injective. In order to conclude, it suffices to show that the image of Φ is dense. Given any $\tilde{f} \in C_b(\mathrm{X} \times \mathrm{Y})$, we have that $\lim_{x' \to x} \int \big|\tilde{f}(x', y) - \tilde{f}(x, y)\big|\,\mathrm{d}\nu(y) = 0$ for every $x \in \mathrm{X}$ by dominated convergence theorem, so that $x \mapsto \tilde{f}(x, \cdot) \in L^1(\nu)$ is continuous and accordingly in $L^1(\mu; L^1(\nu))$. In other words, we proved that any $\tilde{f} \in C_b(\mathrm{X} \times \mathrm{Y})$ belongs to the image of Φ. Since $C_b(\mathrm{X} \times \mathrm{Y})$ is dense in $L^1(\mu \times \nu)$ by Proposition 1.1.18, we thus obtained the statement. □

Proposition 1.3.21 *Let* $(x \mapsto [f_x]) \in L^1(\mu; L^1(\nu))$ *and call* $[\tilde{f}\,]$ *its image under* Φ. *Then*

$$\Big(\int [f_x]\,\mathrm{d}\mu(x)\Big)(y) = \int \tilde{f}(x, y)\,\mathrm{d}\mu(x) \quad \textit{holds for } \nu\textit{-a.e. } y \in \mathrm{Y}. \tag{1.51}$$

Proof First of all, we define the linear and continuous operator $T_1 : L^1(\mu; L^1(\nu)) \to L^1(\nu)$ as $T_1(f) := \int [f_x]\, \mathrm{d}\mu(x) \in L^1(\nu)$ for every $f \in L^1(\mu; L^1(\nu))$. On the other hand, by Fubini theorem it makes sense to define $T_2(\tilde{f}) := \big(y \mapsto \int \tilde{f}(x, y)\, \mathrm{d}\mu(x)\big) \in L^1(\nu)$ for all $\tilde{f} \in L^1(\mu \times \nu)$, so that $T_2 : L^1(\mu \times \nu) \to L^1(\nu)$ is a linear and continuous operator. Therefore the diagram is commutative, because T_1 and $T_2 \circ \Phi$ clearly agree on simple maps $f : \mathrm{X} \to L^1(\nu)$. Hence formula (1.51) is proved, as required. □

Lemma 1.3.22 (Easy Version of Dunford-Pettis) *Let $(f_n)_n \subseteq L^1(\nu)$ be a sequence with the following property: there exists $g \in L^1(\nu)$ such that $|f_n| \leq g$ holds ν-a.e. for every $n \in \mathbb{N}$. Then there exists a subsequence $(n_k)_k$ and some function $f \in L^1(\nu)$ such that $f_{n_k} \rightharpoonup f$ weakly in $L^1(\nu)$ and $|f| \leq g$ holds ν-a.e. in* Y.

Proof For any $k \in \mathbb{N}$, denote $f_n^k := \min\big\{\max\{f_n, -k\}, k\big\}$ and $g_k := \min\big\{\max\{g, -k\}, k\big\}$. The sequence $(f_n^k)_n$ is bounded in $L^2(\nu)$ for any fixed $k \in \mathbb{N}$, thus a diagonalisation argument shows the existence of $(n_i)_i$ and $(h_k)_k \subseteq L^2(\nu)$ such that $f_{n_i}^k \rightharpoonup h_k$ weakly in $L^2(\nu)$ for all k. In particular, $f_{n_i}^k \rightharpoonup h_k$ weakly in $L^1(\nu)$ for all k. Moreover, one can readily check that

$$|f_{n_i}^k - f_{n_i}^{k'}| \leq |g_k - g_{k'}| \text{ holds } \nu\text{-a.e.} \quad \text{for every } i, k, k' \in \mathbb{N}. \tag{1.52}$$

By using (1.52), the lower semicontinuity of $\|\cdot\|_{L^1(\nu)}$ with respect to the weak topology and the dominated convergence theorem, we then deduce that

$$\int |h_k - h_{k'}|\, \mathrm{d}\nu \leq \varliminf_{i\to\infty} \int |f_{n_i}^k - f_{n_i}^{k'}|\, \mathrm{d}\nu \leq \int |g_k - g_{k'}|\, \mathrm{d}\nu \xrightarrow{k,k'} 0, \tag{1.53}$$

which grants that the sequence $(h_k)_k \subseteq L^1(\nu)$ is Cauchy. Call $f \in L^1(\nu)$ its limit. To prove that $f_{n_i} \rightharpoonup f$ weakly in $L^1(\nu)$ as $i \to \infty$, observe that for any $\ell \in L^\infty(\nu)$ it holds that

$$\begin{aligned}
\varlimsup_{i\to\infty} \left| \int (f_{n_i} - f)\, \ell\, \mathrm{d}\nu \right| &\leq \varlimsup_{i\to\infty} \left[\int |f_{n_i} - f_{n_i}^k|\, |\ell|\, \mathrm{d}\nu + \left| \int (f_{n_i}^k - h_k)\, \ell\, \mathrm{d}\nu \right| \right. \\
&\qquad \left. + \int |h_k - f|\, |\ell|\, \mathrm{d}\nu \right] \\
&\leq \Big(\|g - g_k\|_{L^1(\nu)} + \|h_k - f\|_{L^1(\nu)} \Big)\, \|\ell\|_{L^\infty(\nu)} \\
&\leq 2\, \|g - g_k\|_{L^1(\nu)}\, \|\ell\|_{L^\infty(\nu)} \xrightarrow{k} 0,
\end{aligned}$$

where the second inequality stems from (1.52) and the third one from (1.53).

Finally, in order to prove the ν-a.e. inequality $|f| \leq g$ it is clearly sufficient to show that

$$\left| \int f \, \ell \, d\nu \right| \leq \int g \, \ell \, d\nu \quad \text{for every } \ell \in L^\infty(\nu) \text{ with } \ell \geq 0. \tag{1.54}$$

Property (1.54) can be proved by noticing that for any non-negative $\ell \in L^\infty(\nu)$ one has

$$\left| \int f \, \ell \, d\nu \right| = \lim_{i \to \infty} \left| \int f_{n_i} \, \ell \, d\nu \right| \leq \varliminf_{i \to \infty} \int |f_{n_i}| \, \ell \, d\nu \leq \int g \, \ell \, d\nu.$$

Therefore the statement is achieved. □

Hereafter, we shall make use of the following shorthand notation:

$$\mathcal{L}_1 := \mathcal{L}^1|_{[0,1]} \quad \text{and} \quad \Delta := \{(t, s) \in [0, 1]^2 : s \leq t\}. \tag{1.55}$$

Proposition 1.3.23 *Let $f : [0, 1] \to L^1(\nu)$ and $g \in L^1(\mathcal{L}_1; L^1(\nu))$ be given. Suppose that*

$$|f_t(y) - f_s(y)| \leq \int_s^t g_r(y) \, dr \;\; \textit{holds for } \nu\textit{-a.e. } y \in Y, \quad \textit{for every } (t, s) \in \Delta. \tag{1.56}$$

Then f is absolutely continuous and $\mathcal{L}_1$-a.e. differentiable. Moreover, its derivative satisfies

$$|f_t'|(y) \leq g_t(y) \quad \textit{for } (\mathcal{L}_1 \times \nu)\textit{-a.e. } (t, y) \in [0, 1] \times Y. \tag{1.57}$$

Proof By integrating (1.56), we get that $\| f_t - f_s \|_{L^1(\nu)} \leq \int_s^t \| g_r \|_{L^1(\nu)} \, dr$ for every $(t, s) \in \Delta$. This proves that $t \mapsto f_t \in L^1(\nu)$ is AC, but in general this does not grant that $t \mapsto f_t$ is a.e. differentiable, cf. for instance Example 1.3.17. We thus proceed in the following way: let us define $g_t^\varepsilon := \frac{1}{\varepsilon} \int_t^{t+\varepsilon} g_r \, dr$ for every $\varepsilon > 0$ and $t \in [0, 1]$. Observe that

$$\begin{aligned} \| g_\cdot^\varepsilon \|_{L^1(\mathcal{L}_1 \times \nu)} &= \int_0^1 \int |g_t^\varepsilon|(y) \, d\nu(y) \, dt \leq \int_0^1 \int \fint_t^{t+\varepsilon} |g_r|(y) \, dr \, d\nu(y) \, dt \\ &\leq \int_0^1 \int |g_r|(y) \, d\nu(y) \, dr = \| g_\cdot \|_{L^1(\mathcal{L}_1 \times \nu)} \end{aligned} \tag{1.58}$$

is satisfied for every $\varepsilon > 0$. Given any map $h \in C\big([0,1], L^1(\nu)\big)$, it clearly holds that $h^\varepsilon_\cdot \to h_\cdot$ in $L^1(\mathcal{L}_1 \times \nu)$ as $\varepsilon \searrow 0$. Therefore for any such h one has that

$$\begin{aligned}\varlimsup_{\varepsilon\searrow 0} \|g^\varepsilon - g\|_{L^1(\mathcal{L}_1\times\nu)} &\leq \varlimsup_{\varepsilon\searrow 0}\Big[\big\|(g-h)^\varepsilon\big\|_{L^1(\mathcal{L}_1\times\nu)} + \|h^\varepsilon - h\|_{L^1(\mathcal{L}_1\times\nu)}\Big] \\ &\quad + \|h - g\|_{L^1(\mathcal{L}_1\times\nu)} \\ &\leq 2\,\|g-h\|_{L^1(\mathcal{L}_1\times\nu)} + \varlimsup_{\varepsilon\searrow 0} \|h^\varepsilon - h\|_{L^1(\mathcal{L}_1\times\nu)} \\ &= 2\,\|g-h\|_{L^1(\mathcal{L}_1\times\nu)},\end{aligned}$$

where the second inequality follows from (1.58) and the third one from continuity of h. Given that $C\big([0,1], L^1(\nu)\big)$ is dense in $L^1\big(\mathcal{L}_1; L^1(\nu)\big)$, we conclude that $\lim_{\varepsilon\searrow 0} \|g^\varepsilon - g\|_{L^1(\mathcal{L}_1\times\nu)} = 0$.

In particular, there exists a sequence $\varepsilon_n \searrow 0$ and a function $G \in L^1(\mathcal{L}_1 \times \nu)$ such that the inequality $g^{\varepsilon_n} \leq G$ holds $(\mathcal{L}_1 \times \nu)$-a.e. for every $n \in \mathbb{N}$. This grants that

$$\left|\frac{f_{t+\varepsilon_n} - f_t}{\varepsilon_n}\right| \leq \frac{1}{\varepsilon_n}\int_t^{t+\varepsilon_n} g_r \,\mathrm{d}r = g_t^{\varepsilon_n} \leq G_t \text{ holds } \nu\text{-a.e.} \quad \text{for a.e. } t \in [0,1]. \tag{1.59}$$

The bound in (1.59) allows us to apply Lemma 1.3.22: up to a not relabeled subsequence, we have that $(f_{\cdot+\varepsilon_n} - f_\cdot)/\varepsilon_n$ weakly converges in $L^1(\mathcal{L}_1 \times \nu)$ to some function $f' \in L^1(\mathcal{L}_1 \times \nu)$. Moreover, simple computations yield

$$\int_s^t \frac{f_{r+\varepsilon_n} - f_r}{\varepsilon_n}\,\mathrm{d}r = \fint_t^{t+\varepsilon_n} f_r \,\mathrm{d}r - \fint_s^{s+\varepsilon_n} f_r \,\mathrm{d}r \quad \text{for every } (t,s) \in \Delta. \tag{1.60}$$

The continuity of $r \mapsto f_r \in L^1(\nu)$ grants that the right hand side in (1.60) converges to $f_t - f_s$ in $L^1(\nu)$ as $n \to \infty$. On the other hand, for every $\ell \in L^\infty(\nu)$ it holds that

$$\begin{aligned}&\int \ell(y)\bigg(\int_s^t \frac{f_{r+\varepsilon_n} - f_r}{\varepsilon_n}\,\mathrm{d}r\bigg)(y)\,\mathrm{d}\nu(y) \\ &= \int \underbrace{\ell(y)\,\chi_{[s,t]}(r)}_{\in L^\infty(\mathcal{L}_1\times\nu)} \frac{f_{r+\varepsilon_n}(y) - f_r(y)}{\varepsilon_n} \times \mathrm{d}(\mathcal{L}_1 \times \nu)(r,y),\end{aligned}$$

which in turn converges to $\int \ell(y)\big(\int_s^t f'_r\,\mathrm{d}r\big)(y)\,\mathrm{d}\nu(y)$ as $n \to \infty$. In other words, we showed that $\int_s^t (f_{r+\varepsilon_n} - f_r)/\varepsilon_n\,\mathrm{d}r \rightharpoonup \int_s^t f'_r\,\mathrm{d}r$ weakly in $L^1(\nu)$. So by letting $n \to \infty$ in (1.60) we get

$$\int_s^t f'_r\,\mathrm{d}r = f_t - f_s \quad \text{for every } (t,s) \in \Delta.$$

Therefore Proposition 1.3.18 implies that f_t' is the strong derivative in $L^1(\nu)$ of the map $t \mapsto f_t$ for a.e. $t \in [0, 1]$. Finally, by recalling (1.56) we also conclude that (1.57) is verified. □

Lemma 1.3.24 *Let $h \in L^1(0, 1)$ be given. Then $h \in W^{1,1}(0, 1)$ if and only if there exists a function $g \in L^1(0, 1)$ such that*

$$h_t - h_s = \int_s^t g_r \, \mathrm{d}r \quad \textit{holds for } \mathcal{L}^2\textit{-a.e. } (t, s) \in \Delta. \tag{1.61}$$

Moreover, in such case it holds that $h' = g$.

Proof NECESSITY. Fix any family of convolution kernels $\rho_\varepsilon \in C_c^\infty(\mathbb{R})$, i.e. $\int \rho_\varepsilon(x) \, \mathrm{d}x = 1$, the support of ρ_ε is contained in $(-\varepsilon, \varepsilon)$ and $\rho_\varepsilon \geq 0$. Let us define $h^\varepsilon := h * \rho_\varepsilon$ for all $\varepsilon > 0$. Recall that $h^\varepsilon \in C_c^\infty(\mathbb{R})$ and that $(h^\varepsilon)' = (h') * \rho_\varepsilon$. Choose a sequence $\varepsilon_n \searrow 0$ and a negligible Borel set $N \subseteq [0, 1]$ such that $h_t^{\varepsilon_n} \to h_t$ as $n \to \infty$ for every $t \in [0, 1] \setminus N$. Given that we have the equality $h_t^{\varepsilon_n} - h_s^{\varepsilon_n} = \int_s^t (h^{\varepsilon_n})_r' \, \mathrm{d}r$ for every $n \in \mathbb{N}$ and $(t, s) \in \Delta$, we can finally conclude that $h_t - h_s = \int_s^t h_r' \, \mathrm{d}r$ for $\mathcal{L}^2$-a.e. $(t, s) \in \Delta$, proving (1.61) with $g = h'$.
SUFFICIENCY. By Fubini theorem, we see that for a.e. $\varepsilon > 0$ it holds that $h_{t+\varepsilon} - h_t = \int_t^{t+\varepsilon} g_r \, \mathrm{d}r$ for a.e. $t \in [0, 1]$. In particular, there is a sequence $\varepsilon_n \searrow 0$ such that $h_{t+\varepsilon_n} - h_t = \int_t^{t+\varepsilon_n} g_r \, \mathrm{d}r$ for every $n \in \mathbb{N}$ and for a.e. $t \in [0, 1]$. Now fix $\varphi \in C_c^\infty(0, 1)$. Then

$$\int \frac{\varphi_{t-\varepsilon_n} - \varphi_t}{\varepsilon_n} h_t \, \mathrm{d}t = \int \frac{h_{t+\varepsilon_n} - h_t}{\varepsilon_n} \varphi_t \, \mathrm{d}t = \int \left(⨍_t^{t+\varepsilon_n} g_r \, \mathrm{d}r \right) \varphi_t \, \mathrm{d}t. \tag{1.62}$$

By applying the dominated convergence theorem, we finally deduce by letting $n \to \infty$ in the equation (1.62) that $-\int \varphi_t' \, h_t \, \mathrm{d}t = \int g_t \, \varphi_t \, \mathrm{d}t$. Hence $h \in W^{1,1}(0, 1)$ and $h' = g$. □

Bibliographical Remarks

Much of the material of Sect. 1.1 can be found e.g. in the authoritative monograph [11].

The definitions and results about (absolutely) continuous curves presented in Sect. 1.2 are mostly taken from the book [12]; the above proof of Theorem 1.2.5 can be found in [2].

The results in Sect. 1.3 about the Bochner integral are taken from [14].

Chapter 2
Sobolev Calculus on Metric Measure Spaces

Several different approaches to the theory of weakly differentiable functions over abstract metric measure spaces made their appearance in the literature throughout the last twenty years. Amongst them, we shall mainly follow the one (based upon the concept of *test plan*) that has been proposed by Ambrosio, Gigli and Savaré. The whole Sect. 2.1 is devoted to the definition of such notion of *Sobolev space* $W^{1,2}(\mathrm{X})$ and to its most important properties.

Furthermore, in Sect. 2.2 we describe two alternative definitions of Sobolev space, which are both completely equivalent to the previous one: the approach of Cheeger and that of Shanmugalingam, discussed in Sects. 2.2.1 and 2.2.2 respectively. The former is obtained via relaxation of the local Lipschitz constant, while the latter relies upon the potential-theoretic notion of 2-modulus of curves.

2.1 Sobolev Space via Test Plans

2.1.1 *Test Plans*

Let $(\mathrm{X}, \mathsf{d}, \mathfrak{m})$ be a fixed metric measure space.

For every $t \in [0, 1]$, we define the *evaluation map* at time t as follows:

$$\begin{aligned} \mathrm{e}_t : C([0,1], \mathrm{X}) &\longrightarrow \mathrm{X}, \\ \gamma &\longmapsto \gamma_t. \end{aligned} \tag{2.1}$$

It is clear that each map e_t is 1-Lipschitz.

In Sect. 2.1.2, a special role will be played by the class of Borel probability measures that we are now going to describe: the so-called 'test plans'.

N. Gigli, E. Pasqualetto, *Lectures on Nonsmooth Differential Geometry*, SISSA Springer Series 2, https://doi.org/10.1007/978-3-030-38613-9_2

Definition 2.1.1 (Test Plan) A probability measure $\pi \in \mathscr{P}\big(C([0,1],\mathrm{X})\big)$ is said to be a *test plan* on X provided the following two properties are satisfied:

i) There exists a constant $C > 0$ such that $(\mathrm{e}_t)_*\pi \leq C\mathfrak{m}$ for every $t \in [0,1]$.
ii) It holds that $\int \mathrm{KE}(\gamma)\,\mathrm{d}\pi(\gamma) = \int_0^1\int |\dot\gamma_t|^2\,\mathrm{d}\pi(\gamma)\,\mathrm{d}t < +\infty$.

The least constant $C > 0$ that can be chosen in i) is called *compression constant* of π and is denoted by $\mathrm{Comp}(\pi)$.

It follows from $ii)$ that test plans must be concentrated on absolutely continuous curves.

Example 2.1.2 Let us fix a measure $\mu \in \mathscr{P}(\mathrm{X})$ with $\mu \leq C\mathfrak{m}$ for some $C > 0$. Let us denote by $\mathsf{Const} : \mathrm{X} \to C([0,1],\mathrm{X})$ the function sending any point $x \in \mathrm{X}$ to the curve identically equal to x. Then $\mathsf{Const}_*\mu$ turns out to be a test plan on X. ■

Example 2.1.3 Suppose to have a Borel map $F : \mathrm{X} \times [0,1] \to \mathrm{X}$, called *flow*, with the following properties: there exist two constants $L, C > 0$ such that

$$\begin{aligned} F_\cdot(x) : t \mapsto F_t(x) \quad &\text{is } L\text{-Lipschitz for every } x \in \mathrm{X}, \\ (F_t)_*\mathfrak{m} \leq C\mathfrak{m} \quad &\text{for every } t \in [0,1]. \end{aligned} \tag{2.2}$$

The second requirement means, in a sense, that the mass is well-distributed by the flow F.

Now consider any measure $\mu \in \mathscr{P}(\mathrm{X})$ such that $\mu \leq c\,\mathfrak{m}$ for some $c > 0$. Then

$$\pi := (F_\cdot)_*\mu \quad \text{is a test plan on X.} \tag{2.3}$$

Its verification is straightforward: $(\mathrm{e}_t)_*\pi = (\mathrm{e}_t)_*(F_\cdot)_*\mu = (F_t)_*\mu \leq c\,(F_t)_*\mathfrak{m} \leq c\,C\mathfrak{m}$ shows the first property of test plans, while the fact that $\big|\dot F_t(x)\big| \leq L$ holds for every $x \in \mathrm{X}$ and almost every $t \in [0,1]$ grants the second one. Therefore (2.3) is proved. ■

Proposition 2.1.4 *Let π be a test plan on* X *and $p \in [1,\infty)$. Then for every $f \in L^p(\mathfrak{m})$ the map $[0,1] \ni t \mapsto f \circ \mathrm{e}_t \in L^p(\pi)$ is continuous.*

Proof First of all, one has that $\int |f \circ \mathrm{e}_t|^p\,\mathrm{d}\pi \leq \mathrm{Comp}(\pi)\int |f|^p\,\mathrm{d}\mathfrak{m}$ for every $f \in L^p(\mathfrak{m})$. Given any $g \in C_b(\mathrm{X}) \cap L^p(\mathfrak{m})$, it holds that $\big|g(\gamma_s) - g(\gamma_t)\big|^p \to 0$ as $s \to t$ for every $\gamma \in C([0,1],\mathrm{X})$ and $|g \circ \mathrm{e}_s - g \circ \mathrm{e}_t|^p \leq 2\,\|g\|^p_{C_b(\mathrm{X})} \in L^\infty(\pi)$, so that $\lim_{s\to t}\int \big|g \circ \mathrm{e}_s - g \circ \mathrm{e}_t\big|^p\,\mathrm{d}\pi = 0$ by the dominated convergence theorem. This guarantees that

$$\begin{aligned} \varlimsup_{s\to t}\|f \circ \mathrm{e}_s - f \circ \mathrm{e}_t\|_{L^p(\pi)} &\leq \varlimsup_{s\to t}\Big[\|f \circ \mathrm{e}_s - g \circ \mathrm{e}_s\|_{L^p(\pi)} + \|g \circ \mathrm{e}_t - f \circ \mathrm{e}_t\|_{L^p(\pi)}\Big] \\ &\leq 2\,\mathrm{Comp}(\pi)^{1/p}\,\|f - g\|_{L^p(\mathfrak{m})}, \end{aligned}$$

whence $\|f \circ \mathrm{e}_s - f \circ \mathrm{e}_t\|_{L^p(\pi)} \to 0$ as $s \to t$ by density of $C_b(\mathrm{X}) \cap L^p(\mathfrak{m})$ in $L^p(\mathfrak{m})$, which can be proved by suitably adapting the proof of Proposition 1.1.18. □

Let $t, s \in [0, 1]$ be fixed. Then we define the map $\mathsf{Restr}_t^s : C([0, 1], \mathrm{X}) \to C([0, 1], \mathrm{X})$ as

$$\mathsf{Restr}_t^s(\gamma)_r := \gamma_{(1-r)t+rs} \quad \text{for every } \gamma \in C([0,1],\mathrm{X}) \text{ and } r \in [0,1]. \tag{2.4}$$

We call Restr_t^s the *restriction operator* between the times t and s.

Exercise 2.1.5 Prove that the map Restr_t^s is continuous. ■

Lemma 2.1.6 *Let π be a test plan on X. Then:*

i) *For any $\Gamma \subseteq C([0, 1], \mathrm{X})$ Borel with $\pi(\Gamma) > 0$, it holds that $\pi(\Gamma)^{-1}\pi_{|\Gamma}$ is a test plan.*
ii) *For any $t, s \in [0, 1]$, the measure $(\mathsf{Restr}_t^s)_*\pi$ is a test plan on X.*

Proof In order to prove i), just observe that

$$(\mathrm{e}_t)_*\big(\pi(\Gamma)^{-1}\pi_{|\Gamma}\big) \leq \pi(\Gamma)^{-1}(\mathrm{e}_t)_*\pi \leq \mathrm{Comp}(\pi)\,\pi(\Gamma)^{-1}\,\mathfrak{m},$$

$$\int_0^1\!\!\int |\dot\gamma_t|^2\,\mathrm{d}\big(\pi(\Gamma)^{-1}\pi_{|\Gamma}\big)(\gamma)\,\mathrm{d}t = \pi(\Gamma)^{-1}\int_0^1\!\!\int_\Gamma |\dot\gamma_t|^2\,\mathrm{d}\pi(\gamma)\,\mathrm{d}t < +\infty.$$

To prove ii), notice that if $\gamma \in C([0, 1], \mathrm{X})$ is absolutely continuous, then $\sigma := \mathsf{Restr}_t^s(\gamma)$ is absolutely continuous as well and satisfies $|\dot\sigma_r| = |s - t||\dot\gamma_{(1-r)t+rs}|$ for a.e. $r \in [0, 1]$. Hence

$$(\mathrm{e}_r)_*(\mathsf{Restr}_t^s)_*\pi = (\mathrm{e}_r \circ \mathsf{Restr}_t^s)_*\pi = (\mathrm{e}_{(1-r)t+rs})_*\pi \leq \mathrm{Comp}(\pi)\,\mathfrak{m},$$

$$\int_0^1\!\!\int |\dot\sigma_r|^2\,\mathrm{d}\big((\mathsf{Restr}_t^s)_*\pi\big)(\sigma)\,\mathrm{d}r \leq |s-t|\int_0^1\!\!\int |\dot\gamma_r|^2\,\mathrm{d}\pi(\gamma)\,\mathrm{d}r < +\infty,$$

which concludes the proof of the statement. □

2.1.2 *Definition of Sobolev Space*

The definition of Sobolev function (via test plans) is strongly inspired by the following fact:

Remark 2.1.7 Consider $f \in C^1(\mathbb{R}^n)$ and $G \in C(\mathbb{R}^n)$. Then $G \geq |\mathrm{d}f|$ if and only if

$$\big|f(\gamma_1) - f(\gamma_0)\big| \leq \int_0^1 G(\gamma_t)|\gamma_t'|\,\mathrm{d}t \quad \text{for every } \gamma \in C^1([0,1],\mathbb{R}^n). \tag{2.5}$$

This means that the map $|df|$ can be characterised, in a purely variational way, as the least continuous function $G : \mathbb{R}^n \to \mathbb{R}$ for which (2.5) is satisfied. ■

With the previous observation in mind, we can provide the following definition of Sobolev function for general metric measure spaces (by relying upon the notion of test plan):

Definition 2.1.8 (Sobolev Class) The *Sobolev class* $\mathrm{S}^2(\mathrm{X})$ is defined as the space of all Borel functions $f : \mathrm{X} \to \mathbb{R}$ that satisfy the following property: there exists a function $G \in L^2(\mathfrak{m})$ with $G \geq 0$ such that

$$\int \big|f(\gamma_1) - f(\gamma_0)\big| \, \mathrm{d}\boldsymbol{\pi}(\gamma) \leq \int_0^1 \int G(\gamma_t)|\dot{\gamma}_t| \, \mathrm{d}\boldsymbol{\pi}(\gamma) \, \mathrm{d}t \quad \text{for every test plan } \boldsymbol{\pi} \text{ on X.} \tag{2.6}$$

Any such G is said to be a *weak upper gradient* for f.

Remark 2.1.9 In giving Definition 2.1.8 we implicitly used the fact that

$$C([0,1],\mathrm{X}) \times [0,1] \ni (\gamma, t) \longmapsto G(\gamma_t)|\dot{\gamma}_t| \quad \text{is Borel.} \tag{2.7}$$

The map $\mathrm{e} : C([0,1],\mathrm{X}) \times [0,1] \to \mathrm{X}$ sending (γ, t) to γ_t can be easily seen to be continuous, whence $G \circ \mathrm{e}$ is Borel. Since the map in (2.7) is nothing but $G \circ \mathrm{e}\,\mathsf{ms}$—where ms is has been defined and proven to be Borel in Remark 1.2.6—we conclude that (2.7) is satisfied. ■

Remark 2.1.10 We claim that

$$f \circ \mathrm{e}_1 - f \circ \mathrm{e}_0 \in L^1(\boldsymbol{\pi}) \quad \text{for every } f \in \mathrm{S}^2(\mathrm{X}). \tag{2.8}$$

In order to prove (2.8), by (2.6) it suffices to notice that the Hölder inequality gives

$$\begin{aligned} \left(\int_0^1 \int G(\gamma_t)|\dot{\gamma}_t| \, \mathrm{d}\boldsymbol{\pi}(\gamma) \, \mathrm{d}t\right)^2 &\leq \left(\int_0^1 \int G^2 \circ \mathrm{e}_t \, \mathrm{d}\boldsymbol{\pi} \, \mathrm{d}t\right)\left(\int_0^1 \int |\dot{\gamma}_t|^2 \, \mathrm{d}\boldsymbol{\pi}(\gamma) \, \mathrm{d}t\right) \\ &\leq \mathrm{Comp}(\boldsymbol{\pi}) \, \|G\|^2_{L^2(\mathfrak{m})} \int_0^1 \int |\dot{\gamma}_t|^2 \, \mathrm{d}\boldsymbol{\pi}(\gamma) \, \mathrm{d}t < +\infty. \end{aligned}$$

In particular,

$$\text{the map } L^2(\mathfrak{m}) \ni G \mapsto \int_0^1 \int G(\gamma_t)|\dot{\gamma}_t| \, \mathrm{d}\boldsymbol{\pi}(\gamma) \, \mathrm{d}t \text{ is linear and continuous.} \tag{2.9}$$

■

Proposition 2.1.11 *Let $f \in \mathrm{S}^2(\mathrm{X})$ be fixed. Then the set of all weak upper gradients of f is closed and convex in $L^2(\mathfrak{m})$. In particular, there exists a unique weak upper gradient of f having minimal $L^2(\mathfrak{m})$-norm.*

Proof Convexity is trivial. To prove closedness, fix a sequence $(G_n)_n \subseteq L^2(\mathfrak{m})$ of weak upper gradients of f that $L^2(\mathfrak{m})$-converges to some $G \in L^2(\mathfrak{m})$. Hence (2.9) grants that

$$\int \big|f(\gamma_1) - f(\gamma_0)\big| \, \mathrm{d}\pi(\gamma) \leq \int_0^1 \int G_n(\gamma_t)|\dot{\gamma}_t| \, \mathrm{d}\pi(\gamma) \, \mathrm{d}t \overset{n}{\longrightarrow} \int_0^1 \int G(\gamma_t)|\dot{\gamma}_t| \, \mathrm{d}\pi(\gamma) \, \mathrm{d}t,$$

proving that G is a weak upper gradient of f. Hence the set of weak upper gradients of f is closed. Since $L^2(\mathfrak{m})$ is Hilbert, even the last statement follows. □

Definition 2.1.12 (Minimal Weak Upper Gradient) Let $f \in \mathrm{S}^2(\mathrm{X})$. Then the unique weak upper gradient of f having minimal norm is called *minimal weak upper gradient* of f and is denoted by $|Df| \in L^2(\mathfrak{m})$.

An important property of weak upper gradients is given by their lower semicontinuity:

Proposition 2.1.13 *Let the sequence $(f_n)_n \subseteq \mathrm{S}^2(\mathrm{X})$ satisfy $f_n(x) \to f(x)$ for a.e. $x \in \mathrm{X}$, for some Borel map $f : \mathrm{X} \to \mathbb{R}$. Let $G_n \in L^2(\mathfrak{m})$ be a weak upper gradient of f_n for every $n \in \mathbb{N}$. Suppose that $G_n \rightharpoonup G$ weakly in $L^2(\mathfrak{m})$, for some $G \in L^2(\mathfrak{m})$. Then $f \in \mathrm{S}^2(\mathrm{X})$ and G is a weak upper gradient of f.*

Proof First of all, it holds that $f_n(\gamma_1) - f_n(\gamma_0) \overset{n}{\to} f(\gamma_1) - f(\gamma_0)$ for π-a.e. γ. Moreover, the map sending $H \in L^2(\mathfrak{m})$ to $\int_0^1 \int H(\gamma_t)|\dot{\gamma}_t| \, \mathrm{d}\pi(\gamma) \, \mathrm{d}t$ is strongly continuous and linear by Remark 2.1.10, thus it is also weakly continuous. Hence Fatou's lemma yields

$$\begin{aligned}
\int \big|f(\gamma_1) - f(\gamma_0)\big| \, \mathrm{d}\pi(\gamma) &\leq \varliminf_{n\to\infty} \int \big|f_n(\gamma_1) - f(\gamma_0)\big| \, \mathrm{d}\pi(\gamma) \\
&\leq \lim_{n\to\infty} \int_0^1 \int G_n(\gamma_t)|\dot{\gamma}_t| \, \mathrm{d}\pi(\gamma) \, \mathrm{d}t \\
&= \int_0^1 \int G(\gamma_t)|\dot{\gamma}_t| \, \mathrm{d}\pi(\gamma) \, \mathrm{d}t,
\end{aligned}$$

which shows that $f \in \mathrm{S}^2(\mathrm{X})$ and that G is a weak upper gradient for f. □

Exercise 2.1.14 Given a metric space (X, d) and $\alpha \in (0, 1)$, we set the distance d_α on X as

$$\mathsf{d}_\alpha(x, y) := \mathsf{d}(x, y)^\alpha \quad \text{for every } x, y \in \mathrm{X}.$$

Prove that the metric space $(\mathrm{X}, \mathsf{d}_\alpha)$, which is called the *snowflaking* of (X, d), has the following property: if a curve γ is d_α-absolutely continuous, then it is constant.

Now consider any Borel measure $\mathfrak{m}$ on (X, d). Since d and d_α induce the same topology on X, we have that $\mathfrak{m}$ is also a Borel measure on $(\mathrm{X}, \mathsf{d}_\alpha)$. Prove that any Borel map on X belongs to $\mathrm{S}^2(\mathrm{X}, \mathsf{d}_\alpha, \mathfrak{m})$ and has null minimal weak upper gradient. ■

Those elements of the Sobolev class $\mathrm{S}^2(\mathrm{X})$ that are also 2-integrable constitute the Sobolev space $W^{1,2}(\mathrm{X})$, which comes with a natural Banach space structure:

Definition 2.1.15 (Sobolev Space) We define the *Sobolev space* $W^{1,2}(\mathrm{X})$ associated to the metric measure space $(\mathrm{X}, \mathsf{d}, \mathfrak{m})$ as $W^{1,2}(\mathrm{X}) := L^2(\mathfrak{m}) \cap \mathrm{S}^2(\mathrm{X})$. Moreover, we define

$$\|f\|_{W^{1,2}(\mathrm{X})} := \sqrt{\|f\|^2_{L^2(\mathfrak{m})} + \big\||Df|\big\|^2_{L^2(\mathfrak{m})}} \quad \text{for every } f \in W^{1,2}(\mathrm{X}). \tag{2.10}$$

Remark 2.1.16 It is trivial to check that

$$\begin{aligned} \big|D(\lambda f)\big| &= |\lambda||Df| && \text{for every } f \in \mathrm{S}^2(\mathrm{X}) \text{ and } \lambda \in \mathbb{R}, \\ \big|D(f+g)\big| &\le |Df| + |Dg| && \text{for every } f, g \in \mathrm{S}^2(\mathrm{X}). \end{aligned} \tag{2.11}$$

In particular, $\mathrm{S}^2(\mathrm{X})$ is a vector space, so accordingly $W^{1,2}(\mathrm{X})$ is a vector space as well. ■

Theorem 2.1.17 *The space* $\big(W^{1,2}(\mathrm{X}), \|\cdot\|_{W^{1,2}(\mathrm{X})}\big)$ *is a Banach space.*

Proof First of all, we claim that $\mathrm{S}^2(\mathrm{X}) \ni f \mapsto \big\||Df|\big\|_{L^2(\mathfrak{m})} \in \mathbb{R}$ is a seminorm: this follows by taking the $L^2(\mathfrak{m})$-norm in (2.11). Then also $\|\cdot\|_{W^{1,2}(\mathrm{X})}$ is a seminorm. Actually, it is a norm because $\|f\|_{W^{1,2}(\mathrm{X})} = 0$ implies $\|f\|_{L^2(\mathfrak{m})} = 0$ and accordingly $f = 0$. It thus remains to show that $W^{1,2}(\mathrm{X})$ is complete. To this aim, fix a Cauchy sequence $(f_n)_n \subseteq W^{1,2}(\mathrm{X})$. In particular, such sequence is $L^2(\mathfrak{m})$-Cauchy, so that it has an $L^2(\mathfrak{m})$-limit f. Moreover, the sequence $\big(|Df_n|\big)_n$ is bounded in $L^2(\mathfrak{m})$. Hence there exists a subsequence $(f_{n_k})_k$ such that

$$\begin{aligned} |Df_{n_k}| &\rightharpoonup G && \text{weakly in } L^2(\mathfrak{m}), \text{ for some } G \in L^2(\mathfrak{m}), \\ f_{n_k}(x) &\overset{k}{\longrightarrow} f(x) && \text{for } \mathfrak{m}\text{-a.e. } x \in \mathrm{X}. \end{aligned} \tag{2.12}$$

Then Proposition 2.1.13 grants that $f \in W^{1,2}(\mathrm{X})$ and that G is a weak upper gradient for f. Finally, with a similar argument we get $\big\||D(f_{n_k} - f)|\big\|_{L^2(\mathfrak{m})} \le \underline{\lim}_m \big\||D(f_{n_k} - f_{n_m})|\big\|_{L^2(\mathfrak{m})}$ for every $k \in \mathbb{N}$. By recalling that $(f_n)_n$ is $W^{1,2}(\mathrm{X})$-Cauchy, we thus conclude that

$$\overline{\lim_{k\to\infty}} \big\||D(f_{n_k} - f)|\big\|_{L^2(\mathfrak{m})} \le \overline{\lim_{k\to\infty}}\ \underline{\lim_{m\to\infty}} \big\||D(f_{n_k} - f_{n_m})|\big\|_{L^2(\mathfrak{m})} = 0,$$

proving that $f_{n_k} \to f$ in $W^{1,2}(\mathrm{X})$, which in turn grants that $f_n \to f$ in $W^{1,2}(\mathrm{X})$. □

Remark 2.1.18 In general, $W^{1,2}(\mathrm{X})$ is not a Hilbert space. For instance, $W^{1,2}(\mathbb{R}^n, \mathsf{d}, \mathcal{L}^n)$ is not Hilbert for any distance d induced by a norm not coming from a scalar product. ■

Proposition 2.1.19 *Let $(f_n)_n \subseteq \mathrm{S}^2(\mathrm{X})$ be given. Suppose that there exists $f : \mathrm{X} \to \mathbb{R}$ Borel such that $f(x) = \lim_n f_n(x)$ for $\mathfrak{m}$-a.e. $x \in \mathrm{X}$. Then $\big\||Df|\big\|_{L^2(\mathfrak{m})} \leq \underline{\lim}_n \big\||Df_n|\big\|_{L^2(\mathfrak{m})}$, where we adopt the convention that $\big\||Df|\big\|_{L^2(\mathfrak{m})} := +\infty$ whenever $f \notin \mathrm{S}^2(\mathrm{X})$.*

In particular, if a sequence $(g_n)_n \subseteq W^{1,2}(\mathrm{X})$ is $L^2(\mathfrak{m})$-converging to some limit $g \in L^2(\mathfrak{m})$, then it holds that $\big\||Dg|\big\|_{L^2(\mathfrak{m})} \leq \underline{\lim}_n \big\||Dg_n|\big\|_{L^2(\mathfrak{m})}$.

Proof The case $\underline{\lim}_n \big\||Df_n|\big\|_{L^2(\mathfrak{m})} = +\infty$ is trivial, then assume that such liminf is finite. Up to subsequence, we can also assume that such liminf is actually a limit. This grants that the sequence $\big(|Df_n|\big)_n$ is bounded in $L^2(\mathfrak{m})$, thus (up to subsequence) we have that $|Df_n| \rightharpoonup G$ weakly in $L^2(\mathfrak{m})$ for some $G \in L^2(\mathfrak{m})$. Hence Proposition 2.1.13 grants that $f \in \mathrm{S}^2(\mathrm{X})$ and G is a weak upper gradient for f, so that $\big\||Df|\big\|_{L^2(\mathfrak{m})} \leq \|G\|_{L^2(\mathfrak{m})} \leq \underline{\lim}_n \big\||Df_n|\big\|_{L^2(\mathfrak{m})}$.

For the last assertion, first take a subsequence such that $\underline{\lim}_n \big\||Dg_n|\big\|_{L^2(\mathfrak{m})}$ is actually a limit and then note that there is a further subsequence $(g_{n_k})_k$ such that $g(x) = \lim_k g_{n_k}(x)$ holds for $\mathfrak{m}$-a.e. $x \in \mathrm{X}$. To conclude, apply the first part of the statement. □

Proposition 2.1.20 *Let $f \in \mathrm{S}^2(\mathrm{X})$ be given. Consider a weak upper gradient $G \in L^2(\mathfrak{m})$ of f. Then for every test plan $\boldsymbol{\pi}$ on X and for every $t, s \in [0, 1]$ with $s < t$ it holds that*

$$\big|f(\gamma_t) - f(\gamma_s)\big| \leq \int_s^t G(\gamma_r)|\dot{\gamma}_r| \, \mathrm{d}r \quad \text{for } \boldsymbol{\pi}\text{-a.e. } \gamma \in C([0, 1], \mathrm{X}). \tag{2.13}$$

Proof We argue by contradiction: suppose the existence of $t, s \in [0, 1]$ with $s < t$ and of a Borel set $\Gamma \subseteq C([0, 1], \mathrm{X})$ with $\boldsymbol{\pi}(\Gamma) > 0$ such that $\big|f(\gamma_t) - f(\gamma_s)\big| > \int_s^t G(\gamma_r)|\dot{\gamma}_r| \, \mathrm{d}r$ holds for every $\gamma \in \Gamma$. Lemma 2.1.6 grants that the measure $\tilde{\boldsymbol{\pi}} := (\mathsf{Restr}_s^t)_*\big(\boldsymbol{\pi}(\Gamma)^{-1} \boldsymbol{\pi}_{|\Gamma}\big)$ is a test plan on X, thus accordingly

$$\begin{aligned} \boldsymbol{\pi}(\Gamma)^{-1} & \int_\Gamma \big|f(\gamma_t) - f(\gamma_s)\big| \, \mathrm{d}\boldsymbol{\pi}(\gamma) \\ &= \int \big|f(\sigma_1) - f(\sigma_0)\big| \, \mathrm{d}\tilde{\boldsymbol{\pi}}(\sigma) \leq \int_0^1 \int G(\sigma_r)|\dot{\sigma}_r| \, \mathrm{d}\tilde{\boldsymbol{\pi}}(\sigma) \, \mathrm{d}r \\ &= \boldsymbol{\pi}(\Gamma)^{-1} \int_s^t \int_\Gamma G(\gamma_r)|\dot{\gamma}_r| \, \mathrm{d}\boldsymbol{\pi}(\gamma) \, \mathrm{d}r, \end{aligned}$$

which leads to a contradiction. Therefore the statement is achieved. □

We are in a position to prove some alternative characterisations of weak upper gradients:

Theorem 2.1.21 *Let* $(\mathrm{X}, \mathsf{d}, \mathfrak{m})$ *be a metric measure space as in* (1.21)*. Let us fix a Borel function* $f : \mathrm{X} \to \mathbb{R}$*. Let* $G \in L^2(\mathfrak{m})$ *satisfy* $G \geq 0$ $\mathfrak{m}$*-a.e. Then the following are equivalent:*

i) $f \in \mathrm{S}^2(\mathrm{X})$ *and* G *is a weak upper gradient of* f*.*
ii) *For any test plan* $\boldsymbol{\pi}$*, we have that* $t \mapsto f \circ \mathrm{e}_t - f \circ \mathrm{e}_0 \in L^1(\boldsymbol{\pi})$ *is AC. For a.e.* $t \in [0, 1]$*, there exists the strong* $L^1(\boldsymbol{\pi})$*-limit of* $(f \circ \mathrm{e}_{t+h} - f \circ \mathrm{e}_t)/h$ *as* $h \to 0$*. Such limit, denoted by* $\mathsf{Der}_{\boldsymbol{\pi}}(f)_t \in L^1(\boldsymbol{\pi})$*, satisfies* $\left|\mathsf{Der}_{\boldsymbol{\pi}}(f)_t\right|(\gamma) \leq G(\gamma_t)|\dot{\gamma}_t|$ *for* $(\boldsymbol{\pi} \times \mathcal{L}_1)$*-a.e.* (γ, t)*.*
iii) *For every test plan* $\boldsymbol{\pi}$*, we have for* $\boldsymbol{\pi}$*-a.e.* γ *that* $f \circ \gamma$ *belongs to* $W^{1,1}(0, 1)$ *and that the inequality* $\left|(f \circ \gamma)'_t\right| \leq G(\gamma_t)|\dot{\gamma}_t|$ *holds for a.e.* $t \in [0, 1]$*.*

If the above hold, then the equality $\mathsf{Der}_{\boldsymbol{\pi}}(f)_t(\gamma) = (f \circ \gamma)'_t$ *is verified for* $(\boldsymbol{\pi} \times \mathcal{L}_1)$*-a.e.* (γ, t)*.*

Proof The proof goes as follows:

i) $\Longrightarrow$ ii) We have that $\left|f(\gamma_t) - f(\gamma_s)\right| \leq \int_s^t G(\gamma_r)|\dot{\gamma}_r| \, \mathrm{d}r$ is satisfied for every $(t, s) \in \Delta$ and for $\boldsymbol{\pi}$-a.e. γ by Proposition 2.1.20. Since the map $(\gamma, t) \mapsto G(\gamma_t)|\dot{\gamma}_t|$ belongs to $L^1(\boldsymbol{\pi} \times \mathcal{L}_1)$ by Remark 2.1.10 and Remark 1.2.6, we obtain ii) by applying Proposition 1.3.23.

ii) $\Longrightarrow$ iii) By Fubini's theorem, one has for $\boldsymbol{\pi}$-a.e. γ that $f(\gamma_t) - f(\gamma_s) = \int_s^t \mathsf{Der}_{\boldsymbol{\pi}}(f)_r(\gamma) \, \mathrm{d}r$ holds for $\mathcal{L}^2$-a.e. $(t, s) \in \Delta$, whence iii) stems from Lemma 1.3.24. Further, for $\boldsymbol{\pi}$-a.e. γ we have

$$\int_s^t (f \circ \gamma)'_r \, \mathrm{d}r = f(\gamma_t) - f(\gamma_s) = \int_s^t \mathsf{Der}_{\boldsymbol{\pi}}(f)_r(\gamma) \, \mathrm{d}r \quad \text{for } \mathcal{L}^2\text{-a.e. } (t, s) \in \Delta,$$

which in turn implies the last statement of the theorem.

iii) $\Longrightarrow$ i) Fix a test plan $\boldsymbol{\pi}$ on X. Choose a point $\bar{x} \in \mathrm{X}$ and a sequence of 1-Lipschitz functions $(\eta_n)_n \subseteq C_b(\mathrm{X})$ such that $\eta_n = 1$ on $B_n(\bar{x})$ and $\mathrm{spt}(\eta_n) \subseteq B_{n+2}(\bar{x})$. Let us define

$$f^{mn} := \eta_n \min\left\{\max\{f, -m\}, m\right\} \quad \text{for every } m, n \in \mathbb{N}.$$

Fix $m, n \in \mathbb{N}$. Notice that $f^{mn} \circ \gamma \in W^{1,1}(0, 1)$ for $\boldsymbol{\pi}$-a.e. γ, so that Lemma 1.3.24 implies that

$$\int \left|f^{mn}(\gamma_t) - f^{mn}(\gamma_s)\right| \mathrm{d}\boldsymbol{\pi}(\gamma) \leq \iint_s^t \left|(f^{mn} \circ \gamma)'_r\right| \mathrm{d}r \, \mathrm{d}\boldsymbol{\pi}(\gamma) \quad \text{for } \mathcal{L}^2\text{-a.e. } (t, s) \in \Delta. \tag{2.14}$$

The right hand side in (2.14) is clearly continuous in (t, s). Since $f^{mn} \in L^1(\mathfrak{m})$, we deduce from Proposition 2.1.4 that also the left hand side is continuous in (t, s), thus in particular

$$\int \left|f^{mn}(\gamma_1) - f^{mn}(\gamma_0)\right| \mathrm{d}\pi(\gamma) \leq \iint_0^1 \left|(f^{mn} \circ \gamma)'_t\right| \mathrm{d}t \, \mathrm{d}\pi(\gamma). \tag{2.15}$$

Moreover, $\left|(f^{mn} \circ \gamma)'_t\right| \leq m\,|\dot{\gamma}_t|\,\chi_{B_n(\bar{x})^c}(\gamma_t) + \left|(f \circ \gamma)'_t\right|$ is satisfied for $(\pi \times \mathcal{L}_1)$-a.e. (γ, t) as a consequence of the Leibniz rule, whence

$$\begin{aligned}
&\int \left|f(\gamma_1) - f(\gamma_0)\right| \mathrm{d}\pi(\gamma) \\
&\leq \varliminf_{m\to\infty} \varliminf_{n\to\infty} \int \left|f^{mn}(\gamma_1) - f^{mn}(\gamma_0)\right| \mathrm{d}\pi(\gamma) \\
&\leq \varliminf_{m\to\infty} \varliminf_{n\to\infty} \iint_0^1 \left[m\,|\dot{\gamma}_t|\,\chi_{B_n(\bar{x})^c}(\gamma_t) + \left|(f \circ \gamma)'_t\right|\right] \mathrm{d}t \, \mathrm{d}\pi(\gamma) \\
&= \varliminf_{m\to\infty} \iint_0^1 \left|(f \circ \gamma)'_t\right| \mathrm{d}t \, \mathrm{d}\pi(\gamma) \leq \iint_0^1 G(\gamma_t)|\dot{\gamma}_t| \, \mathrm{d}t \, \mathrm{d}\pi(\gamma),
\end{aligned}$$

where the first line follows from Fatou lemma, the second one from (2.15) and the third one from the dominated convergence theorem. Therefore i) is proved. □

Remark 2.1.22 To be more precise, the last statement in Theorem 2.1.21 should be stated as follows: we can choose a Borel representative $F \in L^1(\mathcal{L}_1 \times \pi)$ of $t \mapsto \mathsf{Der}_\pi(f)_t \in L^1(\pi)$ in the sense of Proposition 1.3.19, since such map belongs to $L^1(\mathcal{L}_1; L^1(\pi))$ by ii). Analogously, we can choose a Borel representative $\tilde{F} \in L^1(\pi \times \mathcal{L}_1)$ of $\gamma \mapsto (t \mapsto (f \circ \gamma)'_t \in L^1(0, 1))$, which belongs to $L^1(\pi; L^1(\mathcal{L}_1))$ by iii). Then $F(t, \gamma) = \tilde{F}(\gamma, t)$ holds for $(\pi \times \mathcal{L}_1)$-a.e. (γ, t). ■

We point out some consequences of Theorem 2.1.21:

Proposition 2.1.23 *Let $f \in \mathrm{S}^2(\mathrm{X})$ be given. Consider two weak upper gradients $G_1, G_2 \in L^2(\mathfrak{m})$ of f. Then $G_1 \wedge G_2$ is a weak upper gradient of f.*

Proof By point ii) of Theorem 2.1.21 we have $\left|\mathrm{Der}_\pi(f)_t\right|(\gamma) \leq G_i(\gamma_t)|\dot{\gamma}_t|$ for $i = 1, 2$ and for $(\pi \times \mathcal{L}_1)$-a.e. (γ, t), thus also $\left|\mathrm{Der}_\pi(f)_t\right|(\gamma) \leq (G_1 \wedge G_2)(\gamma_t)|\dot{\gamma}_t|$ for $(\pi \times \mathcal{L}_1)$-a.e. (γ, t). Therefore $G_1 \wedge G_2$ is a weak upper gradient of f, again by Theorem 2.1.21. □

Corollary 2.1.24 *Let $f \in \mathrm{S}^2(\mathrm{X})$ be given. Let $G \in L^2(\mathfrak{m})$ be a weak upper gradient of f. Then it $\mathfrak{m}$-a.e. holds that $|Df| \leq G$. In other words, $|Df|$ is minimal also in the $\mathfrak{m}$-a.e. sense.*

Proof We argue by contradiction: suppose that there exists a weak upper gradient G of f such that $\mathfrak{m}(\{G < |Df|\}) > 0$. Hence the function $G \wedge |Df|$, which has an $L^2(\mathfrak{m})$-norm that is strictly smaller than $\||Df|\|_{L^2(\mathfrak{m})}$, is a weak upper gradient of f by Proposition 2.1.23. This leads to a contradiction, thus proving the statement. □

Given any $f \in \mathrm{LIP}(\mathrm{X})$, we define the *local Lipschitz constant* $\mathrm{lip}(f) : \mathrm{X} \to [0, +\infty)$ as

$$\mathrm{lip}(f)(x) := \varlimsup_{y \to x} \frac{|f(y) - f(x)|}{\mathsf{d}(y, x)} \quad \text{if } x \in \mathrm{X} \text{ is an accumulation point} \tag{2.16}$$

and $\mathrm{lip}(f)(x) := 0$ otherwise.

Remark 2.1.25 Given a Lipschitz function $f \in \mathrm{LIP}(\mathrm{X})$ and an AC curve $\gamma : [0, 1] \to \mathrm{X}$, it holds that $t \mapsto f(\gamma_t) \in \mathbb{R}$ is AC and satisfies

$$\left|(f \circ \gamma)'_t\right| \leq \mathrm{lip}(f)(\gamma_t)\,|\dot{\gamma}_t| \quad \text{for a.e. } t \in [0, 1]. \tag{2.17}$$

Indeed, to check that $f \circ \gamma$ is AC simply notice that $\left|f(\gamma_t) - f(\gamma_s)\right| \leq \mathrm{Lip}(f) \int_s^t |\dot{\gamma}_r|\,\mathrm{d}r$ holds for any $t, s \in [0, 1]$ with $s \leq t$. Now fix $t \in [0, 1]$ such that both $(f \circ \gamma)'_t$ and $|\dot{\gamma}_t|$ exist (which holds for a.e. t). If γ is constant in some neighbourhood of t, then (2.17) is trivially verified (since the left hand side is null). In the remaining case, we have that

$$\begin{aligned}
(f \circ \gamma)'_t &= \lim_{h \to 0} \frac{\left|(f \circ \gamma)_{t+h} - (f \circ \gamma)_t\right|}{|h|} \\
&\leq \varlimsup_{h \to 0} \frac{\left|f(\gamma_{t+h}) - f(\gamma_t)\right|}{\mathsf{d}(\gamma_{t+h}, \gamma_t)} \lim_{h \to 0} \frac{\mathsf{d}(\gamma_{t+h}, \gamma_t)}{|h|} \leq \mathrm{lip}(f)(\gamma_t)\,|\dot{\gamma}_t|,
\end{aligned}$$

thus obtaining (2.17). ■

Proposition 2.1.26 *Let $f \in \mathrm{LIP}_{bs}(\mathrm{X})$ be given. Then $f \in \mathrm{S}^2(\mathrm{X})$ and $|Df| \leq \mathrm{lip}(f) \leq \mathrm{Lip}(f)$ holds $\mathfrak{m}$-a.e. in* X.

Proof For any AC curve γ, we have that $\left|f(\gamma_1) - f(\gamma_0)\right| \leq \int_0^1 \mathrm{lip}(f)(\gamma_t)\,|\dot{\gamma}_t|\,\mathrm{d}t$ by (2.17). By integrating such inequality with respect to any test plan $\boldsymbol{\pi}$, we get the statement. □

We conclude the present subsection by proving that the Sobolev space $W^{1,2}(\mathrm{X})$ is separable whenever it is reflexive:

Theorem 2.1.27 *Let* $(\mathrm{X}, \mathsf{d}, \mathfrak{m})$ *be a metric measure space. Suppose that* $W^{1,2}(\mathrm{X})$ *is reflexive. Then* $W^{1,2}(\mathrm{X})$ *is separable.*

Proof Apply Lemma A.1 to $\mathbb{E}_1 = W^{1,2}(\mathrm{X})$, $\mathbb{E}_2 = L^2(\mathfrak{m})$ and i the inclusion $\mathbb{E}_1 \hookrightarrow \mathbb{E}_2$. □

2.1.3 Calculus Rules

Minimal weak upper gradients satisfy the following calculus rules:

Theorem 2.1.28 *The following properties hold:*

A) LOCALITY. *Let $f, g \in \mathrm{S}^2(\mathrm{X})$ be given. Then $|Df| = |Dg|$ holds $\mathfrak{m}$-a.e. in $\{f = g\}$.*

B) CHAIN RULE. *Let $f \in \mathrm{S}^2(\mathrm{X})$ be given.*

  B1) *If a Borel set $N \subseteq \mathbb{R}$ is $\mathcal{L}^1$-negligible, then $|Df| = 0$ holds $\mathfrak{m}$-a.e. in $f^{-1}(N)$.*

  B2) *If $\varphi : \mathbb{R} \to \mathbb{R}$ is a Lipschitz function, then $\varphi \circ f \in \mathrm{S}^2(\mathrm{X})$ and $|D(\varphi \circ f)| = |\varphi'| \circ f\,|Df|$ holds $\mathfrak{m}$-a.e., where $|\varphi'| \circ f$ is arbitrarily defined on $f^{-1}\big(\{t \in \mathbb{R} : \nexists\, \varphi'(t)\}\big)$.*

C) LEIBNIZ RULE. *Let $f, g \in \mathrm{S}^2(\mathrm{X}) \cap L^\infty(\mathfrak{m})$ be given. Then $fg \in \mathrm{S}^2(\mathrm{X}) \cap L^\infty(\mathfrak{m})$ and the inequality $|D(fg)| \leq |f||Dg| + |g||Df|$ holds $\mathfrak{m}$-a.e. in X.*

Proof We divide the proof into several steps:

STEP 1. First of all, we claim that

$$f \in \mathrm{S}^2(\mathrm{X}),\ \varphi \in \mathrm{LIP}(\mathbb{R}) \quad \Longrightarrow \quad \varphi \circ f \in \mathrm{S}^2(\mathrm{X}),\ |D(\varphi \circ f)| \leq \mathrm{Lip}(\varphi)|Df| \ \ \mathfrak{m}\text{-a.e.} \tag{2.18}$$

Indeed, the inequality $\int \big|(\varphi \circ f)(\gamma_1) - (\varphi \circ f)(\gamma_0)\big| \,\mathrm{d}\boldsymbol{\pi}(\gamma) \leq \mathrm{Lip}(\varphi) \iint_0^1 |Df|(\gamma_t)\,|\dot{\gamma}_t| \,\mathrm{d}t \,\mathrm{d}\boldsymbol{\pi}(\gamma)$ holds for any test plan $\boldsymbol{\pi}$, thus proving (2.18).

STEP 2. Given $h \in W^{1,1}(0,1)$ and $\varphi \in C^1(\mathbb{R}) \cap \mathrm{LIP}(\mathbb{R})$, we have that $\varphi \circ h \in W^{1,1}(0,1)$ and that $(\varphi \circ h)' = \varphi' \circ h\, h'$ holds a.e. in $(0,1)$. In order to prove it, call $h_\varepsilon := h * \rho_\varepsilon$ for all $\varepsilon > 0$, notice that $(\varphi \circ h_\varepsilon)' = \varphi' \circ h_\varepsilon\, h'_\varepsilon$ because h_ε is smooth and finally pass to the limit as $\varepsilon \searrow 0$.

STEP 3. We now claim that

$$f \in \mathrm{S}^2(\mathrm{X}),\ \varphi \in C^1(\mathbb{R}) \cap \mathrm{LIP}(\mathbb{R}) \quad \Longrightarrow \quad \big|D(\varphi \circ f)\big| \leq |\varphi'| \circ f\,|Df| \ \ \mathfrak{m}\text{-a.e.} \tag{2.19}$$

To prove it: fix a test plan $\boldsymbol{\pi}$. For $\boldsymbol{\pi}$-a.e. γ, it holds that $t \mapsto f(\gamma_t)$ belongs to $W^{1,1}(0,1)$ and that $\big|(f \circ \gamma)'_t\big| \leq |Df|(\gamma_t)|\dot{\gamma}_t|$ for a.e. $t \in [0,1]$, by Theorem 2.1.21. Hence STEP 2 grants that the function $t \mapsto (\varphi \circ f)(\gamma_t)$ is in $W^{1,1}(0,1)$ and satisfies

$$\big|(\varphi \circ f \circ \gamma)'_t\big| \leq \big(|\varphi'| \circ f\big)(\gamma_t)\,\big|(f \circ \gamma)'_t\big| \leq \big(|\varphi'| \circ f\big)(\gamma_t)\,|Df|(\gamma_t)\,|\dot{\gamma}_t| \quad \text{for a.e. } t \in [0,1],$$

whence $|D(\varphi \circ f)| \leq |\varphi'| \circ f\,|Df|$ holds $\mathfrak{m}$-a.e. by Theorem 2.1.21, thus proving (2.19).

STEP 4. We want to show that

$$f \in S^2(X),\ K \subseteq \mathbb{R} \text{ compact with } \mathcal{L}^1(K) = 0 \quad \Longrightarrow \quad |Df| = 0 \ \mathfrak{m}\text{-a.e. in } f^{-1}(K). \tag{2.20}$$

For any $n \in \mathbb{N}$, let us call $\psi_n := n\,\mathsf{d}(\cdot, K) \wedge 1$. Since the $\mathcal{L}^1$-measure of the ε-neighbourhood of K converges to 0 as $\varepsilon \searrow 0$, we deduce that $\mathcal{L}^1\big(\{\psi_n < 1\}\big) \to 0$ as $n \to \infty$. Now call φ_n the primitive of ψ_n equal to 0 in 0. Given that ψ_n is continuous and bounded, we have that φ_n is C^1 and Lipschitz. Moreover, it holds that φ_n uniformly converges to $\mathrm{id}_{\mathbb{R}}$ as $n \to \infty$, because

$$\big|\varphi_n(t) - t\big| \leq \int_0^t \big|\psi_n(s) - 1\big|\,\mathrm{d}s \leq \mathcal{L}^1\big(\{\psi_n < 1\}\big) \xrightarrow{n} 0.$$

In particular $\varphi_n \circ f \to f$ pointwise $\mathfrak{m}$-a.e., whence Proposition 2.1.19 gives

$$\begin{aligned}\int |Df|^2\,\mathrm{d}\mathfrak{m} &\leq \varliminf_{n\to\infty} \int |D(\varphi_n \circ f)|^2\,\mathrm{d}\mathfrak{m} \overset{(2.19)}{\leq} \varliminf_{n\to\infty} \int |\varphi_n'|^2 \circ f\,|Df|^2\,\mathrm{d}\mathfrak{m}\\ &\leq \int_{X\setminus f^{-1}(K)} |Df|^2\,\mathrm{d}\mathfrak{m},\end{aligned}$$

where in the last inequality we used the facts that $|\varphi_n'| \leq \|\psi_n\|_{L^\infty(\mathbb{R})} = 1$ and that $\varphi_n' = \psi_n = 0$ on K. This forces $|Df|$ to be $\mathfrak{m}$-a.e. null in the set $f^{-1}(K)$, obtaining (2.20).

STEP 5. We now use STEP 4 to prove B1). Take $f \in S^2(X)$ and $N \subseteq \mathbb{R}$ Borel with $\mathcal{L}^1(N) = 0$. There exists a measure $\tilde{\mathfrak{m}} \in \mathscr{P}(X)$ such that $\mathfrak{m} \ll \tilde{\mathfrak{m}} \ll \mathfrak{m}$, in other words having exactly the same negligible sets as $\mathfrak{m}$. For instance, choose any Borel partition $(B_n)_{n\geq 1}$ of the space X such that $0 < \mathfrak{m}(B_n) < +\infty$ for every $n \in \mathbb{N}$ and define

$$\tilde{\mathfrak{m}} := \sum_{n=1}^{\infty} \frac{1}{2^n\,\mathfrak{m}(B_n)}\,\mathfrak{m}|_{B_n}.$$

Now let us call $\mu := f_*\tilde{\mathfrak{m}}$. Since $\tilde{\mathfrak{m}}$ is finite, we have that μ is a Radon measure on $\mathbb{R}$, in particular μ is inner regular. Then there exists a sequence $(K_n)_n$ of compact subsets of N such that $\mu\big(N \setminus \bigcup_n K_n\big) = 0$, or equivalently $\mathfrak{m}\big(f^{-1}(N \setminus \bigcup_n K_n)\big) = 0$. Given that $|Df| = 0$ is verified $\mathfrak{m}$-a.e. in $\bigcup_n f^{-1}(K_n) = f^{-1}\big(\bigcup_n K_n\big)$ by (2.20), we thus conclude that B1) is satisfied.

STEP 6. We claim that

$$f \in \mathrm{S}^2(\mathrm{X}),\ \varphi \in \mathrm{LIP}(\mathbb{R}) \quad \Longrightarrow \quad |D(\varphi \circ f)| \leq |\varphi'| \circ f\, |Df| \ \ \mathfrak{m}\text{-a.e.} \qquad (2.21)$$

To prove it, call $\varphi_n := \varphi * \rho_{1/n}$. Up to a not relabeled subsequence, we have that $\varphi_n \to \varphi$ pointwise and $\varphi_n' \to \varphi'$ $\mathcal{L}^1$-a.e. Let us denote by N the ($\mathcal{L}^1$-negligible) set of $t \in \mathbb{R}$ such that either φ is not differentiable at t, or $\lim_n \varphi_n'(t)$ does not exist, or $\varphi'(t)$ and $\lim_n \varphi_n'(t)$ exist but are different. We know that $|D(\varphi_n \circ f)| \leq |\varphi_n'| \circ f\, |Df|$ holds $\mathfrak{m}$-a.e. for all $n \in \mathbb{N}$ by (2.19). Given that the inequality $|\varphi_n'| \circ f\, |Df| \leq \mathrm{Lip}(\varphi)|Df|$ is satisfied $\mathfrak{m}$-a.e. for all n, we can thus deduce that $|\varphi_n'| \circ f\, |Df| \to |\varphi'| \circ f\, |Df|$ in $L^2(\mathfrak{m})$ by B1) and dominated convergence theorem. Moreover, one has that $\varphi_n \circ f \to \varphi \circ f$ in the $\mathfrak{m}$-a.e. sense, whence $|D(\varphi \circ f)| \leq |\varphi'| \circ f\, |Df|$ holds $\mathfrak{m}$-a.e. by Proposition 2.1.13 and Corollary 2.1.24. This proves the claim (2.21).

STEP 7. We now deduce property B2) from (2.21). Suppose wlog that $\mathrm{Lip}(\varphi) = 1$. Let us define $\psi^{\pm}(t) := \pm t - \varphi(t)$ for every $t \in \mathbb{R}$. Then $\mathfrak{m}$-a.e. in the set $f^{-1}\big(\{\pm\varphi' \geq 0\}\big)$ we have

$$|Df| = \big|D(\pm f)\big| \leq \big|D(\varphi \circ f)\big| + \big|D(\psi^{\pm} \circ f)\big| \leq \big(|\varphi'| \circ f + |(\psi^{\pm})'| \circ f\big)\, |Df| = |Df|,$$

which forces $|D(\varphi \circ f)| = \pm\varphi' \circ f\, |Df|$ to hold $\mathfrak{m}$-a.e. in $f^{-1}\big(\{\pm\varphi' \geq 0\}\big)$, which is B2).

STEP 8. Property A) readily follows from B1): if $h := f - g$ then $\big||Df| - |Dg|\big| \leq |Dh| = 0$ holds $\mathfrak{m}$-a.e. in $h^{-1}(\{0\}) = \{f = g\}$ by B1).

STEP 9. We conclude by proving item C). Given two functions $h_1, h_2 \in W^{1,1}(0,1)$, we have that $h_1 h_2 \in W^{1,1}(0,1)$ and $(h_1 h_2)' = h_1' h_2 + h_1 h_2'$. Now fix $f, g \in \mathrm{S}^2(\mathrm{X}) \cap L^\infty(\mathfrak{m})$. Given any test plan $\boldsymbol{\pi}$, we have for $\boldsymbol{\pi}$-a.e. γ that $f \circ \gamma, g \circ \gamma \in W^{1,1}(0,1)$, so that $(fg) \circ \gamma \in W^{1,1}(0,1)$ as well. Further, $\big|(f \circ \gamma)_t'\big| \leq |Df|(\gamma_t)|\dot{\gamma}_t|$ and $\big|(g \circ \gamma)_t'\big| \leq |Dg|(\gamma_t)|\dot{\gamma}_t|$ for a.e. $t \in [0,1]$, whence

$$\big|((fg) \circ \gamma)_t'\big| \leq |f|(\gamma_t)\, \big|(g \circ \gamma)_t'\big| + |g|(\gamma_t)\, \big|(f \circ \gamma)_t'\big| \leq \underbrace{\big[|f||Dg| + |g||Df|\big]}_{\in L^2(\mathfrak{m})}(\gamma_t)\, |\dot{\gamma}_t|$$

is satisfied for a.e. $t \in [0,1]$. Therefore $fg \in \mathrm{S}^2(\mathrm{X})$ and $|f||Dg| + |g||Df|$ is a weak upper gradient of fg by Theorem 2.1.21, thus proving C). □

Remark 2.1.29 We present an alternative proof of property C) of Theorem 2.1.28: First of all, suppose that $f, g \geq c$ for some constant $c > 1$. Note that the function log is Lipschitz in $[c, +\infty)$, then choose any Lipschitz function $\varphi : \mathbb{R} \to \mathbb{R}$ that coincides with log in $[c, +\infty)$. Now call $C := \log\big(\|fg\|_{L^\infty(\mathfrak{m})}\big)$ and choose a Lipschitz function $\psi : \mathbb{R} \to \mathbb{R}$ such that $\psi = \exp$ in the interval $\big[\log(c^2), C\big]$. By applying property B2) of Theorem 2.1.28, we see that $\varphi \circ (fg) = \log(fg) = \log(f) + \log(g) = \varphi \circ f + \varphi \circ g$ belongs to $\mathrm{S}^2(\mathrm{X})$ and accordingly that $fg =$

$\exp\big(\log(fg)\big) = \psi \circ \varphi \circ (fg) \in \mathrm{S}^2(\mathrm{X})$. Furthermore, again by B2) we deduce that

$$|D(fg)| = |\psi'| \circ \varphi \circ (fg)\,\big|D\big(\varphi \circ (fg)\big)\big| \leq |fg|\Big[\big|D\log(f)\big| + \big|D\log(g)\big|\Big]$$

$$= |fg|\left[\frac{|Df|}{|f|} + \frac{|Dg|}{|g|}\right] = |f||Dg| + |g||Df| \qquad \mathfrak{m}\text{-a.e. in X.}$$

Now consider the case of general $f, g \in \mathrm{S}^2(\mathrm{X}) \cap L^\infty(\mathfrak{m})$. For any $n \in \mathbb{N}$ and $i \in \mathbb{Z}$, let us denote $I_{ni} := \big[\frac{i}{n}, \frac{i+1}{n}\big[$. Call φ_{ni} the continuous function that is the identity on I_{ni} and constant elsewhere. Let us define

$$f_{ni} := f - \frac{i-1}{n}, \qquad \tilde{f}_{ni} := \varphi_{ni} \circ f - \frac{i-1}{n},$$

$$g_{nj} := g - \frac{j-1}{n}, \qquad \tilde{g}_{nj} := \varphi_{nj} \circ g - \frac{j-1}{n}.$$

Notice that $f_{ni} = \tilde{f}_{ni}$ and $g_{nj} = \tilde{g}_{nj}$ hold $\mathfrak{m}$-a.e. in $f^{-1}(I_{ni})$ and $g^{-1}(I_{nj})$, respectively. Then the equalities $|Df_{ni}| = |D\tilde{f}_{ni}| = |Df|$ and $|Dg_{nj}| = |D\tilde{g}_{nj}| = |Dg|$ hold $\mathfrak{m}$-a.e. in $f^{-1}(I_{ni})$ and in $g^{-1}(I_{nj})$, respectively. We also have that $\big|D(f_{ni}\,g_{nj})\big| = \big|D(\tilde{f}_{ni}\,\tilde{g}_{nj})\big|$ is verified $\mathfrak{m}$-a.e. in $f^{-1}(I_{ni}) \cap g^{-1}(I_{nj})$. Moreover, we have that $1/n \leq \tilde{f}_{ni}, \tilde{g}_{nj} \leq 2/n$ hold $\mathfrak{m}$-a.e. Therefore

$$\begin{aligned}
|D(fg)| &\leq \big|D(f_{ni}\,g_{nj})\big| + \frac{|i-1|}{n}\,|Dg_{nj}| + \frac{|j-1|}{n}\,|Df_{ni}| \\
&\leq |\tilde{g}_{nj}||D\tilde{f}_{ni}| + |\tilde{f}_{ni}||D\tilde{g}_{nj}| + \frac{|i-1|}{n}\,|Dg_{nj}| + \frac{|j-1|}{n}\,|Df_{ni}| \\
&\leq |Df|\Big(|g| + \frac{4}{n}\Big) + |Dg|\Big(|f| + \frac{4}{n}\Big) \qquad \mathfrak{m}\text{-a.e. in } f^{-1}(I_{ni}) \cap g^{-1}(I_{nj}),
\end{aligned}$$

where the second inequality follows from the case $f, g \geq c > 0$ treated above. This implies that the inequality $\big|D(fg)\big| \leq |f||Dg| + |g||Df| + 4\big(|Df| + |Dg|\big)/n$ holds $\mathfrak{m}$-a.e. in X. Given that $n \in \mathbb{N}$ is arbitrary, the Leibniz rule follows. ■

Remark 2.1.30 Property C) of Theorem 2.1.28 can be easily seen to hold for every $f \in W^{1,2}(\mathrm{X})$ and $g \in \mathrm{LIP}_b(\mathrm{X})$. ■

2.1.4 Local Sobolev Space

We can now introduce the *local Sobolev class* associated to $(\mathrm{X}, \mathsf{d}, \mathfrak{m})$:

Definition 2.1.31 We define $\mathrm{S}^2_{loc}(\mathrm{X})$ as the set of all Borel functions $f : \mathrm{X} \to \mathbb{R}$ with the following property: for any bounded Borel set $B \subseteq \mathrm{X}$, there exists a function $f_B \in \mathrm{S}^2(\mathrm{X})$ such that $f_B = f$ holds $\mathfrak{m}$-a.e. in B. Given any $f \in \mathrm{S}^2_{loc}(\mathrm{X})$, we define the function $|Df|$ as

$$|Df| := |Df_B| \quad \mathfrak{m}\text{-a.e. in } B, \qquad \begin{array}{l}\text{for any bounded Borel set } B \subseteq \mathrm{X} \text{ and for} \\ \text{any } f_B \in \mathrm{S}^2(\mathrm{X}) \text{ with } f_B = f \quad \mathfrak{m}\text{-a.e. in } B.\end{array} \tag{2.22}$$

The well-posedness of definition (2.22) stems from the locality property of minimal weak upper gradients, which has been proved in Theorem 2.1.28.

We define $L^2_{loc}(\mathrm{X})$ as the space of all Borel functions $g : \mathrm{X} \to \mathbb{R}$ such that $g_{|B} \in L^2(\mathfrak{m})$ for every bounded Borel subset B of X. It is then clear that $|Df| \in L^2_{loc}(\mathrm{X})$ for any $f \in \mathrm{S}^2_{loc}(\mathrm{X})$.

Proposition 2.1.32 (Alternative Characterisation of $\mathrm{S}^2_{loc}(\mathrm{X})$, pt. 1) *Let $f \in \mathrm{S}^2_{loc}(\mathrm{X})$ be given. Then it holds that*

$$\int \big|f(\gamma_1) - f(\gamma_0)\big| \,\mathrm{d}\boldsymbol{\pi}(\gamma) \le \iint_0^1 |Df|(\gamma_t)|\dot{\gamma}_t| \,\mathrm{d}t \,\mathrm{d}\boldsymbol{\pi}(\gamma) \quad \text{for every } \boldsymbol{\pi} \text{ test plan.} \tag{2.23}$$

Proof Fix a test plan $\boldsymbol{\pi}$ and a point $\bar{x} \in \mathrm{X}$. For any $n \in \mathbb{N}$, let us define

$$\Gamma_n := \left\{\gamma : [0,1] \to \mathrm{X} \text{ AC} \;\middle|\; \mathsf{d}(\gamma_0, \bar{x}) \le n \text{ and } \int_0^1 |\dot{\gamma}_t|^2 \,\mathrm{d}t \le n\right\},$$

which turns out to be a closed subset of $C([0,1],\mathrm{X})$. It is clear that $\boldsymbol{\pi}\big(\bigcup_n \Gamma_n\big) = 1$. Now let us call $\boldsymbol{\pi}_n := \boldsymbol{\pi}(\Gamma_n)^{-1}\boldsymbol{\pi}_{|\Gamma_n}$ for every $n \in \mathbb{N}$ such that $\boldsymbol{\pi}(\Gamma_n) > 0$. For $\boldsymbol{\pi}_n$-a.e. γ it holds that

$$\mathsf{d}(\gamma_t, \bar{x}) \le \int_0^t |\dot{\gamma}_s| \,\mathrm{d}s + \mathsf{d}(\gamma_0, \bar{x}) \le \left(\int_0^1 |\dot{\gamma}_s|^2 \,\mathrm{d}s\right)^{1/2} + n \le \sqrt{n} + n \quad \text{for every } t \in [0,1].$$

Denote by B_n the open ball of radius $\sqrt{n}+n+1$ centered at $\bar{x}$ and take any function $f_n \in \mathrm{S}^2(\mathrm{X})$ such that $f_n = f$ holds $\mathfrak{m}$-a.e. in B_n. Therefore for $\boldsymbol{\pi}_n$-a.e. curve γ one has that

$$\big|f(\gamma_1) - f(\gamma_0)\big| = \big|f_n(\gamma_1) - f_n(\gamma_0)\big| \le \int_0^1 |Df_n|(\gamma_t)|\dot{\gamma}_t| \,\mathrm{d}t = \int_0^1 |Df|(\gamma_t)|\dot{\gamma}_t| \,\mathrm{d}t,$$

whence (2.23) follows by arbitrariness of n. □

Let us fix some notation: given a Polish space X and a (signed) Borel measure μ on X, we define the *support* of μ as

$$\mathrm{spt}(\mu) := \bigcap \big\{C \subseteq \mathrm{X} \text{ closed} \;:\; \mu^+(\mathrm{X} \setminus C) = \mu^-(\mathrm{X} \setminus C) = 0\big\}. \tag{2.24}$$

Clearly $\mathrm{spt}(\mu)$ is a closed subset of X by construction.

Remark 2.1.33 We point out that

$$\mu_{|\mathrm{X}\setminus\mathrm{spt}(\mu)} = 0. \tag{2.25}$$

Indeed, since X is a Lindelöf space (as it is separable), we can choose a sequence $(U_n)_n$ of open sets such that $\bigcup_n U_n = \bigcup\{\mathrm{X} \setminus C \,:\, C \text{ closed}, |\mu|(\mathrm{X} \setminus C) = 0\}$, whence

$$|\mu|(\mathrm{X} \setminus \mathrm{spt}(\mu)) = |\mu|\Big(\bigcup_n U_n\Big) \leq \sum_n |\mu|(U_n) = 0,$$

which is equivalent to (2.25). ■

We can now prove the converse of Proposition 2.1.32 under the additional assumption that the function f belongs to the space $L^2_{loc}(\mathrm{X})$:

Proposition 2.1.34 (Alternative Characterisation of $\mathrm{S}^2_{loc}(\mathrm{X})$, pt. 2) *Let $f \in L^2_{loc}(\mathrm{X})$ be a given function. Suppose that $G \in L^2_{loc}(\mathrm{X})$ is a non-negative function satisfying*

$$\int \big|f(\gamma_1) - f(\gamma_0)\big| \,\mathrm{d}\boldsymbol{\pi}(\gamma) \leq \iint_0^1 G(\gamma_t)|\dot{\gamma}_t| \,\mathrm{d}t \,\mathrm{d}\boldsymbol{\pi}(\gamma) \quad \textit{for every } \boldsymbol{\pi} \textit{ test plan.} \tag{2.26}$$

Then $f \in \mathrm{S}^2_{loc}(\mathrm{X})$ and $|Df| \leq G$ holds $\mathfrak{m}$*-a.e. in* X.

Proof We divide the proof into three steps:

STEP 1. We say that a test plan $\boldsymbol{\pi}$ is bounded provided $\big\{\gamma_t \,:\, \gamma \in \mathrm{spt}(\boldsymbol{\pi}),\ t \in [0,1]\big\}$ is bounded. By arguing as in the proof of Theorem 2.1.21, one can prove the following claim:

Fix $f : \mathrm{X} \to \mathbb{R}$ Borel, $\boldsymbol{\pi}$ bounded test plan and $G \in L^2_{loc}(\mathrm{X})$ with $G \geq 0$. Then the following are equivalent:

A) (2.26) holds for every test plan $\boldsymbol{\pi}'$ of the form $(\mathrm{Restr}_s^t)_*\big(\boldsymbol{\pi}(\Gamma)^{-1}\,\boldsymbol{\pi}_{|\Gamma}\big)$,

B) for $\boldsymbol{\pi}$-a.e. γ we have $f \circ \gamma \in W^{1,1}(0,1)$ and $\big|(f \circ \gamma)'_t\big| \leq G(\gamma_t)|\dot{\gamma}_t|$ for a.e. t.

(2.27)

STEP 2. Fix a function $f \in L^2_{loc}(\mathrm{X})$ satisfying (2.26), a test plan $\boldsymbol{\pi}$ on X and a Lipschitz function $g \in \mathrm{LIP}_{bs}(\mathrm{X})$. Given $\bar{x} \in \mathrm{X}$ and $n \in \mathbb{N}$, let us define

$$\Gamma_n := \left\{\gamma : [0,1] \to \mathrm{X} \text{ AC} \,\middle|\, \mathsf{d}(\gamma_0, \bar{x}) \leq n \text{ and } \int_0^1 |\dot{\gamma}_t|^2 \,\mathrm{d}t \leq n\right\},$$

so that each Γ_n is a Borel set and $\boldsymbol{\pi}\big(\bigcup_n \Gamma_n\big) = 1$, as in the proof of Proposition 2.1.32. Let us fix $n \in \mathbb{N}$ sufficiently big and define $\boldsymbol{\pi}_n :=$

$\pi(\Gamma_n)^{-1}\pi_{|\Gamma_n}$, so that π_n is a bounded test plan on X. Now choose any open bounded set Ω containing $\mathrm{spt}(g)$, whence we have that the inequality $|(g\circ\gamma)'_t| \leq |Dg|\,|\dot\gamma_t|\,\chi_\Omega(\gamma_t)$ holds for $(\pi_n\times\mathcal{L}_1)$-a.e. (γ,t). Thus B) of (2.27) gives

$$\big|((fg)\circ\gamma)'_t\big| \leq |f|(\gamma_t)\,\big|(g\circ\gamma)'_t\big| + |g|(\gamma_t)\,\big|(f\circ\gamma)'_t\big| \leq \big(\chi_\Omega\,|g|\,G + \chi_\Omega\,|f|\,|Dg|\big)(\gamma_t)\,|\dot\gamma_t|$$

for $(\pi_n\times\mathcal{L}_1)$-a.e. (γ,t), so also for $(\pi\times\mathcal{L}_1)$-a.e. (γ,t). Note that $\chi_\Omega\big(|g|G + |f||Dg|\big) \in L^2(\mathfrak{m})$. Therefore Theorem 2.1.21 grants that $fg \in \mathrm{S}^2(\mathrm{X})$ and $|D(fg)| \leq \chi_\Omega\big(|g|G + |f||Dg|\big)$.

STEP 3. To conclude, fix $f \in L^2_{loc}(\mathrm{X})$ satisfying (2.26). Given a bounded Borel set $B \subseteq \mathrm{X}$, pick a function $g \in \mathrm{LIP}_{bs}(\mathrm{X})$ with $g = 1$ on B, thus $|Dg| = 0$ holds $\mathfrak{m}$-a.e. in B by locality. Hence STEP 2 implies that $|Df| = |D(fg)| \leq G$ $\mathfrak{m}$-a.e. in B, yielding the statement. □

Corollary 2.1.35 *Let $f : \mathrm{X} \to \mathbb{R}$ be a Borel map. Then $f \in \mathrm{S}^2(\mathrm{X})$ if and only if $f \in \mathrm{S}^2_{loc}(\mathrm{X})$ and $|Df| \in L^2(\mathfrak{m})$.*

Proof Immediate consequence of Definition 2.1.31 and Proposition 2.1.32. □

2.1.5 Consistency with the Classical Sobolev Space on $\mathbb{R}^n$

In this subsection we aim to prove that the definition of Sobolev space for abstract metric measure spaces is consistent with the classical one when we work in the Euclidean setting, namely if we consider $(\mathrm{X},\mathsf{d},\mathfrak{m}) = (\mathbb{R}^n, \mathsf{d}_{\mathrm{Eucl}}, \mathcal{L}^n)$. To this purpose, let us fix some notation:

$$\begin{aligned} W^{1,2}(\mathbb{R}^n) &= \text{ the classical Sobolev space on } \mathbb{R}^n,\\ |Df| &= \text{ the minimal weak upper gradient of } f \in \mathrm{S}^2_{loc}(\mathbb{R}^n),\\ \mathrm{d}f &= \text{ the distributional differential of } f \in W^{1,2}_{loc}(\mathbb{R}^n),\\ \nabla f &= \text{ the `true' gradient of } f \in C^\infty(\mathbb{R}^n). \end{aligned}$$

The above-mentioned consistency can be readily got as a consequence of the following facts:

Proposition 2.1.36 *The following properties hold:*

A) *If $f \in C^\infty(\mathbb{R}^n) \subseteq W^{1,2}_{loc}(\mathbb{R}^n)$, then the function f belongs to the space $\mathrm{S}^2_{loc} \cap L^2_{loc}(\mathbb{R}^n)$ and the equalities $|\nabla f| = |\mathrm{d}f| = |Df|$ hold $\mathcal{L}^n$-a.e. in $\mathbb{R}^n$.*
B) *If $f \in W^{1,2}(\mathbb{R}^n)$ and $\rho \in C^\infty_c(\mathbb{R}^n)$ is a convolution kernel, then $f * \rho \in W^{1,2}(\mathbb{R}^n)$ and the inequality $\big|\mathrm{d}(f*\rho)\big| \leq |\mathrm{d}f| * \rho$ holds $\mathcal{L}^n$-a.e. in $\mathbb{R}^n$.*
C) *If $f \in \mathrm{S}^2 \cap L^2(\mathbb{R}^n)$ and $\rho \in C^\infty_c(\mathbb{R}^n)$ is a convolution kernel, then $f * \rho \in \mathrm{S}^2 \cap L^2(\mathbb{R}^n)$ and the inequality $\big|D(f*\rho)\big| \leq |Df| * \rho$ holds $\mathcal{L}^n$-a.e. in $\mathbb{R}^n$.*

Proof The proof goes as follows:

A) It is well-known that $|\nabla f| = |\mathrm{d}f|$ holds $\mathcal{L}^n$-a.e. Moreover, $|Df| \leq \mathrm{lip}(f) = |\nabla f|$ is satisfied $\mathcal{L}^n$-a.e., thus to conclude it suffices to show that $\int |Df|\,\mathrm{d}\mathcal{L}^n \geq \int |\nabla f|\,\mathrm{d}\mathcal{L}^n$. By monotone convergence theorem, it is enough to prove that $\int_K |Df|\,\mathrm{d}\mathcal{L}^n \geq \int_K |\nabla f|\,\mathrm{d}\mathcal{L}^n$ is satisfied for any compact subset K of the open set $\{|\nabla f| > 0\}$. Then let us fix such a compact set K and some $\varepsilon > 0$. Call $\lambda := \min_K |\nabla f| > 0$. We can take a Borel partition $(U_i)_{i=1}^k$ of K and vectors $(v_i)_{i=1}^k \subseteq \mathbb{R}^n$ such that $\mathcal{L}^n(U_i) > 0$, $|v_i| \geq \lambda$ and $\big|\nabla f(x) - v_i\big| < \varepsilon$ for every $x \in U_i$. Fix $i = 1, \ldots, k$. Call $\mu := \mathcal{L}^n(U_i)^{-1}\,\mathcal{L}^n|_{U_i}$ and $\boldsymbol{\pi} := F_*\mu$, where $F : \mathbb{R}^n \to C([0,1], \mathbb{R}^n)$ is given by $x \mapsto \big(t \mapsto x + tv_i\big)$, so that $(\mathrm{e}_t)_*\boldsymbol{\pi} \leq \mathcal{L}^n(U_i)^{-1}\,(\cdot + tv_i)_*\mathcal{L}^n \leq \mathcal{L}^n(U_i)^{-1}\,\mathcal{L}^n$ holds for every $t \in [0,1]$ and $\int\int_0^1 |\dot\gamma_t|^2\,\mathrm{d}t\,\mathrm{d}\boldsymbol{\pi}(\gamma) = |v_i|^2 < +\infty$, which means that $\boldsymbol{\pi}$ is a test plan on $\mathbb{R}^n$. It is clear that $f \in \mathrm{S}^2_{loc} \cap L^2_{loc}(\mathbb{R}^n)$, whence for any $t \in [0,1]$ one has

$$
\begin{aligned}
&\int \big|f(\gamma_t) - f(\gamma_0)\big|\,\mathrm{d}\boldsymbol{\pi}(\gamma)\\
&\leq \int\int_0^t |Df|(\gamma_s)|\dot\gamma_s|\,\mathrm{d}s\,\mathrm{d}\boldsymbol{\pi}(\gamma) = |v_i| \int\int_0^t |Df|(\gamma_s)\,\mathrm{d}s\,\mathrm{d}\boldsymbol{\pi}(\gamma)\\
&= |v_i| \int_0^t \int |Df|\,\mathrm{d}(\mathrm{e}_s)_*\boldsymbol{\pi}\,\mathrm{d}s = |v_i| \int_0^t \int |Df|\,\mathrm{d}(\cdot + sv_i)_*\mu\,\mathrm{d}s\\
&= \frac{|v_i|}{\mathcal{L}^n(U_i)} \int_0^t \int \chi_{U_i + sv_i}\,|Df|\,\mathrm{d}\mathcal{L}^n\,\mathrm{d}s.
\end{aligned}
$$

Since $\chi_{U_i + sv_i}$ converges to χ_{U_i} in $L^2(\mathbb{R}^n)$ as $s \to 0$, if we divide the previous formula by t and we let $t \searrow 0$, then we obtain that

$$
\begin{aligned}
|v_i| \fint_{U_i} |Df|\,\mathrm{d}\mathcal{L}^n &\geq \int \big|\langle \nabla f(\gamma_0), \gamma_0' \rangle\big|\,\mathrm{d}\boldsymbol{\pi}(\gamma) = \int \big|\langle \nabla f, v_i \rangle\big|\,\mathrm{d}(\mathrm{e}_0)_*\boldsymbol{\pi}\\
&= \fint_{U_i} \big|\langle \nabla f, v_i \rangle\big|\,\mathrm{d}\mathcal{L}^n \geq \big(|v_i| - 2\,\varepsilon\big) \fint_{U_i} |\nabla f|\,\mathrm{d}\mathcal{L}^n,
\end{aligned}
$$

where the last inequality follows from $\big|\langle \nabla f, v_i \rangle\big| \geq |\nabla f||v_i| - 2\,|\nabla f||\nabla f - v_i|$. Therefore

$$
\begin{aligned}
&\int_K |Df|\,\mathrm{d}\mathcal{L}^n = \sum_{i=1}^k \mathcal{L}^n(U_i) \fint_{U_i} |Df|\,\mathrm{d}\mathcal{L}^n \geq \sum_{i=1}^k \mathcal{L}^n(U_i)\\
&\left[\fint_{U_i} |\nabla f|\,\mathrm{d}\mathcal{L}^n - \frac{2\,\varepsilon}{|v_i|} \fint_{U_i} |\nabla f|\,\mathrm{d}\mathcal{L}^n\right] \geq \int_K |\nabla f|\,\mathrm{d}\mathcal{L}^n - \frac{2\,\varepsilon}{\lambda} \int_K |\nabla f|\,\mathrm{d}\mathcal{L}^n.
\end{aligned}
$$

By letting $\varepsilon \searrow 0$ we thus conclude that $\int_K |Df| \, d\mathcal{L}^n \geq \int_K |\nabla f| \, d\mathcal{L}^n$, as required.

B) It is well-known that $f * \rho \in W^{1,2}(\mathbb{R}^n)$ and $d(f * \rho) = (df) * \rho$. To conclude, it only remains to observe that $|(df) * \rho| \leq |df| * \rho$ by Jensen's inequality. Hence property B) is achieved.

C) Given any $x \in \mathbb{R}^n$, let us define the translation operator $\mathrm{Tr}_x : C([0,1], \mathbb{R}^n) \to C([0,1], \mathbb{R}^n)$ as $\mathrm{Tr}_x(\gamma)_t := \gamma_t - x$. If γ is absolutely continuous, then γ and $\mathrm{Tr}_x(\gamma)$ have the same metric speed. Now fix a test plan $\boldsymbol{\pi}$. Clearly $(\mathrm{Tr}_x)_* \boldsymbol{\pi}$ is a test plan as well. Therefore

$$
\begin{aligned}
&\int \left|(f * \rho)(\gamma_1) - (f * \rho)(\gamma_0)\right| d\boldsymbol{\pi}(\gamma) \\
&\leq \int \rho(x) \int \left|f(\gamma_1 - x) - f(\gamma_0 - x)\right| d\boldsymbol{\pi}(\gamma) \, dx \\
&= \int \rho(x) \int \left|f(\sigma_1) - f(\sigma_0)\right| d(\mathrm{Tr}_x)_* \boldsymbol{\pi}(\sigma) \, dx \\
&\leq \int \rho(x) \int \int_0^1 |Df|(\sigma_t) \, |\dot{\sigma}_t| \, dt \, d(\mathrm{Tr}_x)_* \boldsymbol{\pi}(\sigma) \, dx \\
&= \iiint_0^1 \rho(x) \, |Df|(\gamma_t - x) \, |\dot{\gamma}_t| \, dt \, d\boldsymbol{\pi}(\gamma) \, dx \\
&= \iint_0^1 \left(\int |Df|(\gamma_t - x) \, \rho(x) \, dx \right) |\dot{\gamma}_t| \, dt \, d\boldsymbol{\pi}(\gamma) \\
&= \iint_0^1 (|Df| * \rho)(\gamma_t) \, |\dot{\gamma}_t| \, dt \, d\boldsymbol{\pi}(\gamma),
\end{aligned}
$$

which grants that $f * \rho \in \mathrm{S}^2 \cap L^2(\mathbb{R}^n)$ and $|D(f * \rho)| \leq |Df| * \rho$ holds $\mathcal{L}^n$-a.e. in $\mathbb{R}^n$. □

With this said, we are in a position to prove the main result:

Theorem 2.1.37 *Let $f : \mathbb{R}^n \to \mathbb{R}$ be a given Borel function. Then $f \in \mathrm{S}^2 \cap L^2(\mathbb{R}^n)$ if and only if $f \in W^{1,2}(\mathbb{R}^n)$. In this case, the equality $|Df| = |df|$ holds $\mathcal{L}^n$-a.e. in $\mathbb{R}^n$.*

Proof Let us fix a family of convolution kernels $(\rho_\varepsilon)_{\varepsilon > 0}$. Given any $f \in W^{1,2}(\mathbb{R}^n)$, we deduce from properties A) and B) of Proposition 2.1.36 that $f * \rho_\varepsilon \in \mathrm{S}^2 \cap L^2(\mathbb{R}^n)$ and that

$$
\left|D(f * \rho_\varepsilon)\right| = \left|d(f * \rho_\varepsilon)\right| \leq |df| * \rho_\varepsilon \longrightarrow |df| \quad \text{in } L^2(\mathbb{R}^n) \quad \text{as } \varepsilon \searrow 0.
$$

Since also $f * \rho_\varepsilon \to f$ in $L^2(\mathbb{R}^n)$ as $\varepsilon \searrow 0$, we have that $f \in \mathrm{S}^2 \cap L^2(\mathbb{R}^n)$ and that $|Df| \leq |df|$ holds $\mathcal{L}^n$-a.e. in $\mathbb{R}^n$, as a consequence of Proposition 2.1.13.

On the other hand, given any function $f \in \mathrm{S}^2 \cap L^2(\mathbb{R}^n)$, we have that $f * \rho_\varepsilon \in \mathrm{S}^2 \cap L^2(\mathbb{R}^n)$ and that $|\mathrm{d}(f * \rho_\varepsilon)| = |D(f * \rho_\varepsilon)| \leq |Df| * \rho_\varepsilon$ holds $\mathcal{L}^n$-a.e. by properties A) and C) of Proposition 2.1.36. Since $|Df| * \rho_\varepsilon \to |Df|$ in $L^2(\mathbb{R}^n)$ as $\varepsilon \searrow 0$, there exist a sequence $\varepsilon_k \searrow 0$ and $w \in L^2(\mathbb{R}^n)$ such that $\mathrm{d}(f * \rho_{\varepsilon_k}) \rightharpoonup w$ weakly in $L^2(\mathbb{R}^n)$, thus necessarily $w = \mathrm{d}f$. In particular, it holds that $\int |\mathrm{d}f|^2 \,\mathrm{d}\mathcal{L}^n \leq \underline{\lim}_k \int |\mathrm{d}(f * \rho_{\varepsilon_k})|^2 \,\mathrm{d}\mathcal{L}^n = \int |Df|^2 \,\mathrm{d}\mathcal{L}^n$, which forces the $\mathcal{L}^n$-a.e. equality $|Df| = |\mathrm{d}f|$, proving the statement. □

2.2 Alternative Notions of Sobolev Space

We now introduce some alternative definitions of Sobolev space on a general metric measure space $(\mathrm{X}, \mathsf{d}, \mathfrak{m})$, which a posteriori turn out to be equivalent to the one (via weak upper gradients) we gave in Definition 2.1.15.

2.2.1 *Approach à la Cheeger*

The rough idea behind this approach is the following; we need an $L^2(\mathfrak{m})$-lower semicontinuous energy functional of the form $\frac{1}{2}\int |\mathrm{d}f|^2 \,\mathrm{d}\mathfrak{m}$, where the function $|\mathrm{d}f|$ is an object which is 'local' and satisfies some sort of chain rule. Given any Lipschitz function $f \in \mathrm{LIP}(\mathrm{X})$, some (seemingly) good candidates for $|\mathrm{d}f|$ could be given by

$$\mathrm{lip}(f)(x) := \overline{\lim_{y \to x}} \frac{|f(y) - f(x)|}{\mathsf{d}(y, x)} \quad \text{(local Lipschitz constant)},$$

$$\mathrm{lip}_a(f)(x) := \overline{\lim_{y,z \to x}} \frac{|f(y) - f(z)|}{\mathsf{d}(y, z)} \quad \text{(asymptotic Lipschitz constant)},$$

for $x \in \mathrm{X}$ accumulation point and $\mathrm{lip}(f)(x), \mathrm{lip}_a(f)(x) := 0$ otherwise. The local Lipschitz constant has been previously introduced in (2.16). Observe that $\mathrm{lip}(f) \leq \mathrm{lip}_a(f) \leq \mathrm{Lip}(f)$ and that the equalities $\mathrm{lip}_a(f)(x) = \lim_{r \searrow 0} \mathrm{Lip}\big(f|_{B_r(x)}\big) = \inf_{r>0} \mathrm{Lip}\big(f|_{B_r(x)}\big)$ hold for every accumulation point $x \in \mathrm{X}$. Moreover, we shall make use of the following property of lip_a:

$$\mathrm{lip}_a(fg) \leq |f| \,\mathrm{lip}_a(g) + |g| \,\mathrm{lip}_a(f) \quad \text{for every } f, g \in \mathrm{LIP}(\mathrm{X}), \tag{2.28}$$

which is the *Leibniz rule* for the asymptotic Lipschitz constant.

Exercise 2.2.1 Prove that $\mathrm{lip}_a(f)$ is an upper semicontinuous function. ■

Another ingredient we need is the notion of upper gradient:

Definition 2.2.2 (Upper Gradient) Consider two functions $f, g : \mathsf{X} \to \mathbb{R}$, with $g \geq 0$. Then we say that g is an *upper gradient* of f provided for any AC curve $\gamma : [0, 1] \to \mathsf{X}$ one has that the curve $f \circ \gamma$ is AC and satisfies $|(f \circ \gamma)'_t| \leq g(\gamma_t)|\dot{\gamma}_t|$ for a.e. $t \in [0, 1]$.

Notice that $\mathrm{lip}(f)$—thus accordingly also $\mathrm{lip}_a(f)$—is an upper gradient of f for any Lipschitz function $f \in \mathrm{LIP}(\mathsf{X})$, as already shown in Remark 2.1.25. Given that, in general, the functionals $f \mapsto \frac{1}{2}\int \mathrm{lip}^2(f)\,\mathrm{dm}$ and $f \mapsto \frac{1}{2}\int \mathrm{lip}_a^2(f)\,\mathrm{dm}$ are not lower semicontinuous, we introduce our energy functionals by means of a relaxation procedure:

Definition 2.2.3 Let us give the following definitions:

i) The functional $\mathsf{E}_{*,a} : L^2(\mathfrak{m}) \to [0, +\infty]$ is given by

$$\mathsf{E}_{*,a}(f) := \inf \varliminf_{n\to\infty} \frac{1}{2}\int \mathrm{lip}_a^2(f_n)\,\mathrm{dm},$$

where the infimum is taken among all sequences $(f_n)_n \subseteq \mathrm{LIP}(\mathsf{X})$ with $f_n \to f$ in $L^2(\mathfrak{m})$.

ii) The functional $\mathsf{E}_* : L^2(\mathfrak{m}) \to [0, +\infty]$ is given by

$$\mathsf{E}_*(f) := \inf \varliminf_{n\to\infty} \frac{1}{2}\int \mathrm{lip}^2(f_n)\,\mathrm{dm},$$

where the infimum is taken among all sequences $(f_n)_n \subseteq \mathrm{LIP}(\mathsf{X})$ with $f_n \to f$ in $L^2(\mathfrak{m})$.

iii) The functional $\mathsf{E}_{\mathrm{Ch}} : L^2(\mathfrak{m}) \to [0, +\infty]$ is given by

$$\mathsf{E}_{\mathrm{Ch}}(f) := \inf \varliminf_{n\to\infty} \frac{1}{2}\int G_n^2\,\mathrm{dm},$$

where the infimum is taken among all sequences $(f_n)_n \subseteq C(\mathsf{X})$ and $(G_n)_n$ such that G_n is an upper gradient of f_n for every $n \in \mathbb{N}$ and $f_n \to f$ in $L^2(\mathfrak{m})$.

Exercise 2.2.4 Prove that $\mathsf{E}_{*,a}$ is $L^2(\mathfrak{m})$-lower semicontinuous and is the maximal $L^2(\mathfrak{m})$-lower semicontinuous functional E such that $\mathsf{E}(f) \leq \frac{1}{2}\int \mathrm{lip}_a^2(f)\,\mathrm{dm}$ holds for every $f \in \mathrm{LIP}(\mathsf{X})$. Actually, the same properties are verified by E_* if we replace $\mathrm{lip}_a(f)$ with $\mathrm{lip}(f)$. ■

Definition 2.2.5 We define the Banach spaces $W^{1,2}_{*,a}(\mathrm{X})$, $W^{1,2}_{*}(\mathrm{X})$ and $W^{1,2}_{\mathrm{Ch}}(\mathrm{X})$ as follows:

$$\begin{aligned} W^{1,2}_{*,a}(\mathrm{X}) &:= \big\{ f \in L^2(\mathfrak{m}) \,:\, \mathsf{E}_{*,a}(f) < +\infty \big\}, \\ W^{1,2}_{*}(\mathrm{X}) &:= \big\{ f \in L^2(\mathfrak{m}) \,:\, \mathsf{E}_{*}(f) < +\infty \big\}, \\ W^{1,2}_{\mathrm{Ch}}(\mathrm{X}) &:= \big\{ f \in L^2(\mathfrak{m}) \,:\, \mathsf{E}_{\mathrm{Ch}}(f) < +\infty \big\}. \end{aligned} \tag{2.29}$$

Any upper gradient is a weak upper gradient, so $W^{1,2}_{*,a}(\mathrm{X}) \subseteq W^{1,2}_{*}(\mathrm{X}) \subseteq W^{1,2}_{\mathrm{Ch}}(\mathrm{X}) \subseteq W^{1,2}(\mathrm{X})$.

Hereafter, we shall mainly focus our attention on the space $W^{1,2}_{*,a}(\mathrm{X})$. Analogous statements for the other two spaces in (2.29) can be shown to hold.

Remark 2.2.6 The fact that the set $W^{1,2}_{*,a}(\mathrm{X})$ is a vector space follows from this observation: the asymptotic Lipschitz constant satisfies $\mathrm{lip}_a(f+g) \leq \mathrm{lip}_a(f) + \mathrm{lip}_a(g)$ for all $f, g \in \mathrm{LIP}(\mathrm{X})$. Given any $f, g \in W^{1,2}_{*,a}(\mathrm{X})$ and $\alpha, \beta \in \mathbb{R}$, we can choose two sequences $(f_n)_n, (g_n)_n \subseteq \mathrm{LIP}(\mathrm{X})$ such that $\lim_n \|f_n - f\|_{L^2(\mathfrak{m})} = \lim_n \|g_n - g\|_{L^2(\mathfrak{m})} = 0$ and $\overline{\lim}_n \int \mathrm{lip}_a^2(f_n) + \mathrm{lip}_a^2(g_n)\,\mathrm{d}\mathfrak{m}$ is finite. Since $\alpha f_n + \beta g_n \to \alpha f + \beta g$ in $L^2(\mathfrak{m})$, we thus deduce that

$$\begin{aligned} 2\,\mathsf{E}_{*,a}(\alpha f + \beta g) &\leq \overline{\lim_n} \int \mathrm{lip}_a^2(\alpha f_n + \beta g_n)\,\mathrm{d}\mathfrak{m} \\ &\leq 2\,\overline{\lim_n} \int \alpha^2\,\mathrm{lip}_a^2(f_n) + \beta^2\,\mathrm{lip}_a^2(g_n)\,\mathrm{d}\mathfrak{m} < +\infty, \end{aligned}$$

which shows that $\alpha f + \beta g \in W^{1,2}_{*,a}(\mathrm{X})$, as required. ■

Definition 2.2.7 (Asymptotic Relaxed Slope) Let $f \in W^{1,2}_{*,a}(\mathrm{X})$ be a given function. Then an element $G \in L^2(\mathfrak{m})$ with $G \geq 0$ is said to be an *asymptotic relaxed slope* for f provided there exists a sequence $(f_n)_n \subseteq \mathrm{LIP}(\mathrm{X})$ such that $f_n \to f$ strongly in $L^2(\mathfrak{m})$ and $\mathrm{lip}_a(f_n) \rightharpoonup G'$ weakly in $L^2(\mathfrak{m})$, for some $G' \in L^2(\mathfrak{m})$ with $G' \leq G$.

Proposition 2.2.8 *Let $f \in W^{1,2}_{*,a}(\mathrm{X})$ be given. Then the set of all asymptotic relaxed slopes for f is a non-empty closed convex subset of $L^2(\mathfrak{m})$. Its element of minimal $L^2(\mathfrak{m})$-norm, denoted by $|Df|_{*,a}$ and called* minimal asymptotic relaxed slope, *satisfies the equality*

$$\mathsf{E}_{*,a}(f) = \frac{1}{2} \int |Df|^2_{*,a}\,\mathrm{d}\mathfrak{m}. \tag{2.30}$$

Proof The proof goes as follows:
EXISTENCE OF ASYMPTOTIC RELAXED SLOPES. Given that $\mathsf{E}_{*,a}(f) < +\infty$, we can find a sequence $(f_n)_n \subseteq \mathrm{LIP}(\mathrm{X})$ such that $f_n \to f$ strongly in $L^2(\mathfrak{m})$ and $\sup_n \int \mathrm{lip}_a^2(f_n)\,\mathrm{d}\mathfrak{m} < +\infty$. Then (up to a not relabeled subsequence) we have that $\mathrm{lip}_a(f_n) \rightharpoonup G$ weakly in $L^2(\mathfrak{m})$ for some $G \in L^2(\mathfrak{m})$, whence G is an asymptotic relaxed slope for f.
CONVEXITY. Let us fix two asymptotic relaxed slopes G_1, G_2 for f and a constant $\alpha \in [0, 1]$. For $i = 1, 2$, choose $(f_n^i)_n \subseteq \mathrm{LIP}(\mathrm{X})$ such that $f_n^i \to f$ and $\mathrm{lip}_a(f_n^i) \rightharpoonup G_i' \leq G_i$. We then claim that $\alpha G_1 + (1 - \alpha)G_2$ is an asymptotic relaxed slope for f. In order to prove it, notice that $\alpha f_n^1 + (1 - \alpha) f_n^2 \to f$ in $L^2(\mathfrak{m})$ and that

$$\begin{aligned}\mathrm{lip}_a\big(\alpha f_n^1 + (1-\alpha) f_n^2\big) &\leq \alpha\, \mathrm{lip}_a(f_n^1) + (1-\alpha)\, \mathrm{lip}_a(f_n^2) \rightharpoonup \alpha G_1' \\ &+ (1-\alpha)G_2' \leq \alpha G_1 + (1-\alpha)G_2.\end{aligned}$$

Up to subsequence, we thus have that $\mathrm{lip}_a\big(\alpha f_n^1 + (1-\alpha) f_n^2\big)$ weakly converges to some limit function $\tilde{G} \leq \alpha G_1 + (1-\alpha)G_2$, proving the claim.
CLOSEDNESS. Fix a sequence $(G_n)_n \subseteq L^2(\mathfrak{m})$ of asymptotic relaxed slopes for f that strongly converges to some $G \in L^2(\mathfrak{m})$. Given any $n \in \mathbb{N}$, we can pick a sequence $(f_{n,m})_m \subseteq \mathrm{LIP}(\mathrm{X})$ with $f_{n,m} \overset{m}{\to} f$ and $\mathrm{lip}_a(f_{n,m}) \overset{m}{\rightharpoonup} G_n' \leq G_n$. Up to subsequence, we have that $G_n' \rightharpoonup G'$ for some $G' \in L^2(\mathfrak{m})$ with $G' \leq G$. Then we can assume without loss of generality that the sequence $\big(\mathrm{lip}_a(f_{n,m})\big)_{n,m}$ is bounded in the space $L^2(\mathfrak{m})$. Since the restriction of the weak topology to any closed ball of $L^2(\mathfrak{m})$ is metrizable, by a diagonalisation argument we can extract a subsequence $(m_n)_n$ for which we have $f_{n,m_n} \overset{n}{\to} f$ and $\mathrm{lip}_a(f_{n,m_n}) \overset{n}{\rightharpoonup} G' \leq G$, i.e. G is an asymptotic relaxed slope for f.
FORMULA (2.30). Call $|Df|_{*,a}$ the asymptotic relaxed slope for f of minimal $L^2(\mathfrak{m})$-norm. By a diagonalisation argument, there exists some $(h_n)_n \subseteq \mathrm{LIP}(\mathrm{X})$ such that $h_n \to f$ in $L^2(\mathfrak{m})$ and $\mathsf{E}_{*,a}(f) = \lim_n \frac{1}{2}\int \mathrm{lip}_a^2(h_n)\,\mathrm{d}\mathfrak{m}$. Up to subsequence, it holds that $\mathrm{lip}_a(h_n) \rightharpoonup H$ weakly for some $H \in L^2(\mathfrak{m})$, thus H is an asymptotic relaxed slope for f and accordingly

$$\frac{1}{2}\int |Df|_{*,a}^2\,\mathrm{d}\mathfrak{m} \leq \frac{1}{2}\int H^2\,\mathrm{d}\mathfrak{m} \leq \mathsf{E}_{*,a}(f). \tag{2.31}$$

Now choose any sequence $(\tilde{f}_n)_n \subseteq \mathrm{LIP}(\mathrm{X})$ such that $\tilde{f}_n \to f$ in $L^2(\mathfrak{m})$ and $\mathrm{lip}_a(\tilde{f}_n) \rightharpoonup |Df|_{*,a}$ weakly in $L^2(\mathfrak{m})$. By Theorem A.2, for any $n \in \mathbb{N}$ there exist $N_n \geq n$ and $(\alpha_{n,i})_{i=n}^{N_n} \subseteq [0, 1]$ in such a way that $\sum_{i=n}^{N_n} \alpha_{n,i} = 1$ and $\sum_{i=n}^{N_n} \alpha_{n,i}\, \mathrm{lip}_a(\tilde{f}_i) \overset{n}{\to} |Df|_{*,a}$ in $L^2(\mathfrak{m})$. Let us now define

$$f_n := \sum_{i=n}^{N_n} \alpha_{n,i}\, \tilde{f}_i \quad \text{for every } n \in \mathbb{N}.$$

It is clear that $f_n \to f$ in $L^2(\mathfrak{m})$: given any $\varepsilon > 0$, there is $\bar{n} \in \mathbb{N}$ such that $\|\tilde{f}_n - f\|_{L^2(\mathfrak{m})} \leq \varepsilon$ for all $n \geq \bar{n}$, so that accordingly one has

$$\|f_n - f\|_{L^2(\mathfrak{m})} \leq \sum_{i=n}^{N_n} \alpha_{n,i} \, \|\tilde{f}_i - f\|_{L^2(\mathfrak{m})} \leq \varepsilon \sum_{i=n}^{N_n} \alpha_{n,i} = \varepsilon \quad \text{for every } n \geq \bar{n}.$$

Note that one has $\mathrm{lip}_a(f_n) \leq \sum_{i=n}^{N_n} \alpha_{n,i} \, \mathrm{lip}_a(\tilde{f}_i) \to |Df|_{*,a}$ in $L^2(\mathfrak{m})$, whence (up to a not relabeled subsequence) it holds that $\mathrm{lip}_a(f_n) \rightharpoonup G$ weakly in $L^2(\mathfrak{m})$ for some $G \leq |Df|_{*,a}$. Therefore G is an asymptotic relaxed slope for f, so that $\int |Df|_{*,a}^2 \, \mathrm{d}\mathfrak{m} \leq \int G^2 \, \mathrm{d}\mathfrak{m}$, which forces the $\mathfrak{m}$-a.e. equality $G = |Df|_{*,a}$. Moreover, it holds that

$$\begin{aligned} \mathsf{E}_{*,a}(f) &\leq \varliminf_{n\to\infty} \frac{1}{2} \int \mathrm{lip}_a^2(f_n) \, \mathrm{d}\mathfrak{m} \leq \varlimsup_{n\to\infty} \frac{1}{2} \int \mathrm{lip}_a^2(f_n) \, \mathrm{d}\mathfrak{m} \\ &\leq \varlimsup_{n\to\infty} \frac{1}{2} \int \Big(\sum_{i=n}^{N_n} \alpha_{n,i} \, \mathrm{lip}_a(\tilde{f}_i) \Big)^2 \, \mathrm{d}\mathfrak{m} = \frac{1}{2} \int |Df|_{*,a}^2 \, \mathrm{d}\mathfrak{m} \overset{(2.31)}{\leq} \mathsf{E}_{*,a}(f). \end{aligned}$$

This ensures that $\frac{1}{2} \int |Df|_{*,a}^2 \, \mathrm{d}\mathfrak{m} = \mathsf{E}_{*,a}(f)$, thus proving (2.30). □

Proposition 2.2.9 (Cheeger) *Let $f \in W_{*,a}^{1,2}(\mathrm{X})$ be given. Let G_1, G_2 be asymptotic relaxed slopes for f. Then $G_1 \wedge G_2$ is an asymptotic relaxed slope for f as well.*

Proof Notice that $G_1 \wedge G_2 = \chi_E \, G_1 + \chi_{E^c} \, G_2$, where $E := \{G_1 < G_2\}$. By inner regularity of the measure $\mathfrak{m}$, it thus suffices to show that $\chi_K \, G_1 + \chi_{K^c} \, G_2$ is an asymptotic relaxed slope for f, for any compact $K \subseteq \mathrm{X}$. Fix any $r > 0$. Let us define the cut-off function $\eta_r \in L^2(\mathfrak{m})$ as $\eta_r := \big(1 - \mathsf{d}(\cdot, K)/r\big)^+$. For any $i = 1, 2$, we can choose $(f_n^i)_n \subseteq \mathrm{LIP}(\mathrm{X})$ such that $f_n^i \to f$ and $\mathrm{lip}_a(f_n^i) \rightharpoonup G_i' \leq G_i$. Now call $h_n^r := \eta_r f_n^1 + (1 - \eta_r) f_n^2 \in \mathrm{LIP}(\mathrm{X})$ for every $n \in \mathbb{N}$. One clearly has that $h_n^r \overset{n}{\to} f$ strongly in $L^2(\mathfrak{m})$. Moreover, given that

$$h_n^r = f_n^1 + (1 - \eta_r)(f_n^2 - f_n^1) = f_n^2 + \eta_r (f_n^1 - f_n^2),$$

we infer from the Leibniz rule (2.28) that

$$\begin{aligned} \mathrm{lip}_a(h_n^r) &\leq \mathrm{lip}_a(f_n^1) + (1 - \eta_r)\big(\mathrm{lip}_a(f_n^1) + \mathrm{lip}_a(f_n^2)\big) + |f_n^1 - f_n^2| \, \mathrm{lip}_a(1 - \eta_r), \\ \mathrm{lip}_a(h_n^r) &\leq \mathrm{lip}_a(f_n^2) + \eta_r \big(\mathrm{lip}_a(f_n^1) + \mathrm{lip}_a(f_n^2)\big) + |f_n^1 - f_n^2| \, \mathrm{lip}_a(\eta_r). \end{aligned} \tag{2.32}$$

Up to subsequence, we obtain from (2.32) that $\mathrm{lip}_a(h_n^r) \overset{n}{\rightharpoonup} G_r$ for some $G_r \in L^2(\mathfrak{m})$ with

$$G_r \leq \min\big\{G_1' + (1 - \eta_r)(G_1' + G_2'), G_2' + \eta_r (G_1' + G_2')\big\}. \tag{2.33}$$

Since $\eta_r = 1$ on K and $\eta_r = 0$ on $X \setminus K^r$, where $K^r := \{x \in X : \mathsf{d}(x, K) < r\}$, we deduce from the inequality (2.33) that

$$G_r \leq \chi_K \, G'_1 + \chi_{X \setminus K^r} \, G'_2 + 2 \, \chi_{K^r \setminus K} \, (G'_1 + G'_2). \tag{2.34}$$

The right hand side in (2.34) converges in $L^2(\mathfrak{m})$ to the function $\chi_K \, G'_1 + \chi_{K^c} \, G'_2$ as $r \searrow 0$, which grants that $\chi_K \, G_1 + \chi_{K^c} \, G_2$ is an asymptotic relaxed slope for f, as required. □

It immediately follows from Proposition 2.2.9 that:

Corollary 2.2.10 *Let $f \in W^{1,2}_{*,a}(X)$. Take any asymptotic relaxed slope G for f. Then the inequality $|Df|_{*,a} \leq G$ holds $\mathfrak{m}$-a.e. in X.*

Proof We argue by contradiction: suppose that there exists a Borel set $P \subseteq X$ with $\mathfrak{m}(P) > 0$ such that $G < |Df|_{*,a}$ holds $\mathfrak{m}$-a.e. on P. Then the function $G' := G \wedge |Df|_{*,a} \in L^2(\mathfrak{m})$ satisfies the inequality $\int (G')^2 \, d\mathfrak{m} < \int |Df|^2_{*,a} \, d\mathfrak{m}$. This contradicts the minimality of $|Df|_{*,a}$, as G' is an asymptotic relaxed slope for f by Proposition 2.2.9. □

Proposition 2.2.11 (Chain Rule) *Let $f \in W^{1,2}_{*,a}(X)$ be fixed. Let $\varphi \in C^1(\mathbb{R}) \cap \mathrm{LIP}(\mathbb{R})$ be such that $\varphi(0) = 0$, which grants that $\varphi \circ f \in L^2(\mathfrak{m})$. Then $\varphi \circ f \in W^{1,2}_{*,a}(X)$ and*

$$\big|D(\varphi \circ f)\big|_{*,a} \leq |\varphi'| \circ f \, |Df|_{*,a} \quad \textit{holds } \mathfrak{m}\textit{-a.e. in } X. \tag{2.35}$$

Proof Pick $(f_n)_n \subseteq \mathrm{LIP}(X)$ such that $f_n \to f$ and $\mathrm{lip}_a(f_n) \to |Df|_{*,a}$ in $L^2(\mathfrak{m})$. It holds that

$$\mathrm{lip}_a(\varphi \circ f_n) \leq |\varphi'| \circ f_n \, \mathrm{lip}_a(f_n) \longrightarrow |\varphi'| \circ f \, |Df|_{*,a} \quad \text{strongly in } L^2(\mathfrak{m}). \tag{2.36}$$

Then there exists $G \in L^2(\mathfrak{m})$ such that, possibly passing to a subsequence, $\mathrm{lip}_a(\varphi \circ f_n) \rightharpoonup G$. In particular $G \leq |\varphi'| \circ f \, |Df|_{*,a}$ by (2.36), while the inequality $\big|D(\varphi \circ f)\big|_{*,a} \leq G$ is granted by the minimality of $\big|D(\varphi \circ f)\big|_{*,a}$. This proves the statement. □

Remark 2.2.12 Analogous properties to the ones described in Theorem 2.1.28 can be shown to hold for the minimal asymptotic relaxed slope $|Df|_{*,a}$. This follows from Proposition 2.2.9 and Proposition 2.2.11 by suitably adapting the proof of Theorem 2.1.28. ■

The vector space $W^{1,2}_{*,a}(X)$ can be endowed with the norm

$$\|f\|^2_{W^{1,2}_{*,a}(X)} := \|f\|^2_{L^2(\mathfrak{m})} + \big\||Df|_{*,a}\big\|^2_{L^2(\mathfrak{m})} \quad \text{for every } f \in W^{1,2}_{*,a}(X). \tag{2.37}$$

Then $\left(W^{1,2}_{*,a}(\mathrm{X}), \|\cdot\|_{W^{1,2}_{*,a}(\mathrm{X})}\right)$ turns out to be a Banach space. Completeness stems from the lower semicontinuity of the energy functional $\mathsf{E}_{*,a}$.

Remark 2.2.13 Similarly to what done so far, one can define the objects $|Df|_*$ and $|Df|_{\mathrm{Ch}}$ associated to the energies E_* and E_{Ch}, respectively. It can be readily checked that

$$|Df| \leq |Df|_{\mathrm{Ch}} \leq |Df|_* \leq |Df|_{*,a} \quad \text{in the } \mathfrak{m}\text{-a.e. sense}$$

for every $f \in W^{1,2}_{*,a}(\mathrm{X})$. ■

Besides the fact of granting completeness of $W^{1,2}_{*,a}(\mathrm{X})$, the relaxation procedure we used to define the energy functional $\mathsf{E}_{*,a}$ is also motivated by the following observation:

Remark 2.2.14 Suppose that X is compact. Define

$$\|f\|^2_{\widetilde{W}} := \|f\|^2_{L^2(\mathfrak{m})} + \left\|\mathrm{lip}_a(f)\right\|^2_{L^2(\mathfrak{m})} \quad \text{for every } f \in \mathrm{LIP}(\mathrm{X}).$$

Hence $\|\cdot\|_{\widetilde{W}}$ is a seminorm on the vector space LIP(X). Now let us denote by $\widetilde{W}$ the completion of the quotient space of $\left(\mathrm{LIP}(\mathrm{X}), \|\cdot\|_{\widetilde{W}}\right)$. The problem is that in general the elements of $\widetilde{W}$ 'are not functions', in the sense that we are going to explain. The natural inclusion $i : \mathrm{LIP}(\mathrm{X}) \to L^2(\mathfrak{m})$ uniquely extends to a linear continuous map $i : \widetilde{W} \to L^2(\mathfrak{m})$, but such map is not necessarily injective, as shown by the following example. ■

Example 2.2.15 Take $\mathrm{X} := [-1, 1]$ with the Euclidean distance and $\mathfrak{m} := \delta_0$. Consider the functions $f_1, f_2 \in \mathrm{LIP}(\mathrm{X})$ given by $f_1(x) := 0$ and $f_2(x) := x$, respectively. Then f_1 and f_2 coincide as elements of $L^2(\mathfrak{m})$, but $\|f_1 - f_2\|_{\widetilde{W}} = \|f_2\|_{\widetilde{W}} = 1$. ■

2.2.2 Approach à la Shanmugalingam

Here we present a further notion of Sobolev space on metric measure spaces, which will turn out to be equivalent to all of the other ones discussed so far.

Given a metric measure space $(\mathrm{X}, \mathsf{d}, \mathfrak{m})$, let us define

$$\Gamma(\mathrm{X}) := \left\{\gamma : J \to \mathrm{X} \;\middle|\; J \subseteq \mathbb{R} \text{ non-trivial interval, } \gamma \text{ is AC}\right\}. \tag{2.38}$$

Given any curve $\gamma \in \Gamma(\mathrm{X})$, we will denote by $\mathrm{Dom}(\gamma)$ the interval where γ is defined and we will typically call $I \in \mathbb{R}$ and $F \in \mathbb{R}$ the infimum and the supremum of $\mathrm{Dom}(\gamma)$, respectively.

If $G : \mathrm{X} \to [0, +\infty]$ is a Borel function and $\gamma \in \Gamma(\mathrm{X})$, then we define

$$\int_\gamma G := \int_I^F G(\gamma_t)|\dot{\gamma}_t|\,\mathrm{d}t, \tag{2.39}$$

with the convention that $\int_\gamma G := +\infty$ in the case in which $\{t \in \mathrm{Dom}(\gamma) : G(\gamma_t) = +\infty\}$ has positive $\mathcal{L}^1$-measure. We call $\int_\gamma G$ the *line integral* of G along the curve γ.

Definition 2.2.16 (2-Modulus of a Curve Family) Let Γ be any subset of $\Gamma(\mathrm{X})$. Then we define the quantity $\mathrm{Mod}_2(\Gamma) \in [0, +\infty]$ as

$$\mathrm{Mod}_2(\Gamma) := \inf\left\{ \int \rho^2\,\mathrm{dm} \;\middle|\; \rho : \mathrm{X} \to [0, +\infty] \text{ Borel}, \int_\gamma \rho \geq 1 \text{ for all } \gamma \in \Gamma \right\}. \tag{2.40}$$

We call $\mathrm{Mod}_2(\Gamma)$ the *2-modulus* of Γ. Moreover, a property is said to *hold 2-a.e.* provided it is satisfied for every γ belonging to some set $\Gamma \subseteq \Gamma(\mathrm{X})$ such that $\mathrm{Mod}_2(\Gamma^c) = 0$.

The 2-modulus Mod_2 is an outer measure on $\Gamma(\mathrm{X})$, in particular it holds that

$$\Gamma \subseteq \Gamma' \subseteq \Gamma(\mathrm{X}) \implies \mathrm{Mod}_2(\Gamma) \leq \mathrm{Mod}_2(\Gamma'),$$

$$\Gamma_n \subseteq \Gamma(\mathrm{X}), \mathrm{Mod}_2(\Gamma_n) = 0 \text{ for all } n \in \mathbb{N} \implies \mathrm{Mod}_2(\Gamma) = 0, \text{ where } \Gamma := \bigcup_{n\in\mathbb{N}} \Gamma_n.$$

To prove the above claim, fix a sequence $(\Gamma_n)_n$ of subsets of $\Gamma(\mathrm{X})$ and some constant $\varepsilon > 0$. For any $n \in \mathbb{N}$, choose a function ρ_n that is admissible for Γ_n in the definition of $\mathrm{Mod}_2(\Gamma_n)$ and such that $\int \rho_n^2\,\mathrm{dm} \leq \mathrm{Mod}_2(\Gamma_n) + \varepsilon/2^n$. Now call $\rho := \sup_n \rho_n$. Clearly ρ is admissible for $\Gamma := \bigcup_n \Gamma_n$ and it holds that

$$\mathrm{Mod}_2(\Gamma) \leq \int \rho^2\,\mathrm{dm} \leq \sum_{n\in\mathbb{N}} \int \rho_n^2\,\mathrm{dm} \leq \sum_{n\in\mathbb{N}} \mathrm{Mod}_2(\Gamma_n) + 2\,\varepsilon,$$

whence $\mathrm{Mod}_2(\Gamma) \leq \sum_{n\in\mathbb{N}} \mathrm{Mod}_2(\Gamma_n)$ by arbitrariness of ε. Hence Mod_2 is an outer measure.

Remark 2.2.17 Let us fix a Borel function $G : \mathrm{X} \to [0, +\infty)$ such that $G \in L^2(\mathrm{m})$. We stress that G is everywhere defined, not an equivalence class. Then $\int_\gamma G < +\infty$ for 2-a.e. γ.

Indeed, call $\Gamma := \{\gamma \in \Gamma(\mathrm{X}) : \int_\gamma G = +\infty\}$. Given any $\varepsilon > 0$, we have that $\rho := \varepsilon\, G$ is admissible for Γ, so that $\mathrm{Mod}_2(\Gamma) \leq \varepsilon^2 \int G^2\,\mathrm{dm}$. By letting $\varepsilon \searrow 0$, we thus finally conclude that $\mathrm{Mod}_2(\Gamma) = 0$, as required. ■

Definition 2.2.18 (2-Weak Upper Gradient) Let $f : \mathrm{X} \to \mathbb{R} \cup \{\pm\infty\}$ and $G : \mathrm{X} \to [0, +\infty]$ be Borel functions, with $G \in L^2(\mathfrak{m})$. Then we say that G is a 2-*weak upper gradient* for f if

$$\big|f(\gamma_F) - f(\gamma_I)\big| \le \int_\gamma G \quad \text{holds for 2-a.e. } \gamma, \tag{2.41}$$

meaning also that $\int_\gamma G$ must equal $+\infty$ as soon as either $\big|f(\gamma_I)\big| = +\infty$ or $\big|f(\gamma_F)\big| = +\infty$.

Remark 2.2.19 Consider two sets $\Gamma, \Gamma' \subseteq \Gamma(\mathrm{X})$ with the following property: for every $\gamma \in \Gamma$, there exists a subcurve of γ that belongs to Γ'. Then $\mathrm{Mod}_2(\Gamma) \le \mathrm{Mod}_2(\Gamma')$.

The validity of such fact easily follows from the observation that any function ρ that is admissible for Γ' is admissible even for Γ. ■

Lemma 2.2.20 *Let G be a 2-weak upper gradient for f. Then for 2-a.e. curve $\gamma \in \Gamma(\mathrm{X})$ it holds that* $\mathrm{Dom}(\gamma) \ni t \mapsto f(\gamma_t)$ *is AC and* $\big|\partial_t(f \circ \gamma)_t\big| \le G(\gamma_t)|\dot\gamma_t|$ *for a.e.* $t \in \mathrm{Dom}(\gamma)$.

Proof Let us denote by Γ the set of curves γ for which the statement fails. Moreover, call

$$\Gamma' := \left\{\gamma \in \Gamma(\mathrm{X}) \;\middle|\; \big|f(\gamma_F) - f(\gamma_I)\big| > \int_\gamma G\right\},$$

$$\widetilde\Gamma := \left\{\gamma \in \Gamma(\mathrm{X}) \;\middle|\; \int_\gamma G = +\infty\right\}.$$

Notice that $\mathrm{Mod}_2(\Gamma') = 0$ because G is a 2-weak upper gradient for f, while $\mathrm{Mod}_2(\widetilde\Gamma) = 0$ by Remark 2.2.17. Now fix $\gamma \in \Gamma \setminus \widetilde\Gamma$, in particular $t \mapsto G(\gamma_t)|\dot\gamma_t|$ belongs to $L^1(I, F)$. Then there exists $t, s \in \mathrm{Dom}(\gamma)$, $s < t$ such that $\big|f(\gamma_t) - f(\gamma_s)\big| > \int_s^t G(\gamma_r)|\dot\gamma_r|\,\mathrm{d}r$: if not, then γ would satisfy the statement of the lemma. Therefore $\gamma_{|[s,t]} \in \Gamma'$, whence $\mathrm{Mod}_2(\Gamma \setminus \widetilde\Gamma) \le \mathrm{Mod}_2(\Gamma')$ by Remark 2.2.19. This yields $\mathrm{Mod}_2(\Gamma) \le \mathrm{Mod}_2(\Gamma') + \mathrm{Mod}_2(\Gamma \cap \widetilde\Gamma) = 0$, as desired. □

We thus deduce from the previous lemma the following locality property:

Proposition 2.2.21 *Let G_1, G_2 be 2-weak upper gradients of f. Then* $\min\{G_1, G_2\}$ *is a 2-weak upper gradient of f as well.*

Proof For $i = 1, 2$, call Γ_i the set of $\gamma \in \Gamma(\mathrm{X})$ such that $f \circ \gamma$ is AC and $\big|\partial_t(f \circ \gamma)\big| \le G_i(\gamma_t)|\dot\gamma_t|$ holds for a.e. $t \in \mathrm{Dom}(\gamma)$. Then for every curve $\gamma \in \Gamma_1 \cap \Gamma_2$ we have that $f \circ \gamma$ is AC and that $\big|\partial_t(f \circ \gamma)\big| \le \min\big\{G_1(\gamma_t), G_2(\gamma_t)\big\}|\dot\gamma_t|$ holds for a.e. $t \in \mathrm{Dom}(\gamma)$. By integrating such inequality over $\mathrm{Dom}(\gamma)$ we get

$$\big|f(\gamma_F) - f(\gamma_I)\big| \le \int_\gamma \min\{G_1, G_2\} \quad \text{for every } \gamma \in \Gamma_1 \cap \Gamma_2.$$

Then the claim follows by simply noticing that $\mathrm{Mod}_2\big(\Gamma(\mathrm{X}) \setminus (\Gamma_1 \cap \Gamma_2)\big) = 0$. □

Theorem 2.2.22 (Fuglede's Lemma) *Let $G, G_n : \mathrm{X} \to [0, +\infty]$, $n \in \mathbb{N}$ be Borel functions that belong to $L^2(\mathfrak{m})$ and satisfy $\lim_n \|G_n - G\|_{L^2(\mathfrak{m})} = 0$. Then there is a subsequence $(n_k)_k$ such that $\int_\gamma |G_{n_k} - G| \overset{k}{\to} 0$ holds for 2-a.e. γ. In particular, $\int_\gamma G_{n_k} \overset{k}{\to} \int_\gamma G$ for 2-a.e. γ.*

Proof Up to subsequence, assume that $\|G_n - G\|_{L^2(\mathfrak{m})} \leq 1/2^n$ for every $n \in \mathbb{N}$. Let us define

$$\Gamma_k := \left\{ \gamma \in \Gamma(\mathrm{X}) \;\middle|\; \varlimsup_{n\to\infty} \int_\gamma |G_n - G| > \frac{1}{k} \right\} \quad \text{for every } k \in \mathbb{N} \setminus \{0\}.$$

Observe that $\int_\gamma |G_n - G| \to 0$ as $n \to \infty$ for every $\gamma \notin \bigcup_k \Gamma_k$, thus to prove the statement it is sufficient to show that $\mathrm{Mod}_2(\Gamma_k) = 0$ holds for any $k \geq 1$. Let $k \geq 1$ be fixed. For any $m \in \mathbb{N}$ we define $\rho_m := k \sum_{n \geq m} |G_n - G|$. For every curve $\gamma \in \Gamma_k$ there is $n \geq m$ such that $\int_\gamma |G_n - G| \geq 1/k$, whence $\int_\gamma \rho_m \geq 1$, in other words ρ_m is admissible for Γ_k. Moreover, one has that $\|\rho_m\|_{L^2(\mathfrak{m})} \leq k \sum_{n \geq m} \|G_n - G\|_{L^2(\mathfrak{m})} \leq k/2^{m-1}$ for every $m \in \mathbb{N}$. Hence $\mathrm{Mod}_2(\Gamma_k) \leq \|\rho_m\|^2_{L^2(\mathfrak{m})} \overset{m}{\to} 0$, getting the statement. □

Theorem 2.2.23 *Given any $n \in \mathbb{N}$, let G_n be a 2-weak upper gradient for some function f_n. Suppose further that $G_n \to G$ and $f_n \to f$ in $L^2(\mathfrak{m})$, for suitable Borel functions $f : \mathrm{X} \to \mathbb{R}$ and $G : \mathrm{X} \to [0, +\infty]$. Then there is a Borel function $\bar{f} : \mathrm{X} \to \mathbb{R}$ such that $\bar{f}(x) = f(x)$ holds for $\mathfrak{m}$-a.e. $x \in \mathrm{X}$ and G is a 2-weak upper gradient for $\bar{f}$.*

Proof Possibly passing to a not relabeled subsequence, we can assume without loss of generality that $f_n \to f$ in the $\mathfrak{m}$-a.e. sense. In addition, we can also suppose that $\int_\gamma |G_n - G| \to 0$ holds for 2-a.e. γ by Theorem 2.2.22. Call $\tilde{f}(x) := \varlimsup_n f_n(x)$ for every $x \in \mathrm{X}$. Then $\tilde{f} = f$ holds $\mathfrak{m}$-a.e. in X, thus accordingly $\tilde{f} \in L^2(\mathfrak{m})$. Let us define

$$\begin{aligned}
\Gamma &:= \left\{ \gamma \in \Gamma(\mathrm{X}) \;\middle|\; \int_\gamma |G_n - G| \overset{n}{\to} 0,\ f_n \circ \gamma \text{ is AC, } \left|(f_n \circ \gamma)'\right| \right.\\
&\qquad \left. \leq G_n \circ \gamma\, |\dot{\gamma}| \text{ for all } n \in \mathbb{N} \right\},\\
\Gamma' &:= \left\{ \gamma \in \Gamma(\mathrm{X}) \;\middle|\; \text{either } \left|\tilde{f}(\gamma_I)\right| < +\infty \text{ or } \left|\tilde{f}(\gamma_F)\right| < +\infty \right\},\\
\mathcal{N} &:= \left\{ \gamma \in \Gamma(\mathrm{X}) \;\middle|\; \left|\tilde{f}(\gamma_t)\right| = +\infty \text{ for every } t \in \mathrm{Dom}(\gamma) \right\}.
\end{aligned}$$

Note that $\mathrm{Mod}_2(\Gamma^c) = 0$ because G_n is a 2-weak upper gradient of f_n for any $n \in \mathbb{N}$. Furthermore, we have that $\mathrm{Mod}_2(\mathcal{N}) = 0$: indeed, for every $\varepsilon > 0$ the function $\rho := \varepsilon|\tilde{f}|$ is admissible for $\mathcal{N}$ and $\|\rho\|_{L^2(\mathfrak{m})} \leq \varepsilon \|f\|_{L^2(\mathfrak{m})}$. We now claim that

$$\big|\tilde{f}(\gamma_F) - \tilde{f}(\gamma_I)\big| \leq \int_\gamma G \quad \text{for every } \gamma \in \Gamma \cap \Gamma'. \tag{2.42}$$

To prove it, just observe that $\big|\tilde{f}(\gamma_F) - \tilde{f}(\gamma_I)\big| \leq \overline{\lim}_n \big|f_n(\gamma_F) - f_n(\gamma_I)\big| \leq \lim_n \int_\gamma G_n = \int_\gamma G$ for every $\gamma \in \Gamma \cap \Gamma'$. We can use (2.42) to prove that

$$\big|\tilde{f}(\gamma_F) - \tilde{f}(\gamma_I)\big| \leq \int_\gamma G \quad \text{for every } \gamma \in \Gamma \setminus \mathcal{N}. \tag{2.43}$$

Indeed: fix $\gamma \in \Gamma \setminus \mathcal{N}$. There exists $t_0 \in \mathrm{Dom}(\gamma)$ such that $\big|\tilde{f}(\gamma_{t_0})\big| < +\infty$. Call $\gamma^1 := \gamma_{|[I,t_0]}$ and $\gamma^2 := \gamma_{|[t_0,F]}$. We have that $\gamma^1, \gamma^2 \in \Gamma \cap \Gamma'$, so that (2.42) yields

$$\big|\tilde{f}(\gamma_F) - \tilde{f}(\gamma_I)\big| \leq \big|\tilde{f}(\gamma_F) - \tilde{f}(\gamma_{t_0})\big| + \big|\tilde{f}(\gamma_{t_0}) - \tilde{f}(\gamma_{t_0})\big| \leq \int_{\gamma^1} G + \int_{\gamma^2} G = \int_\gamma G.$$

Since $\mathrm{Mod}_2\big(\Gamma(\mathrm{X}) \setminus (\Gamma \setminus \mathcal{N})\big) = 0$, we deduce from (2.43) that G is a 2-weak upper gradient of the function $\bar{f} : \mathrm{X} \to \mathbb{R}$, defined by $\bar{f} := \chi_{\{\tilde{f} < +\infty\}} \tilde{f}$, which $\mathfrak{m}$-a.e. coincides with f. □

We now define the Sobolev space $W^{1,2}_{\mathrm{Sh}}(\mathrm{X})$, where 'Sh' stays for Shanmugalingam, who first introduced such object.

Definition 2.2.24 We define the Sobolev space $W^{1,2}_{\mathrm{Sh}}(\mathrm{X})$ as the set of all $f \in L^2(\mathfrak{m})$ such that there exist two Borel functions $\bar{f} : \mathrm{X} \to \mathbb{R}$ and $G : \mathrm{X} \to [0, +\infty]$ in $L^2(\mathfrak{m})$ satisfying these properties: $\bar{f}(x) = f(x)$ for $\mathfrak{m}$-a.e. $x \in \mathrm{X}$ and G is a 2-weak upper gradient for $\bar{f}$.

We endow the vector space $W^{1,2}_{\mathrm{Sh}}(\mathrm{X})$ with the norm given by

$$\|f\|^2_{W^{1,2}_{\mathrm{Sh}}(\mathrm{X})} := \|f\|^2_{L^2(\mathfrak{m})} + \inf \|G\|^2_{L^2(\mathfrak{m})} \quad \text{for every } f \in W^{1,2}_{\mathrm{Sh}}(\mathrm{X}), \tag{2.44}$$

where the infimum is taken among all Borel functions $G : \mathrm{X} \to [0, +\infty]$ that are 2-weak upper gradients of some Borel representative of f.

Remark 2.2.25 (Minimal 2-Weak Upper Gradient) Given any $f \in W^{1,2}_{\mathrm{Sh}}(\mathrm{X})$, there exists a *minimal 2-weak upper gradient* $|Df|_{\mathrm{Sh}}$, where minimality has to be intended in the $\mathfrak{m}$-a.e. sense. In other words, if $\bar{f}$ is a Borel representative of f and G is a 2-weak upper gradient for $\bar{f}$, then $|Df|_{\mathrm{Sh}} \leq G$ holds $\mathfrak{m}$-a.e. in X. It thus holds that

$$\|f\|^2_{W^{1,2}_{\mathrm{Sh}}(\mathrm{X})} = \|f\|^2_{L^2(\mathfrak{m})} + \big\||Df|_{\mathrm{Sh}}\big\|^2_{L^2(\mathfrak{m})} \quad \text{for every } f \in W^{1,2}_{\mathrm{Sh}}(\mathrm{X}). \tag{2.45}$$

These statements follow from Proposition 2.2.21 and Theorem 2.2.23. ■

Lemma 2.2.26 *Let* Γ *be a subset of* $AC([0,1],\mathrm{X})$ *such that* $\mathrm{Mod}_2(\Gamma) = 0$. *Then* $\pi^*(\Gamma) = 0$ *for every test plan* π *on* X, *where* π^* *denotes the outer measure induced by* π.

Proof Take ρ admissible for Γ. The function $(\gamma, t) \mapsto \rho(\gamma_t)|\dot{\gamma}_t|$ is Borel, hence $\{\gamma : \int_\gamma \rho \geq 1\}$ is a π-measurable set by Fubini theorem. Observe that such set contains Γ, so that

$$\begin{aligned}\pi^*(\Gamma) &\leq \iint_\gamma \rho \,\mathrm{d}\pi(\gamma) = \int_0^1 \int \rho(\gamma_t)|\dot{\gamma}_t| \,\mathrm{d}\pi(\gamma)\,\mathrm{d}t \\ &\leq \left(\int_0^1 \int \rho^2(\gamma_t)\,\mathrm{d}\pi(\gamma)\,\mathrm{d}t\right)^{1/2} \left(\int_0^1 \int |\dot{\gamma}_t|^2 \,\mathrm{d}\pi(\gamma)\,\mathrm{d}t\right)^{1/2} \\ &\leq \sqrt{\mathrm{Comp}(\pi)} \left(\int_0^1 \int |\dot{\gamma}_t|^2 \,\mathrm{d}\pi(\gamma)\,\mathrm{d}t\right)^{1/2} \left(\int \rho^2 \,\mathrm{dm}\right)^{1/2}.\end{aligned}$$

By arbitrariness of ρ, we conclude that $\pi^*(\Gamma) = 0$. □

Remark 2.2.27 It holds that

$$\begin{aligned}&|Df|_{*,a} \geq |Df|_* \geq |Df|_{\mathrm{Ch}} \geq |Df|_{\mathrm{Sh}} \geq |Df|, \\ &W^{1,2}_{*,a}(\mathrm{X}) \subseteq W^{1,2}_*(\mathrm{X}) \subseteq W^{1,2}_{\mathrm{Ch}}(\mathrm{X}) \subseteq W^{1,2}_{\mathrm{Sh}}(\mathrm{X}) \subseteq W^{1,2}(\mathrm{X}).\end{aligned} \tag{2.46}$$

To prove $|Df|_{\mathrm{Ch}} \geq |Df|_{\mathrm{Sh}}$, observe that any upper gradient is a 2-weak upper gradient. On the other hand, to show $|Df|_{\mathrm{Sh}} \geq |Df|$ it suffices to apply Lemma 2.2.26. ■

To prove the equivalence of all the notions of Sobolev function on metric measure spaces described so far, we need the following deep approximation result, whose proof we omit:

Theorem 2.2.28 (Ambrosio-Gigli-Savaré) *Let* $(\mathrm{X}, \mathsf{d}, \mathfrak{m})$ *be any metric measure space. Then Lipschitz functions in* X *are* dense in energy *in* $W^{1,2}(\mathrm{X})$, *namely for every* $f \in W^{1,2}(\mathrm{X})$ *there exists a sequence* $(f_n)_n \subseteq \mathrm{LIP}(\mathrm{X}) \cap L^2(\mathfrak{m})$ *such that* $f_n \to f$ *and* $\mathrm{lip}_a(f_n) \to |Df|$ *in* $L^2(\mathfrak{m})$, *thus accordingly also* $\mathrm{lip}(f_n) \to |Df|$ *and* $|Df_n| \to |Df|$ *in* $L^2(\mathfrak{m})$.

In particular, we have that $W^{1,2}_{*,a}(\mathrm{X}) = W^{1,2}(\mathrm{X})$ *and that the equality* $|Df|_{*,a} = |Df|$ *is satisfied* $\mathfrak{m}$*-a.e. for every* $f \in W^{1,2}(\mathrm{X})$.

We directly deduce from Theorem 2.2.28 that all inequalities and inclusions in (2.46) are actually equalities. In other words, all the several approaches we saw are in fact equivalent.

Remark 2.2.29 In order to prove that $|Df|_{\mathrm{Ch}} = |Df|_{\mathrm{Sh}}$, the following fact is sufficient:

$$\begin{aligned}&\text{Let } G \text{ be a 2-weak upper gradient for } f \text{ and let } \varepsilon > 0. \text{ Then there exists}\\&\text{an upper gradient } \widetilde{G} \text{ for } f \text{ such that } \|\widetilde{G}\|_{L^2(\mathrm{m})} \leq \|G\|_{L^2(\mathrm{m})} + \varepsilon.\end{aligned} \tag{2.47}$$

To prove it: call Γ the set of $\gamma \in \Gamma(\mathrm{X})$ such that $\big|f(\gamma_F) - f(\gamma_I)\big| > \int_\gamma G$, so that $\mathrm{Mod}_2(\Gamma) = 0$. We first claim that

$$\exists\, \rho : \mathrm{X} \to [0, +\infty] \text{ Borel such that } \int_\gamma \rho = +\infty \text{ for all } \gamma \in \Gamma \text{ and } \|\rho\|_{L^2(\mathrm{m})} \leq \varepsilon. \tag{2.48}$$

Indeed, there is $(\rho_n)_n$ such that $\int_\gamma \rho_n \geq 1$ and $\|\rho_n\|_{L^2(\mathrm{m})} \leq \varepsilon/2^n$ for all $n \in \mathbb{N}$ and $\gamma \in \Gamma$. Thus it can be easily seen that the function $\rho := \sum_{n\geq 1} \rho_n$ satisfies (2.48): for every $\gamma \in \Gamma$ we have that $\int_\gamma \rho = \lim_{m\to\infty} \sum_{n=1}^m \int_\gamma \rho_n \geq \lim_{m\to\infty} m = +\infty$, while $\|\rho\|_{L^2(\mathrm{m})} \leq \sum_{n\geq 1} \|\rho_n\|_{L^2(\mathrm{m})} \leq \varepsilon$.

Finally, let us call $\widetilde{G} := G + \rho$. Clearly $\widetilde{G}$ satisfies (2.47): if $\gamma \in \Gamma$ then $\int_\gamma \widetilde{G} = +\infty$, while if $\gamma \notin \Gamma$ then $\big|f(\gamma_F) - f(\gamma_I)\big| \leq \int_\gamma G \leq \int_\gamma \widetilde{G}$, i.e. $\widetilde{G}$ is an upper gradient of f; moreover, one has $\|\widetilde{G}\|_{L^2(\mathrm{m})} \leq \|G\|_{L^2(\mathrm{m})} + \|\rho\|_{L^2(\mathrm{m})} \leq \|G\|_{L^2(\mathrm{m})} + \varepsilon$. This concludes the proof. ∎

Bibliographical Remarks

The first definition of Sobolev space on a metric measure space has been proposed by Hajłasz in [21]. The notion that in [21] is analogous to that of minimal weak upper gradient discussed here is non-local in nature; as such, the definition in [21] lacks one of the key properties that Sobolev functions have in the classical smooth setting and is not suitable to the discussion we intend to pursue here, where locality of minimal weak upper gradients plays a pivotal role.

The paper which introduced the by-now most widely used notion of Sobolev spaces on metric measure spaces is the seminal work of Cheeger [13], of which we gave an account in Sect. 2.2.1. Cheeger's approach was at least in part inspired by Koskela and MacManus, who in [22] introduced the notion of upper gradient in a metric setting.

Soon after Cheeger's contribution, Shanmugalingam proposed in [28] the alternative definition we recalled in Sect. 2.2.2, and proved the equivalence with Cheeger's one: her theory is an adaptation to the metric setting of the results contained in [16], which are in turn inspired by the ideas of [23].

Finally, the approach to Sobolev functions by duality with the concept of test plan has been proposed in [4], where also the equivalence with Cheeger's and Shanmugalingam's approach has been proved. The presentation we gave here also takes into account some ideas contained in [18]. Theorem 2.1.21, constitutes a (partially) new result, inspired by the study of test plans carried out in [17]. The formulation of the density in energy of Lipschitz functions given here, namely Theorem 2.2.28, comes from [3], but the argument was in fact mostly contained in [4].

Chapter 3
The Theory of Normed Modules

This chapter is devoted to the study of the so-called *normed modules* over metric measure spaces. These represent a tool that has been introduced by Gigli in order to build up a differential structure on nonsmooth spaces. In a few words, an $L^2(\mathfrak{m})$-normed $L^\infty(\mathfrak{m})$-module is a generalisation of the concept of 'space of 2-integrable sections of some measurable bundle'; it is an algebraic module over the commutative ring $L^\infty(\mathfrak{m})$ that is additionally endowed with a pointwise norm operator. This notion, its basic properties and some of its technical variants constitute the topics of Sect. 3.1.

Many constructions are available in the framework of normed modules. For instance, it is possible to take duals, tensor products and pullbacks of normed modules. Furthermore, there is a special class of normed modules, called *Hilbert modules*, which have nicer functional analytic properties. All these objects are described in detail in Sect. 3.2.

3.1 Definition of Normed Module and Basic Properties

3.1.1 L^2-Normed L^∞-Modules

Let $(\mathrm{X}, \mathsf{d}, \mathfrak{m})$ be a fixed metric measure space.

Definition 3.1.1 (L^2-Normed L^∞-Module) We define an $L^2(\mathfrak{m})$-*normed* $L^\infty(\mathfrak{m})$-*module*, or briefly *module*, as a quadruplet $\big(\mathscr{M}, \|\cdot\|_{\mathscr{M}}, \cdot, |\cdot|\big)$ with the following properties:

N. Gigli, E. Pasqualetto, *Lectures on Nonsmooth Differential Geometry*,
SISSA Springer Series 2, https://doi.org/10.1007/978-3-030-38613-9_3

i) $(\mathscr{M}, \|\cdot\|_{\mathscr{M}})$ is a Banach space.

ii) The *multiplication by L^∞-functions* $\cdot : L^\infty(\mathfrak{m}) \times \mathscr{M} \to \mathscr{M}$ is a bilinear map satisfying

$$\begin{aligned} f\cdot(g\cdot v) &= (fg)\cdot v && \text{for every } f,g \in L^\infty(\mathfrak{m}) \text{ and } v \in \mathscr{M}, \\ \hat{1}\cdot v &= v && \text{for every } v \in \mathscr{M}, \end{aligned} \tag{3.1}$$

where $\hat{1}$ denotes the (equivalence class of the) function on X identically equal to 1.

iii) The *pointwise norm* $|\cdot| : \mathscr{M} \to L^2(\mathfrak{m})$ satisfies

$$\begin{aligned} |v| &\geq 0 \ \ \mathfrak{m}\text{-a.e.} && \text{for every } v \in \mathscr{M}, \\ |f\cdot v| &= |f||v| \ \ \mathfrak{m}\text{-a.e.} && \text{for every } f \in L^\infty(\mathfrak{m}) \text{ and } v \in \mathscr{M}, \\ \|v\|_{\mathscr{M}} &= \big\||v|\big\|_{L^2(\mathfrak{m})} && \text{for every } v \in \mathscr{M}. \end{aligned} \tag{3.2}$$

For the sake of brevity, we shall often write fv instead of $f\cdot v$.

Proposition 3.1.2 *Let $\mathscr{M}$ be a module. Then:*

i) $\|fv\|_{\mathscr{M}} \leq \|f\|_{L^\infty(\mathfrak{m})}\|v\|_{\mathscr{M}}$ *for every $f \in L^\infty(\mathfrak{m})$ and $v \in \mathscr{M}$.*

ii) $\lambda v = \hat{\lambda} v$ *for every $\lambda \in \mathbb{R}$, where $\hat{\lambda}$ denotes the (equivalence class of the) function on* X *identically equal to λ.*

iii) *It holds that*

$$\left.\begin{aligned} |v+w| &\leq |v| + |w| \\ |\lambda v| &= |\lambda||v| \end{aligned}\right\} \quad \mathfrak{m}\text{-a.e.} \quad \text{for every } v,w \in \mathscr{M} \text{ and } \lambda \in \mathbb{R}. \tag{3.3}$$

Proof The proof goes as follows:

i) Simply notice that

$$\|fv\|_{\mathscr{M}} = \big\||f||v|\big\|_{L^2(\mathfrak{m})} \leq \|f\|_{L^\infty(\mathfrak{m})}\big\||v|\big\|_{L^2(\mathfrak{m})} = \|f\|_{L^\infty(\mathfrak{m})}\|v\|_{\mathscr{M}}$$

is verified for every $f \in L^\infty(\mathfrak{m})$ and $v \in \mathscr{M}$ by (3.2) and by Hölder inequality.

ii) Given any $\lambda \in \mathbb{R}$ and $v \in \mathscr{M}$, we have that $\hat{\lambda}v = (\lambda\hat{1})v = \lambda(\hat{1}v) = \lambda v$ by (3.1) and by bilinearity of the multiplication by L^∞-functions.

iii) Fix $\lambda \in \mathbb{R}$ and $v, w \in \mathscr{M}$. Clearly $|\lambda v| = |\hat{\lambda}v| = |\hat{\lambda}||v| = |\lambda||v|$ holds $\mathfrak{m}$-a.e. in X as a consequence of ii). On the other hand, in order to prove that $|v+w| \leq |v|+|w|$ holds $\mathfrak{m}$-a.e. we argue by contradiction: suppose the contrary, thus there exist $a, b, c \in \mathbb{R}$ with $a+b<c$ and $E \subseteq$ X Borel with $\mathfrak{m}(E) > 0$ such that

$$\begin{cases} |v| \leq a \\ |w| \leq b \\ |v+w| \geq c \end{cases} \quad \text{holds } \mathfrak{m}\text{-a.e. in } E. \tag{3.4}$$

Hence we deduce from (3.4) that

$$\begin{aligned}\|\chi_E(v+w)\|_{\mathscr{M}} &= \left(\int_E |v+w|^2\,\mathrm{dm}\right)^{1/2} \geq c\,\mathrm{m}(E)^{1/2} > (a+b)\,\mathrm{m}(E)^{1/2} \\ &\geq \left(\int_E |v|^2\,\mathrm{dm}\right)^{1/2} + \left(\int_E |w|^2\,\mathrm{dm}\right)^{1/2} = \|\chi_E\, v\|_{\mathscr{M}} + \|\chi_E\, w\|_{\mathscr{M}},\end{aligned}$$

which contradicts the fact the $\|\cdot\|_{\mathscr{M}}$ is a norm. Therefore (3.3) is proved. □

Exercise 3.1.3 Let V, W, Z be normed spaces. Let $B : V \times W \to Z$ be a bilinear operator.

i) Suppose V is Banach. Show that B is continuous if and only if both $B(v, \cdot)$ and $B(\cdot, w)$ are continuous for every $v \in V$ and $w \in W$.
ii) Prove that B is continuous if and only if there exists a constant $C > 0$ such that the inequality $\|B(v, w)\|_Z \leq C\,\|v\|_V\,\|w\|_W$ holds for every $(v, w) \in V \times W$. ■

Remark 3.1.4 It directly follows from property i) of Proposition 3.1.2 and from Exercise 3.1.3 that the multiplication by L^∞-functions is a continuous operator. ■

Example 3.1.5 We provide some examples of $L^2(\mathfrak{m})$-normed $L^\infty(\mathfrak{m})$-modules:

i) The space $L^2(\mathfrak{m})$ itself can be viewed as a module.
ii) More in general, the space $L^2(\mathrm{X}, \mathbb{B})$ is a module for every Banach space $\mathbb{B}$. (In the case in which $\mathfrak{m}$ is a finite measure, the space $L^2(\mathrm{X}, \mathbb{B})$ is defined as the set of all elements v of $L^1(\mathrm{X}, \mathbb{B})$ for which the quantity $\int \|v(x)\|_{\mathbb{B}}^2\,\mathrm{d}\mathfrak{m}(x)$ is finite.)
iii) The space of L^2-vector fields on a Riemannian manifold is a module with respect to the pointwise operations. Actually, the same holds true even for a Finsler manifold (i.e., roughly speaking, a manifold endowed with a norm on each tangent space).
iv) The space of L^2-sections of a 'measurable bundle' over X (whose fibers are Banach spaces) has a natural structure of L^2-normed L^∞-module. For instance, consider the spaces of covector fields or higher dimensional tensors with pointwise norm in L^2. ■

Remark 3.1.6 One can imagine a module $\mathscr{M}$, in a sense, as the space of L^2-sections of some measurable Banach bundle over X; cf. the Serre-Swan theorem. ■

Definition 3.1.7 Let $\mathscr{M}$ be a module and $v \in \mathscr{M}$. Then let us define

$$\{v = 0\} := \big\{|v| = 0\big\}. \tag{3.5}$$

Notice that $\{v = 0\}$ is a Borel set in X, defined up to $\mathfrak{m}$-a.e. equality. Similarly, one can define $\{v \neq 0\}$, $\{v = w\}$ for $w \in \mathscr{M}$ and so on.

It is trivial to check that for any $E \subseteq \mathrm{X}$ Borel one has

$$\chi_E\, v = 0 \iff |v| = 0 \quad \mathfrak{m}\text{-a.e. in } E. \tag{3.6}$$

Indeed, $\chi_E\, v = 0$ if and only if $\|\chi_E\, v\|_{\mathscr{M}} = 0$ if and only if $\int_E |v|^2\, \mathrm{d}\mathfrak{m} = 0$ if and only if $|v| = 0$ holds $\mathfrak{m}$-a.e. in E. If the two conditions in (3.6) hold, we say that v *is* $\mathfrak{m}$*-a.e. null in* E.

Remark 3.1.8 Let $\mathscr{M}$ be a module. Let $v \in \mathscr{M}$. Suppose to have a sequence $(E_n)_n$ of Borel subsets of X such that $\chi_{E_n} v = 0$ for every $n \in \mathbb{N}$. Then v is $\mathfrak{m}$-a.e. null in $\bigcup_n E_n$, as one can readily deduce from the characterisation (3.6). ■

Proposition 3.1.9 ($\mathfrak{m}$-Essential Union) *Let $\{E_i\}_{i\in I}$ be a (not necessarily countable) family of Borel subsets of* X*. Then there exists a Borel set* $E \subseteq$ X *such that:*

i) $\mathfrak{m}(E_i \setminus E) = 0$ *for every* $i \in I$.
ii) *If* $F \subseteq$ X *Borel satisfies* $\mathfrak{m}(E_i \setminus F) = 0$ *for all* $i \in I$, *then* $\mathfrak{m}(E \setminus F) = 0$.

Such set E, *which is called the* $\mathfrak{m}$-essential union *of* $\{E_i\}_{i\in I}$, *is* $\mathfrak{m}$*-a.e. unique, in the sense that any other Borel set* $\widetilde{E}$ *with the same properties must satisfy* $\mathfrak{m}(E\Delta\widetilde{E}) = 0$.

Proof Uniqueness follows from condition ii). To prove existence, assume without loss of generality that $\mathfrak{m} \in \mathscr{P}(\mathrm{X})$ (otherwise, we can replace $\mathfrak{m}$ with a Borel probability measure $\tilde{\mathfrak{m}}$ such that $\tilde{\mathfrak{m}} \ll \mathfrak{m} \ll \tilde{\mathfrak{m}}$, which can be built as in the proof of STEP 5 of Theorem 2.1.28). Let us denote by $\mathcal{A}$ the family of all finite unions of the E_i's and call $S := \sup\big\{\mathfrak{m}(A) \,:\, A \in \mathcal{A}\big\}$. Hence there is an increasing sequence of sets $(A_n)_n \subseteq \mathcal{A}$ with $\mathfrak{m}(A_n) \nearrow S$. Define $E := \bigcup_n A_n$. Clearly E satisfies i): if not, there exists some $i \in I$ such that $\mathfrak{m}(E_i \setminus E) > 0$, whence

$$S = \mathfrak{m}(E) < \mathfrak{m}(E \cup E_i) = \lim_{n\to\infty} \mathfrak{m}(A_n \cup E_i) \leq S,$$

which leads to a contradiction. Moreover, the set E can be clearly written as countable union of elements in $\{E_i\}_{i\in I}$, say $E = \bigcup_{j\in J} E_j$ for some $J \subseteq I$ countable. Hence for any $F \subseteq$ X Borel with $\mathfrak{m}(E_i \setminus F) = 0$ for each $i \in I$, it holds that

$$\mathfrak{m}(E \setminus F) \leq \sum_{j\in J} \mathfrak{m}(E_j \setminus F) = 0,$$

proving ii) and accordingly the existence part of the statement. □

Given any $v \in \mathscr{M}$, it holds that $\{v = 0\}$ can equivalently described as the $\mathfrak{m}$-essential union of all Borel sets $E \subseteq$ X such that $\chi_E\, v = 0$.

Example 3.1.10 Define $E_i := \{i\}$ for every $i \in \mathbb{R}$. Then the set-theoretic union of $\{E_i\}_{i\in\mathbb{R}}$ is the whole real line $\mathbb{R}$, while its $\mathcal{L}^1$-essential union is given by the empty set. ■

Definition 3.1.11 (Localisation of a Module) Let $\mathscr{M}$ be a module. Let E be any Borel subset of X. Then we define

$$\mathscr{M}_{|E} := \big\{\chi_E\, v \,:\, v \in \mathscr{M}\big\} \subseteq \mathscr{M}. \tag{3.7}$$

It turns out that the space $\mathscr{M}_{|E}$ is stable under all module operations and is complete, thus it is a submodule of $\mathscr{M}$.

Proposition 3.1.12 *Let S be any subset of $\mathscr{M}$. Let us define*

$$\mathscr{M}(S) := \mathscr{M}\text{-closure of } \mathcal{S} := \left\{ \sum_{i=1}^{n} f_i\, v_i \;\middle|\; n \in \mathbb{N},\ (f_i)_{i=1}^{n} \subseteq L^\infty(\mathfrak{m}),\ (v_i)_{i=1}^{n} \subseteq S \right\}. \tag{3.8}$$

Then $\mathscr{M}(S)$ is the smallest submodule of $\mathscr{M}$ containing S.

Proof We omit the simple proof of the fact that $\mathscr{M}(S)$ inherits from $\mathscr{M}$ a module structure. Moreover, any module containing the set S must contain also $\mathcal{S}$ and must be closed, whence the required minimality. □

Definition 3.1.13 (Generators) The module $\mathscr{M}(S)$ that we defined in Proposition 3.1.12 is called the *module generated by* S. Moreover, if $E \subseteq \mathrm{X}$ is Borel and $\mathscr{M}(S)_{|E} = \mathscr{M}_{|E}$, then we say that S *generates* $\mathscr{M}$ *on* E

Remark 3.1.14 The space $L^2(\mathfrak{m})$, viewed as a module, can be generated by a single element, namely by any $L^2(\mathfrak{m})$-function which is $\mathfrak{m}$-a.e. different from 0. ■

Proposition 3.1.15 *Let V be a vector subspace of $\mathscr{M}$. Then $\mathscr{M}(V)$ is the $\mathscr{M}$-closure of*

$$\mathcal{V} := \left\{ \sum_{i=1}^{n} \chi_{E_i} v_i \;\middle|\; n \in \mathbb{N},\ (E_i)_{i=1}^{n} \text{ Borel partition of } \mathrm{X},\ (v_i)_{i=1}^{n} \subseteq V \right\}. \tag{3.9}$$

Proof The inclusion $\mathrm{cl}_{\mathscr{M}}(\mathcal{V}) \subseteq \mathscr{M}(V)$ is trivial. To prove the converse inclusion, since $\mathcal{V}$ and accordingly also $\mathrm{cl}_{\mathscr{M}}(\mathcal{V})$ are vector spaces, it suffices to show that $f\, v \in \mathrm{cl}_{\mathscr{M}}(\mathcal{V})$ whenever we have $f \in L^\infty(\mathfrak{m})$ and $v \in V \setminus \{0\}$. Given any $\varepsilon > 0$, pick a simple function $g = \sum_{i=1}^{n} \alpha_i\, \chi_{E_i}$ such that $\|f - g\|_{L^\infty(\mathfrak{m})} \leq \varepsilon / \|v\|_{\mathscr{M}}$. Then $\|f\, v - g\, v\|_{\mathscr{M}} \leq \varepsilon$ and $g\, v = \sum_{i=1}^{n} \chi_{E_i}(\alpha_i\, v) \in \mathcal{V}$, as required. Hence the statement is achieved. □

Remark 3.1.16 Let $\mathscr{M}$ be a module. Then the pointwise norm $|\cdot| : \mathscr{M} \to L^2(\mathfrak{m})$ is continuous.

Indeed, since $\big||v| - |w|\big| \leq |v - w|$ holds $\mathfrak{m}$-a.e. for any $v, w \in \mathscr{M}$ by (3.3), one immediately deduces that $\big\||v| - |w|\big\|_{L^2(\mathfrak{m})} \leq \|v - w\|_{\mathscr{M}}$ for every $v, w \in \mathscr{M}$. ■

Lemma 3.1.17 *Let $S \subseteq \mathscr{M}$ be a separable subset with the following property: the $L^\infty(\mathfrak{m})$-linear combinations of elements of S are dense in $\mathscr{M}$. Then the space $\mathscr{M}$ is separable.*

Proof Pick a countable dense subset $(v_n)_n$ of S. It is then clear that the $L^\infty(\mathfrak{m})$-linear combinations of the v_n's are dense in $\mathscr{M}$. It only remains to show that the family of such combinations is separable. Now fix a Borel probability measure $\mathfrak{m}'$ on

X with $\mathfrak{m} \ll \mathfrak{m}' \ll \mathfrak{m}$. Then there exists a countable family $\mathcal{A}$ of Borel subsets of X such that for any $E \subseteq$ X Borel there is a sequence $(E_i)_i \subseteq \mathcal{A}$ with $\mathfrak{m}'(E_i \Delta E) \to 0$. For instance, define $\mathcal{A}$ as the set of all open balls with rational radii that are centered at some fixed countable dense subset of X. Hence let us define the separable set D as

$$D := \left\{ \sum_{n=0}^{N} \alpha_n \chi_{E_n} v_n \,\middle|\, N \in \mathbb{N},\ (\alpha_n)_{n=0}^{N} \subseteq \mathbb{Q},\ (E_n)_{n=0}^{N} \subseteq \mathcal{A} \right\}.$$

It can be readily proved that the set of all $L^\infty(\mathfrak{m})$-linear combinations of the v_n's is contained in the closure of D. Therefore the statement is achieved. □

3.1.2 L^0-Normed L^0-Modules

We introduce an alternative notion of normed module over (X, d, $\mathfrak{m}$), for which no integrability assumption is required:

Definition 3.1.18 (L^0-Normed L^0-Module) Let (X, d, $\mathfrak{m}$) be a metric measure space. We define an $L^0(\mathfrak{m})$-*normed* $L^0(\mathfrak{m})$-*module* as any quadruple $(\mathscr{M}^0, \tau, \cdot, |\cdot|)$, where:

i) $(\mathscr{M}^0, \tau)$ is a topological vector space.

ii) The bilinear map $\cdot : L^0(\mathfrak{m}) \times \mathscr{M}^0 \to \mathscr{M}^0$ satisfies $f \cdot (g \cdot v) = (fg) \cdot v$ and $\hat{1} \cdot v = v$ for every $f, g \in L^0(\mathfrak{m})$ and $v \in \mathscr{M}^0$.

iii) The map $|\cdot| : \mathscr{M}^0 \to L^0(\mathfrak{m})$, which satisfies both $|v| \geq 0$ and $|f \cdot v| = |f||v|$ $\mathfrak{m}$-a.e. for every $v \in \mathscr{M}^0$ and $f \in L^0(\mathfrak{m})$, is such that the function $\mathsf{d}_{\mathscr{M}^0} : \mathscr{M}^0 \times \mathscr{M}^0 \to [0, +\infty)$, defined by

$$\mathsf{d}_{\mathscr{M}^0}(v, w) := \int |v - w| \wedge 1 \, d\mathfrak{m}' \quad \text{for some } \mathfrak{m}' \in \mathscr{P}(\mathrm{X}) \text{ with } \mathfrak{m} \ll \mathfrak{m}' \ll \mathfrak{m}, \tag{3.10}$$

is a complete distance on $\mathscr{M}^0$ that induces the topology τ.

Remark 3.1.19 The topology τ in the definition of an L^0-normed module does not depend on the particular choice of the measure $\mathfrak{m}'$. Indeed, it holds that a given sequence $(v_n)_n \subseteq \mathscr{M}^0$ is $\mathsf{d}_{\mathscr{M}^0}$-Cauchy if and only if

$$\varlimsup_{n,m\to\infty} \mathfrak{m}\Big(E \cap \big\{|v_n - v_m| > \varepsilon\big\}\Big) = 0 \quad \begin{array}{l}\text{for every } \varepsilon > 0 \text{ and } E \subseteq \mathrm{X} \\ \text{Borel with } \mathfrak{m}(E) < +\infty.\end{array}$$

Such statement can be achieved by arguing as in the proof of Proposition 1.1.19. ■

Definition 3.1.20 (L^0-Completion) Let $\mathscr{M}$ be an $L^2(\mathfrak{m})$-normed module. Then we define an $L^0(\mathfrak{m})$-*completion* of $\mathscr{M}$ as any couple $(\mathscr{M}^0, i)$, where $\mathscr{M}^0$ is an $L^0(\mathfrak{m})$-

normed module and the map $i : \mathscr{M} \to \mathscr{M}^0$ is a linear operator with dense image that preserves the pointwise norm, i.e. such that the equality $|i(v)| = |v|$ holds $\mathfrak{m}$-a.e. for every $v \in \mathscr{M}$.

Remark 3.1.21 Let $\mathscr{M}^0$ be an $L^0(\mathfrak{m})$-normed module. Then

$$\begin{aligned} |\cdot| : \mathscr{M}^0 \to L^0(\mathfrak{m}) &\quad \text{is continuous,} \\ \cdot : L^0(\mathfrak{m}) \times \mathscr{M}^0 \to \mathscr{M}^0 &\quad \text{is continuous.} \end{aligned} \tag{3.11}$$

To prove the first in (3.11), we begin by observing that $|v + w| \leq |v| + |w|$ holds $\mathfrak{m}$-a.e. for any $v, w \in \mathscr{M}^0$: if not, we can find constants $a, b, c > 0$ with $a + b < c$ and a Borel set $P \subseteq \mathrm{X}$ with $\mathfrak{m}(P) > 0$ such that $|v| < a$, $|w| < b$ and $|v + w| > c$ hold $\mathfrak{m}$-a.e. on P, so that

$$\begin{aligned} \mathsf{d}_{\mathscr{M}^0}(c^{-1}\chi_P\, v, 0) + \mathsf{d}_{\mathscr{M}^0}(c^{-1}\chi_P\, w, 0) &= \int_P \frac{|v|}{c} \wedge 1\, \mathrm{d}\mathfrak{m}' + \int_P \frac{|w|}{c} \wedge 1\, \mathrm{d}\mathfrak{m}' \\ &= \int_P \frac{|v| + |w|}{c}\, \mathrm{d}\mathfrak{m}' \\ &< \int_P \frac{a+b}{c}\, \mathrm{d}\mathfrak{m}' < \int_P \frac{|v+w|}{c}\, \mathrm{d}\mathfrak{m}' \\ &= \mathsf{d}_{\mathscr{M}^0}(c^{-1}\chi_P\,(v+w), 0), \end{aligned}$$

which contradicts the fact that $\mathsf{d}_{\mathscr{M}^0}$ is a distance. Therefore

$$\mathsf{d}_{L^0}(|v|, |w|) = \int \big||v| - |w|\big| \wedge 1\, \mathrm{d}\mathfrak{m}' \leq \int |v - w| \wedge 1\, \mathrm{d}\mathfrak{m}' = \mathsf{d}_{\mathscr{M}^0}(v, w).$$

To prove the second in (3.11), suppose that $f_n \to f$ and $v_n \to v$ in $L^0(\mathfrak{m})$ and $\mathscr{M}^0$, respectively. We aim to show that $f_n v_n \to f v$ in $\mathscr{M}^0$. First of all, observe that

$$|f_n v_n - f v| \leq |f_n||v_n - v| + |v||f_n - f| \quad \text{holds } \mathfrak{m}\text{-a.e. in X.} \tag{3.12}$$

We claim that

$$\forall \delta > 0 \quad \exists M > 0 : \quad \varlimsup_{n\to\infty} \mathfrak{m}'\big(\{|f_n| > M\}\big) < \delta. \tag{3.13}$$

Clearly, given any $\delta > 0$ there exists $M > 1$ such that $\mathfrak{m}'\big(\{|f| > M - 1\}\big) < \delta$. Hence

$$\varlimsup_{n\to\infty} \mathfrak{m}'\big(\{|f_n| > M\}\big) \leq \mathfrak{m}'\big(\{|f| > M - 1\}\big) + \varlimsup_{n\to\infty} \mathfrak{m}'\big(\{|f_n - f| > 1\}\big) < \delta,$$

which proves (3.13). Now let $\varepsilon > 0$ be fixed. Given any $\delta > 0$, take $M > 0$ as in (3.13), so

$$\overline{\lim_n}\, \mathfrak{m}'\big(\{|f_n||v_n - v| > \varepsilon/2\}\big) \leq \overline{\lim_n}\, \mathfrak{m}'\big(\{|f_n| > M\}\big) + \overline{\lim_n}\, \mathfrak{m}'\big(\{|v_n - v| > \varepsilon/(2M)\}\big) < \delta.$$

Hence $\overline{\lim}_n\, \mathfrak{m}'\big(\{|f_n||v_n - v| > \varepsilon/2\}\big) = 0$ by letting $\delta \searrow 0$. In an analogous way, we can see that also $\overline{\lim}_n\, \mathfrak{m}'\big(\{|v||f_n - f| > \varepsilon/2\}\big) = 0$. Therefore (3.12) yields

$$\overline{\lim_n}\, \mathfrak{m}'\big(\{|f_n v_n - f v| > \varepsilon\}\big) \leq \overline{\lim_n}\, \mathfrak{m}'\big(\{|f_n||v_n - v| > \varepsilon/2\}\big) + \overline{\lim_n}\, \mathfrak{m}'\big(\{|v||f_n - f| > \varepsilon/2\}\big) = 0,$$

which proves that $f_n v_n \to f v$ in $\mathscr{M}^0$, as desired. ■

Proposition 3.1.22 (Existence and Uniqueness of the L^0-Completion) *Let $\mathscr{M}$ be any given $L^2(\mathfrak{m})$-normed module. Then there exists a unique $L^0(\mathfrak{m})$-completion $(\mathscr{M}^0, i)$ of $\mathscr{M}$.*

Uniqueness has to be intended up to unique isomorphism, in the following sense: given any other $L^0(\mathfrak{m})$-completion $(\widetilde{\mathscr{M}^0}, \tilde{i}\,)$ of $\mathscr{M}$, there is a unique module isomorphism $\Psi : \mathscr{M}^0 \to \widetilde{\mathscr{M}^0}$ such that

$$\begin{array}{ccc} \mathscr{M} & \xrightarrow{\;i\;} & \mathscr{M}^0 \\ & \searrow^{\tilde{i}} & \downarrow \Psi \\ & & \widetilde{\mathscr{M}^0} \end{array} \tag{3.14}$$

is a commutative diagram. Moreover, it holds that:

i) *The map $i : \mathscr{M} \to \mathscr{M}^0$ is continuous and $i(fv) = f\, i(v)$ for all $f \in L^\infty(\mathfrak{m})$ and $v \in \mathscr{M}$.*
ii) *$i(\mathscr{M})$ coincides with the set of all $v \in \mathscr{M}^0$ such that $|v| \in L^2(\mathfrak{m})$.*

Proof The proof goes as follows:

i) Since $\big|i(v)\big| = |v|$ holds $\mathfrak{m}$-a.e. for every $v \in \mathscr{M}$, we deduce that $\big\||i(v)|\big\|_{L^2(\mathfrak{m})} = \|v\|_{\mathscr{M}}$ for every $v \in \mathscr{M}$. Hence if $(v_n)_n \subseteq \mathscr{M}$ converges to $v \in \mathscr{M}$ then $\big\||i(v_n - v)|\big\|_{L^2(\mathfrak{m})} \to 0$, so that we have $\mathsf{d}_{\mathscr{M}^0}\big(i(v_n), i(v)\big) = \mathsf{d}_{L^0}\big(|i(v_n - v)|, 0\big) \to 0$ by Remark 1.1.22.

Moreover, we have that $\chi_E\, i(v) = i(\chi_E\, v)$ for every $E \subseteq \mathrm{X}$ Borel, indeed

$$|\chi_E\, i(v) - i(\chi_E\, v)| = \begin{cases} |i(v) - i(\chi_E\, v)| = |i((1-\chi_E)v)| = \chi_{E^c}|v| = 0 & \mathfrak{m}\text{-a.e. on } E, \\ |i(\chi_E\, v)| = |\chi_E\, v| = \chi_E |v| = 0 & \mathfrak{m}\text{-a.e. on } E^c. \end{cases}$$

By linearity of i, we immediately see that $f\, i(v) = i(fv)$ for any simple function $f : \mathrm{X} \to \mathbb{R}$, thus also for every $f \in L^\infty(\mathfrak{m})$ by continuity of i and Remark 3.1.21.

UNIQUENESS. The choice $\Psi\big(i(v)\big) := \tilde{i}(v)$ for every $v \in \mathscr{M}$ is obliged. Moreover, we have that the equalities $|i(v)| = |v| = |\tilde{i}(v)|$ hold $\mathfrak{m}$-a.e. in X for every $v \in \mathscr{M}$. Hence

$$\begin{aligned} \mathsf{d}_{\widetilde{\mathscr{M}}^0}\big(\Psi(i(v)), \Psi(i(w))\big) &= \int \big|\tilde{i}(v) - \tilde{i}(w)\big| \wedge 1 \, \mathrm{d}\mathfrak{m}' = \int |v - w| \wedge 1 \, \mathrm{d}\mathfrak{m}' \\ &= \int \big|i(v) - i(w)\big| \wedge 1 \, \mathrm{d}\mathfrak{m}' = \mathsf{d}_{\mathscr{M}^0}\big(i(v), i(w)\big) \end{aligned}$$

is satisfied for every $v, w \in \mathscr{M}$, which shows that $\Psi : i(\mathscr{M}) \to \tilde{i}(\mathscr{M})$ is an isometry, in particular it is continuous. Since $i(\mathscr{M})$ is dense in $\mathscr{M}^0$, we can uniquely extend Ψ to some map $\Psi : \mathscr{M}^0 \to \widetilde{\mathscr{M}}^0$, which is a linear isometry. Furthermore, Ψ preserves the pointwise norm and the multiplication by $L^0(\mathfrak{m})$-functions by i) and Remark 3.1.21, while it is surjective by density of $\tilde{i}(\mathscr{M})$ in $\widetilde{\mathscr{M}}^0$. Therefore this (uniquely determined) map Ψ is a module isomorphism satisfying property (3.14).

EXISTENCE. Define the distance d_0 on $\mathscr{M}$ as $\mathsf{d}_0(v, w) := \int |v - w| \wedge 1 \, \mathrm{d}\mathfrak{m}'$ and denote by $\mathscr{M}^0$ the completion of $(\mathscr{M}, \mathsf{d}_0)$. It can be readily proved that

$$\begin{aligned} \mathsf{d}_0(v_1 + w_1, v_2 + w_2) &\le \mathsf{d}_0(v_1, v_2) + \mathsf{d}_0(w_1, w_2), \\ \mathsf{d}_0(\lambda\, v, \lambda\, w) &\le \big(|\lambda| \vee 1\big)\, \mathsf{d}_0(v, w), \\ \mathsf{d}_{L^0}\big(|v|, |w|\big) &\le \mathsf{d}_0(v, w), \\ (f_n)_n \ L^0(\mathfrak{m})\text{-Cauchy}, \ (v_n)_n \ \mathsf{d}_0\text{-Cauchy} &\Longrightarrow (f_n v_n)_n \ \mathsf{d}_0\text{-Cauchy}. \end{aligned} \tag{3.15}$$

The first two properties in (3.15) grant that the vector space structure of $\mathscr{M}$ can be carried over to $\mathscr{M}^0$, while the third one and the fourth one show that we can extend to $\mathscr{M}^0$ the pointwise norm and the multiplication by $L^0(\mathfrak{m})$-functions, respectively.

ii) It clearly suffices to prove that $i(\mathscr{M}) \supseteq \big\{v \in \mathscr{M}^0 : |v| \in L^2(\mathfrak{m})\big\}$. To this aim, let us fix any $v \in \mathscr{M}^0$ with $|v| \in L^2(\mathfrak{m})$. There exists $(v_n)_n \subseteq \mathscr{M}$ such that $i(v_n) \to v$ in $\mathscr{M}^0$. Define

$$w_n := \chi_{\{|i(v_n)|>0\}} \frac{|v|}{|i(v_n)|}\, i(v_n) \in \mathscr{M}^0 \quad \text{for every } n \in \mathbb{N}.$$

Notice that $|w_n| = \chi_{\{|i(v_n)|>0\}} |v| \in L^2(\mathfrak{m})$ for every $n \in \mathbb{N}$. Moreover, one can easily prove that $(w_n)_n \subseteq i(\mathscr{M})$. Since $|w_n - v| \to 0$ in $L^2(\mathfrak{m})$ by dominated convergence theorem, we thus conclude that $v \in i(\mathscr{M})$ as well.

□

3.2 Operations on Normed Modules

3.2.1 Dual Normed Module

In order to define the dual of a normed module, we need to introduce the following concept:

Lemma 3.2.1 (Essential Supremum) *Let* $f_i : \mathrm{X} \to \mathbb{R} \cup \{\pm\infty\}$ *be given Borel functions, with* $i \in I$*. Then there is a unique (up to equality* $\mathfrak{m}$*-a.e.) Borel function* $g : \mathrm{X} \to \mathbb{R} \cup \{\pm\infty\}$ *such that the following conditions holds:*

i) $g \geq f_i$ *holds* $\mathfrak{m}$*-a.e. for every* $i \in I$*.*
ii) If $h \geq f_i$ *holds* $\mathfrak{m}$*-a.e. for every* $i \in I$*, then* $h \geq g$ *in the* $\mathfrak{m}$*-a.e. sense.*

Moreover, there exists an at most countable subfamily $(f_{i_n})_n$ *of* $(f_i)_{i\in I}$ *such that* $g = \sup_n f_{i_n}$*. Such function* g *is called* essential supremum *of the family* $(f_i)_{i\in I}$*.*

Proof The $\mathfrak{m}$-a.e. uniqueness of g follows trivially from ii), so we pass to existence. Replacing if necessary the f_i's with $\varphi \circ f_i$—where $\varphi : \mathbb{R} \cup \{\pm\infty\} \to [0, 1]$ is monotone and injective—we can assume that the given functions are bounded. Similarly, replacing $\mathfrak{m}$ with a Borel probability measure with the same negligible sets we can assume that $\mathfrak{m}$ is a probability measure. Now let

$$\mathcal{A} := \big\{ f_{i_1} \vee \ldots \vee f_{i_n} \, : \, n \in \mathbb{N}, \, i_j \in I \text{ for all } j = 1, \ldots, n \big\},$$

set $S := \sup_{\tilde{f} \in \mathcal{A}} \int \tilde{f} \, \mathrm{d}\mathfrak{m}$ and notice that—since the f_i's are uniformly bounded and $\mathfrak{m}(\mathrm{X}) < \infty$—we have $S < +\infty$. Let $(\tilde{f}_n)_n \subseteq \mathcal{A}$ be such that $S = \sup_n \int \tilde{f}_n \, \mathrm{d}\mathfrak{m}$. Let us set $g := \sup_n \tilde{f}_n$, so that by construction we have $S = \int g \, \mathrm{d}\mathfrak{m}$ and by definition there must exist a countable family $(f_{i_n})_n$, with $i_n \in I$, such that $g = \sup_{n\in\mathbb{N}} f_{i_n}$. We claim that g satisfies i) and ii). Indeed, suppose i) does not hold, i.e. for some $\bar{i} \in I$ it holds that $f_{\bar{i}} > g$ on a set of positive $\mathfrak{m}$-measure. Then

$$S = \int g \, \mathrm{d}\mathfrak{m} < \int g \vee f_{\bar{i}} \, \mathrm{d}\mathfrak{m} = \lim_{n\to\infty} \int f_{i_1} \vee \ldots \vee f_{i_n} \vee f_{\bar{i}} \, \mathrm{d}\mathfrak{m},$$

contradicting the definition of S. To get ii), simply notice that if $h \geq f_{i_n}$ holds $\mathfrak{m}$-a.e. for every n, then $h \geq g$ is verified in the $\mathfrak{m}$-a.e. sense. □

We are ready to define the concept of *dual* $\mathscr{M}^*$ of an $L^2(\mathfrak{m})$-normed $L^\infty(\mathfrak{m})$-module $\mathscr{M}$. As a set we define

$$\mathscr{M}^* := \Big\{ L : \mathscr{M} \to L^1(\mathfrak{m}) \,\Big|\, L \text{ linear continuous, } L(fv) = f L(v) \text{ for all } v \in \mathscr{M},\ f \in L^\infty(\mathfrak{m}) \Big\}$$

and we endow it with the operator norm, i.e. $\|L\|_* := \sup_{\|v\|\le 1} \|L(v)\|_{L^1(\mathfrak{m})}$. The product between a function $f \in L^\infty(\mathfrak{m})$ and an element $L \in \mathscr{M}^*$ is defined as

$$(fL)(v) := fL(v) \quad \text{for every } v \in \mathscr{M},$$

while the pointwise norm of L is given by

$$|L|_* := \operatorname*{ess\,sup}_{v \in \mathscr{M},\, |v| \le 1\ \mathfrak{m}-a.e.} L(v).$$

Proposition 3.2.2 *The space $\mathscr{M}^*$ is an $L^2(\mathfrak{m})$-normed $L^\infty(\mathfrak{m})$-module. Moreover, it holds*

$$|L|_* = \operatorname*{ess\,sup}_{v \in \mathscr{M},\, |v| \le 1\ \mathfrak{m}-a.e.} \big|L(v)\big| \quad \textit{for every } L \in \mathscr{M}^*, \tag{3.16a}$$

$$\big|L(v)\big| \le |v||L|_* \quad \mathfrak{m}\text{-}a.e. \quad \textit{for every } v \in \mathscr{M} \textit{ and } L \in \mathscr{M}^*. \tag{3.16b}$$

Proof The fact that $(\mathscr{M}^*, \|\cdot\|_*)$ is a Banach space is obvious. The fact that $fL \in \mathscr{M}^*$ for any $f \in L^\infty(\mathfrak{m})$ and $L \in \mathscr{M}^*$ follows from the commutativity of $L^\infty(\mathfrak{m})$: indeed, the fact that the operator fL is linear continuous is obvious and moreover we have

$$(fL)(gv) = fL(gv) = fgL(v) = gfL(v) = g(fL)(v).$$

The required properties of the multiplication by L^∞-functions are easily derived, as for $v \in \mathscr{M}$ we have that

$$\big(f(gL)\big)(v) = f\big((gL)(v)\big) = f\big(gL(v)\big) = fgL(v) = (fgL)(v)$$

and $(\hat{1}L)(v) = L(\hat{1}v) = L(v)$. We come to the pointwise norm. To check that $|L|_* \ge 0$, let us pick $v = 0$ in the definition. Inequality $\le$ in (3.16a) is obvious, for the converse let $v \in \mathscr{M}$ be with $|v| \le 1$ $\mathfrak{m}$-a.e. and set $\tilde{v} := \chi_{\{L(v)\ge 0\}}v - \chi_{\{L(v)<0\}}v$, so that $|\tilde{v}| = |v|$ and $L(\tilde{v}) = |L(v)|$. Then it holds that $|L|_* \ge L(\tilde{v}) = \big|L(v)\big|$, thus getting (3.16a).

We pass to (3.16b) and observe that $\chi_{\{v=0\}}L(v) = L(\chi_{\{v=0\}}v) = 0$, so that (3.16b) holds $\mathfrak{m}$-a.e. on $\{v = 0\}$. Hence it is sufficient to prove that for any $c \in (0, 1)$ the same inequality holds $\mathfrak{m}$-a.e. on $S_c := \{c \le |v| \le c^{-1}\}$. To see this, notice that on S_c the functions $|v|, |v|^{-1}$ are in $L^\infty(\mathfrak{m})$, hence we can write $\chi_{S_c}v = \chi_{S_c}|v|\frac{v}{|v|}$ and since $\big|\chi_{S_c}\frac{v}{|v|}\big| \le 1$ $\mathfrak{m}$-a.e. we obtain

$$\chi_{S_c}\big|L(v)\big| = \chi_{S_c}\Big|L\Big(|v|\frac{v}{|v|}\Big)\Big| = \chi_{S_c}|v|\Big|L\Big(\frac{v}{|v|}\Big)\Big| \le \chi_{S_c}|v|\,|L|_*.$$

We now observe that for every $f \in L^\infty(\mathfrak{m})$ and $L \in \mathscr{M}^*$ we have

$$|fL|_* = \operatorname{ess\,sup} \big| fL(v) \big| = \operatorname{ess\,sup} |f| \big| L(v) \big| = |f| \operatorname{ess\,sup} \big| L(v) \big| = |f||L|_*,$$

where each essential supremum is taken among all $v \in \mathscr{M}$ with $|v| \leq 1$ $\mathfrak{m}$-a.e. Hence to conclude we need to prove that

$$\|L\|_* = \sqrt{\int |L|_*^2 \,\mathrm{d}\mathfrak{m}}. \tag{3.17}$$

The inequality

$$\int \big| L(v) \big| \,\mathrm{d}\mathfrak{m} \leq \int |v||L|_* \,\mathrm{d}\mathfrak{m} \leq \sqrt{\int |v|^2 \,\mathrm{d}\mathfrak{m}} \sqrt{\int |L|_*^2 \,\mathrm{d}\mathfrak{m}} = \|v\|_{\mathscr{M}} \sqrt{\int |L|_*^2 \,\mathrm{d}\mathfrak{m}},$$

valid for any $v \in \mathscr{M}$ and $L \in \mathscr{M}^*$, shows that $\leq$ holds in (3.17). For the converse inequality, recall that the properties of the essential supremum ensure that there is a sequence $(v_n)_n \subseteq \mathscr{M}$ with $|v_n| \leq 1$ $\mathfrak{m}$-a.e. for every $n \in \mathbb{N}$ such that $|L|_* = \sup_n L(v_n)$. Define recursively the sequence $(\tilde{v}_n)_n \subseteq \mathscr{M}$ by setting $\tilde{v}_0 := v_0$ and

$$\tilde{v}_{n+1} := \chi_{\{L(v_{n+1}) \geq L(\tilde{v}_n)\}} v_{n+1} + \chi_{\{L(v_{n+1}) < L(\tilde{v}_n)\}} \tilde{v}_n.$$

Notice that $L(\tilde{v}_n) = \sup_{i \leq n} L(v_i)$, so that $L(\tilde{v}_n)$ increases monotonically to $|L|_*$. Moreover, we have $|\tilde{v}_n| \leq 1$ $\mathfrak{m}$-a.e. for every $n \in \mathbb{N}$. Given any function $f \in L^\infty(\mathfrak{m}) \cap L^2(\mathfrak{m})$ with $f \geq 0$, we also have that $\|f\tilde{v}_n\|_{\mathscr{M}} = \big\| |f\tilde{v}_n| \big\|_{L^2(\mathfrak{m})} \leq \|f\|_{L^2(\mathfrak{m})}$ and thus

$$\int fL(v_n) \,\mathrm{d}\mathfrak{m} = \int L(fv_n) \,\mathrm{d}\mathfrak{m} \leq \|L\|_{\mathscr{M}^*} \|f\tilde{v}_n\|_{\mathscr{M}} \leq \|L\|_{\mathscr{M}^*} \|f\|_{L^2(\mathfrak{m})},$$

so that—by letting $n \to \infty$ and using the monotone convergence theorem to pass to the limit in the left hand side—we obtain

$$\int f|L|_* \,\mathrm{d}\mathfrak{m} \leq \|L\|_{\mathscr{M}^*} \|f\|_{L^2(\mathfrak{m})}.$$

By arbitrariness of f, we thus get (3.17). □

Proposition 3.2.3 *Let $L : \mathscr{M} \to L^1(\mathfrak{m})$ be linear, continuous and satisfying*

$$L(\chi_E v) = \chi_E L(v)$$

for every $v \in \mathscr{M}$ and $E \subseteq \mathrm{X}$ Borel. Then $L \in \mathscr{M}^*$.

Proof We need to prove that

$$L(fv) = fL(v) \quad \text{for every } v \in \mathscr{M} \text{ and } f \in L^\infty(\mathfrak{m}). \tag{3.18}$$

By assumption and taking into account the linearity of L, we see that (3.18) is true for every simple function f. The claim then follows by continuity of both sides of (3.18) with respect to $f \in L^\infty(\mathfrak{m})$. □

Exercise 3.2.4 Assume that $\mathfrak{m}$ has no atoms and let $L : \mathscr{M} \to L^\infty(\mathfrak{m})$ be linear, continuous and satisfying $L(fv) = fL(v)$ for every $v \in \mathscr{M}$ and $f \in L^\infty(\mathfrak{m})$. Prove that $L = 0$. ■

We now study the relation between the dual module and the dual in the sense of Banach spaces. Thus let $\mathscr{M}'$ be the dual of $\mathscr{M}$ seen as a Banach space. Integration provides a natural map $\mathrm{Int}_{\mathscr{M}} : \mathscr{M}^* \to \mathscr{M}'$, sending $L \in \mathscr{M}^*$ to the operator $\mathrm{Int}_{\mathscr{M}}(L) \in \mathscr{M}'$ defined as

$$\mathrm{Int}_{\mathscr{M}}(L)(v) := \int L(v)\,\mathrm{d}\mathfrak{m} \quad \text{for every } v \in \mathscr{M}.$$

Proposition 3.2.5 *The map* $\mathrm{Int}_{\mathscr{M}}$ *is a bijective isometry, i.e. it holds that*

$$\|L\|_{\mathscr{M}^*} = \big\|\mathrm{Int}_{\mathscr{M}}(L)\big\|_{\mathscr{M}'} \quad \textit{for every } L \in \mathscr{M}^*.$$

Proof From the inequality

$$\big|\mathrm{Int}_{\mathscr{M}}(L)(v)\big| = \left|\int L(v)\,\mathrm{d}\mathfrak{m}\right| \le \big\|L(v)\big\|_{L^1(\mathfrak{m})} \le \|v\|_{\mathscr{M}}\|L\|_{\mathscr{M}^*}$$

we see that $\big\|\mathrm{Int}_{\mathscr{M}}(L)\big\|_{\mathscr{M}'} \le \|L\|_{\mathscr{M}^*}$. For the converse inequality, let $L \in \mathscr{M}^*$, fix $\varepsilon > 0$ and find $v \in \mathscr{M}$ such that $\big\|L(v)\big\|_{L^1(\mathfrak{m})} \ge \|v\|_{\mathscr{M}}\big(\|L\|_{\mathscr{M}^*} - \varepsilon\big)$. Set $\tilde{v} := \chi_{\{L(v)\ge 0\}}v - \chi_{\{L(v)<0\}}v$, notice that $|\tilde{v}| = |v|$ and $L(\tilde{v}) = \big|L(v)\big|$ $\mathfrak{m}$-a.e. and conclude by

$$\begin{aligned}\big\|\mathrm{Int}_{\mathscr{M}}(L)\big\|_{\mathscr{M}'}\|\tilde{v}\|_{\mathscr{M}} \ge \big|\mathrm{Int}_{\mathscr{M}}(L)(\tilde{v})\big| &= \left|\int L(\tilde{v})\,\mathrm{d}\mathfrak{m}\right| \\ &= \big\|L(v)\big\|_{L^1(\mathfrak{m})} \ge \|v\|_{\mathscr{M}}\big(\|L\|_{\mathscr{M}^*} - \varepsilon\big) \\ &= \|\tilde{v}\|_{\mathscr{M}}\big(\|L\|_{\mathscr{M}^*} - \varepsilon\big)\end{aligned}$$

and the arbitrariness of $\varepsilon > 0$. It remains to prove that $\mathrm{Int}_{\mathscr{M}}$ is surjective. Fix $\ell \in \mathscr{M}'$ and for any $v \in \mathscr{M}$ consider the function sending a Borel set $E \subseteq \mathrm{X}$ to $\mu_v(E) := \ell(\chi_E\, v) \in \mathbb{R}$. Clearly μ_v is additive and—given a disjoint sequence

$(E_i)_i$ of Borel sets—we have that

$$\left|\mu_v\Big(\bigcup_n E_n\Big)-\mu_v\Big(\bigcup_{n=1}^N E_n\Big)\right|=\left|\mu_v\Big(\bigcup_{n>N} E_n\Big)\right|=\left|\ell\big(\chi_{\bigcup_{n>N}E_n}v\big)\right|\leq\|\ell\|_{\mathscr{M}'}\big\|\chi_{\bigcup_{n>N}E_n}v\big\|_{\mathscr{M}}.$$

Since $\big\|\chi_{\bigcup_{n>N}E_n}v\big\|_{\mathscr{M}}^2=\int_{\bigcup_{n>N}E_n}|v|^2\,\mathrm{dm}\to 0$ by the dominated convergence theorem, we see that μ_v is a Borel measure. By construction, it is also absolutely continuous with respect to the measure $\mathfrak{m}$ and thus it has a Radon-Nikodým derivative: call it $L(v)\in L^1(\mathfrak{m})$.

By construction we clearly have that the mapping $v\mapsto L(v)$ is linear. Moreover, since for every $E,F\subseteq \mathrm{X}$ Borel the identities $\mu_{\chi_E v}(F)=\ell(\chi_F\chi_E v)=\ell(\chi_{E\cap F}v)=\mu_v(E\cap F)$ grant that the equality $\int_F L(\chi_E v)\,\mathrm{dm}=\int_{E\cap F}L(v)\,\mathrm{dm}$ is satisfied, we see that

$$L(\chi_E v)=\chi_E L(v)\quad\text{for every } v\in\mathscr{M}\text{ and } E\subseteq\mathrm{X}\text{ Borel.}\tag{3.19}$$

Now let us prove that the map $v\mapsto L(v)\in L^1(\mathfrak{m})$ is continuous. For a given $v\in\mathscr{M}$, let us set $\tilde{v}:=\chi_{\{L(v)\geq 0\}}v-\chi_{\{L(v)<0\}}v$, so that $|\tilde{v}|=|v|$ and—by (3.19) and the linearity of L—we have $\big|L(v)\big|=L(\tilde{v})$ in the $\mathfrak{m}$-a.e. sense. Then

$$\big\|L(v)\big\|_{L^1(\mathfrak{m})}=\int L(\tilde{v})\,\mathrm{dm}=\mu_{\tilde{v}}(\mathrm{X})=\ell(\tilde{v})\leq\|\ell\|_{\mathscr{M}'}\|\tilde{v}\|_{\mathscr{M}}=\|\ell\|_{\mathscr{M}'}\|v\|_{\mathscr{M}},$$

which was the claim. The fact that $L\in\mathscr{M}^*$ follows from (3.19) and Proposition 3.2.3. □

Remark 3.2.6 We point out that the map

$$I_{\mathscr{M}}:\mathscr{M}\hookrightarrow\mathscr{M}^{**},\quad \mathscr{M}\ni v\mapsto\Big(I_{\mathscr{M}}(v):\mathscr{M}^*\ni L\mapsto L(v)\in L^1(\mathfrak{m})\Big)\in\mathscr{M}^{**}\tag{3.20}$$

is an isometric embedding. Indeed, its $L^\infty(\mathfrak{m})$-linearity can be easily proved, while to prove that it preserves the pointwise norm observe that

$$\big|I_{\mathscr{M}}(v)\big|=\operatorname*{ess\,sup}_{|L|_*\leq 1}\big|I_{\mathscr{M}}(v)(L)\big|=\operatorname*{ess\,sup}_{|L|_*\leq 1}\big|L(v)\big|\leq|v|\quad\mathfrak{m}\text{-a.e.}\qquad\text{for every } v\in\mathscr{M}$$

and that for any $v\in\mathscr{M}$ there exists $L\in\mathscr{M}^*$ such that $L(v)=|v|^2=|L|_*^2$ holds $\mathfrak{m}$-a.e., namely choose $\ell\in\mathscr{M}'$ such that $\ell(v)=\|v\|_{\mathscr{M}}^2=\|\ell\|_{\mathscr{M}'}^2$ and set $L:=\mathrm{Int}_{\mathscr{M}}^{-1}(\ell)$. Then one has that $\big|I_{\mathscr{M}}(v)\big|=|v|$ holds $\mathfrak{m}$-a.e. for all $v\in\mathscr{M}$, whence $I_{\mathscr{M}}$ is an isometric embedding. ■

Definition 3.2.7 The $L^2(\mathfrak{m})$-normed module $\mathscr{M}$ is said to be *reflexive as module* provided the embedding $I_{\mathscr{M}}$ is surjective.

Proposition 3.2.8 *The $L^2(\mathfrak{m})$-normed module $\mathscr{M}$ is reflexive as module if and only if it is reflexive as Banach space.*

Proof The map $\mathrm{Int}_{\mathscr{M}} : \mathscr{M}^* \to \mathscr{M}'$ induces an isomorphism $\mathrm{Int}^{\mathrm{tr}}_{\mathscr{M}} : \mathscr{M}'' \to (\mathscr{M}^*)'$. Let us denote by $J : \mathscr{M} \hookrightarrow \mathscr{M}''$ the canonical embedding. We have that

$$\mathrm{Int}_{\mathscr{M}^*}\big(I_{\mathscr{M}}(v)\big)(L) = \int I_{\mathscr{M}}(v)(L)\,\mathrm{d}\mathfrak{m} = \int L(v)\,\mathrm{d}\mathfrak{m},$$

$$\mathrm{Int}^{\mathrm{tr}}_{\mathscr{M}}\big(J(v)\big)(L) = J(v)\big(\mathrm{Int}_{\mathscr{M}}(L)\big) = \mathrm{Int}_{\mathscr{M}}(L)(v) = \int L(v)\,\mathrm{d}\mathfrak{m}$$

for every $v \in \mathscr{M}$ and $L \in \mathscr{M}^*$, whence we deduce that the diagram

$$\begin{array}{ccc} \mathscr{M} & \xrightarrow{\;I_{\mathscr{M}}\;} & \mathscr{M}^{**} \\ {\scriptstyle J}\downarrow & & \downarrow{\scriptstyle \mathrm{Int}_{\mathscr{M}^*}} \\ \mathscr{M}'' & \xrightarrow[\mathrm{Int}^{\mathrm{tr}}_{\mathscr{M}}]{} & (\mathscr{M}^*)' \end{array}$$

commutes. Since $I_{\mathscr{M}}$, J are injective and $\mathrm{Int}^{\mathrm{tr}}_{\mathscr{M}}$, $\mathrm{Int}_{\mathscr{M}^*}$ are bijective, we thus conclude that $I_{\mathscr{M}}$ is surjective if and only if J is surjective. □

Proposition 3.2.9 *Let V be a generating linear subspace of $\mathscr{M}$. Suppose that $L : V \to L^1(\mathfrak{m})$ is a linear map such that for some $g \in L^2(\mathfrak{m})$ it holds*

$$\big|L(v)\big| \leq g\,|v| \quad \mathfrak{m}\text{-a.e.} \quad \text{for every } v \in V. \tag{3.21}$$

Then there exists a unique $\widetilde{L} \in \mathscr{M}^*$ such that $\widetilde{L}_{|V} = L$ Moreover, the inequality $|L|_* \leq g$ holds $\mathfrak{m}$-a.e. in X.

Proof We claim that for any $v, w \in V$ and $E \subseteq \mathrm{X}$ Borel we have that

$$v = w \quad \mathfrak{m}\text{-a.e. on } E \quad \Longrightarrow \quad L(v) = L(w) \quad \mathfrak{m}\text{-a.e. on } E. \tag{3.22}$$

Indeed, note that (3.21) yields $\big|L(v) - L(w)\big| = \big|L(v - w)\big| \leq g\,|v - w| = 0$ $\mathfrak{m}$-a.e. on E. Now call $\widetilde{V}$ the set of all elements $\sum_{i=1}^n \chi_{E_i} v_i$, with $(E_i)_{i=1}^n$ Borel partition of X and $v_1, \ldots, v_n \in V$. The vector space $\widetilde{V}$ is dense in $\mathscr{M}$ by hypothesis. We are forced to define $\widetilde{L} : \widetilde{V} \to L^1(\mathfrak{m})$ as follows: $\widetilde{L}(\widetilde{v}) := \sum_{i=1}^n \chi_{E_i} L(v_i)$ for every $\widetilde{v} = \sum_{i=1}^n \chi_{E_i} v_i \in \widetilde{V}$, which is well-posed by (3.22) and linear by construction. Given that for every $\widetilde{v} = \sum_{i=1}^n \chi_{E_i} v_i \in \widetilde{V}$ we have

$$\big|\widetilde{L}(\widetilde{v})\big| = \sum_{i=1}^n \chi_{E_i} \big|L(v_i)\big| \leq g \sum_{i=1}^n \chi_{E_i} |v_i| = g\,|\widetilde{v}| \quad \mathfrak{m}\text{-a.e.}, \tag{3.23}$$

we deduce that $\|\widetilde{L}(\widetilde{v})\|_{L^1(\mathfrak{m})} \leq \|g\|_{L^2(\mathfrak{m})} \|\widetilde{v}\|_{\mathscr{M}}$ for every $\widetilde{v} \in \widetilde{V}$. In particular $\widetilde{L}$ is continuous, whence it can be uniquely extended to a linear and continuous map $\widetilde{L} : \mathscr{M} \to L^1(\mathfrak{m})$. It is easy to see that $\widetilde{L}$ is $L^\infty(\mathfrak{m})$-linear, so that $\widetilde{L} \in \mathscr{M}^*$. To conclude, the fact that the $\mathfrak{m}$-a.e. inequality $|\widetilde{L}(v)| \leq g\,|v|$ holds for every $v \in \mathscr{M}$ follows from (3.23) via an approximation argument. Hence $|L|_* \leq g$ holds $\mathfrak{m}$-a.e., as required. □

3.2.2 *Hilbert Modules and Tensor Products*

We now focus our attention on a special class of normed modules:

Definition 3.2.10 (Hilbert Module) An $L^2(\mathfrak{m})$-normed $L^\infty(\mathfrak{m})$-module $\mathscr{H}$ is said to be a *Hilbert module* provided $(\mathscr{H}, \|\cdot\|_{\mathscr{H}})$ is a Hilbert space.

Proposition 3.2.11 *Every Hilbert module is reflexive.*

Proof Any Hilbert module is clearly reflexive when viewed as a Banach space, thus also in the sense of modules by Proposition 3.2.8. □

Proposition 3.2.12 *Let $\mathscr{H}$ be a Hilbert module. Then the formula*

$$\langle v, w \rangle := \frac{1}{2}\big(|v+w|^2 - |v|^2 - |w|^2\big) \in L^1(\mathfrak{m}) \tag{3.24}$$

defines an $L^\infty(\mathfrak{m})$-bilinear map $\langle\cdot,\cdot\rangle : \mathscr{H} \times \mathscr{H} \to L^1(\mathfrak{m})$, called pointwise scalar product, *which satisfies*

$$\begin{aligned} \langle v, w\rangle &= \langle w, v\rangle \\ \big|\langle v, w\rangle\big| &\leq |v||w| \quad \text{in the } \mathfrak{m}\text{-a.e. sense} \quad \text{for every } v, w \in \mathscr{H}. \\ \langle v, v\rangle &= |v|^2 \end{aligned} \tag{3.25}$$

Moreover, the pointwise parallelogram rule *is satisfied, i.e.*

$$2\big(|v|^2 + |w|^2\big) = |v+w|^2 + |v-w|^2 \quad \mathfrak{m}\text{-a.e.} \quad \text{for every } v, w \in \mathscr{H}. \tag{3.26}$$

Proof We only prove the validity of formula (3.26). The other properties can be obtained by suitably adapting the proof of the analogous statements for Hilbert spaces, apart from the $L^\infty(\mathfrak{m})$-bilinearity of $\langle\cdot,\cdot\rangle$, which can be shown by using the fact that $\langle\cdot,\cdot\rangle$ is local and continuous with respect to both entries by its very construction. Then let $v, w \in \mathscr{H}$ be fixed. Since the norm $\|\cdot\|_{\mathscr{H}}$ satisfies the parallelogram rule, we have that for any Borel set $E \subseteq \mathrm{X}$ it holds

$$2\int_E |v|^2 + |w|^2 \, \mathrm{dm} = 2\,\|\chi_E\, v\|^2_{\mathscr{H}} + 2\,\|\chi_E\, w\|^2_{\mathscr{H}}$$
$$= \|\chi_E\, v + \chi_E\, w\|^2_{\mathscr{H}} + \|\chi_E\, v - \chi_E\, w\|^2_{\mathscr{H}}$$
$$= \int_E |v + w|^2 + |v - w|^2 \, \mathrm{dm},$$

which yields (3.26) by arbitrariness of E. □

Given any Hilbert module $\mathscr{H}$, it holds that

$$\int \langle v, w \rangle \, \mathrm{dm} = \langle v, w \rangle_{\mathscr{H}} \quad \text{for every } v, w \in \mathscr{H}, \tag{3.27}$$

as one can immediately see by recalling that $\int |v|^2 \, \mathrm{dm} = \|v\|^2_{\mathscr{H}}$.

Remark 3.2.13 Actually the pointwise parallelogram rule characterises the Hilbert modules: any $L^2(\mathrm{m})$-normed module is a Hilbert module if and only if (3.26) is satisfied. ■

Theorem 3.2.14 (Riesz) *Let $\mathscr{H}$ be a Hilbert module. Then for every $L \in \mathscr{H}^*$ there exists a unique element $v \in \mathscr{H}$ such that*

$$L(w) = \langle v, w \rangle \quad \textit{for every } w \in \mathscr{H}. \tag{3.28}$$

Moreover, the equality $|v| = |L|_*$ holds m-a.e. in X.

Proof Consider $\mathrm{Int}_{\mathscr{H}}(L) \in \mathscr{H}'$. By the classical Riesz theorem, there is (a unique) $v \in \mathscr{H}$ such that $\langle v, w \rangle_{\mathscr{H}} = \mathrm{Int}_{\mathscr{H}}(L)(w)$ for every $w \in \mathscr{H}$. Hence for any $w \in \mathscr{H}$ we have that

$$\int_E \langle v, w \rangle \, \mathrm{dm} = \langle v, \chi_E\, w \rangle_{\mathscr{H}} = \mathrm{Int}_{\mathscr{H}}(L)(\chi_E\, w) = \int_E L(w) \, \mathrm{dm} \quad \text{for every } E \subseteq \mathrm{X} \text{ Borel},$$

so that (3.28) is satisfied. Finally, it is easy to show that $|v| = \operatorname{ess\,sup}_{|w| \le 1} \langle v, w \rangle$. Recall that also $|L|_* = \operatorname{ess\,sup}_{|w| \le 1} L(w)$, therefore the m-a.e. equality $|v| = |L|_*$ follows. □

It immediately follows from Theorem 3.2.14 that the map $\mathscr{H} \ni v \mapsto \langle v, \cdot \rangle \in \mathscr{H}^*$ is an isometric isomorphism of modules.

Example 3.2.15 We compare the Riesz theorem for Hilbert spaces and Theorem 3.2.14 in the special case in which $\mathscr{H} = L^2(\mathrm{m})$.

The former grants that for any linear and continuous map $\ell : L^2(\mathrm{m}) \to \mathbb{R}$ there exists a unique g in $L^2(\mathrm{m})$ such that $\ell(f) = \int fg \, \mathrm{dm}$ for every $f \in L^2(\mathrm{m})$, thus $\|g\|_{L^2(\mathrm{m})} = \|\ell\|_{L^2(\mathrm{m})'}$.

The latter grants that for any $L^\infty(\mathfrak{m})$-linear and continuous map $L : L^2(\mathfrak{m}) \to L^1(\mathfrak{m})$ there exists a unique g in $L^2(\mathfrak{m})$ such that $L(f) = fg$ holds $\mathfrak{m}$-a.e. for every $f \in L^2(\mathfrak{m})$, thus accordingly $|g| = |L|_*$ holds $\mathfrak{m}$-a.e. in X. ■

In order to introduce the notion of tensor product of Hilbert modules, we first recall what is the tensor product of two Hilbert spaces. Fix H_1, H_2 Hilbert spaces. We call $H_1 \otimes_{\mathrm{Alg}} H_2$ their tensor product as vector spaces, namely the space of formal finite sums $\sum_{i=1}^n v_i \otimes w_i$, with $(v, w) \mapsto v \otimes w$ bilinear. The space $H_1 \otimes_{\mathrm{Alg}} H_2$ satisfies the following universal property: given any vector space V and any bilinear map $B : H_1 \times H_2 \to V$, there exists a unique linear map $T : H_1 \otimes_{\mathrm{Alg}} H_2 \to V$ such that the diagram

$$\begin{array}{ccc} H_1 \times H_2 & \xhookrightarrow{\;\otimes\;} & H_1 \otimes_{\mathrm{Alg}} H_2 \\ & {\scriptstyle B}\searrow & \downarrow{\scriptstyle T} \\ & & V \end{array} \tag{3.29}$$

commutes, where $\otimes : H_1 \times H_2 \hookrightarrow H_1 \otimes_{\mathrm{Alg}} H_2$ denotes the map $(v, w) \mapsto v \otimes w$. Hence we can define a scalar product on $H_1 \otimes_{\mathrm{Alg}} H_2$ in the following way: first we declare

$$\langle v \otimes w, v' \otimes w' \rangle := \langle v, v' \rangle_{H_1} \langle w, w' \rangle_{H_2} \quad \text{for every } v, v' \in H_1 \text{ and } w, w' \in H_2,$$

then we can uniquely extend it to a bilinear operator $\langle \cdot, \cdot \rangle : \left[H_1 \otimes_{\mathrm{Alg}} H_2\right]^2 \to \mathbb{R}$, which is a scalar product as a consequence of the lemma below.

Lemma 3.2.16 *Let $v_1, \ldots, v_n \in H_1$ and $w_1, \ldots, w_n \in H_2$ be given. Then*

$$\left\langle \sum_{i=1}^n v_i \otimes w_i, \sum_{i=1}^n v_i \otimes w_i \right\rangle \geq 0,$$

with equality if and only if $\sum_{i=1}^n v_i \otimes w_i = 0$.

Proof We can suppose with no loss of generality that H_1 and H_2 are finite-dimensional. Choose orthonormal bases $\mathsf{e}_1, \ldots, \mathsf{e}_k$ and $\mathsf{f}_1, \ldots, \mathsf{f}_h$ of H_1 and H_2, respectively. Therefore a basis of $H_1 \otimes_{\mathrm{Alg}} H_2$ is given by $(\mathsf{e}_i \otimes \mathsf{f}_j)_{i,j}$. Now notice that for any $(a_{ij})_{i,j} \subseteq \mathbb{R}$ it holds

$$\left\langle \sum_{i,j} a_{ij}\, \mathsf{e}_i \otimes \mathsf{f}_j, \sum_{i,j} a_{ij}\, \mathsf{e}_i \otimes \mathsf{f}_j \right\rangle = \sum_{i,i',j,j'} a_{ij}\, a_{i'j'} \underbrace{\langle \mathsf{e}_i \otimes \mathsf{f}_j, \mathsf{e}_{i'} \otimes \mathsf{f}_{j'} \rangle}_{=\delta_{(i,j)(i',j')}} = \sum_{i,j} a_{ij}^2,$$

whence the statement follows. □

Then we define the tensor product $H_1 \otimes H_2$ of Hilbert spaces as the completion of $H_1 \otimes_{\mathrm{Alg}} H_2$ with the respect to the distance coming from $\langle \cdot, \cdot \rangle$.

Now consider two Hilbert modules $\mathscr{H}_1, \mathscr{H}_2$ over a metric measure space $(\mathrm{X}, \mathsf{d}, \mathfrak{m})$. Denote by $\mathscr{H}_1^0, \mathscr{H}_2^0$ the L^0-completions of $\mathscr{H}_1, \mathscr{H}_2$, respectively. Since $\mathscr{H}_1^0, \mathscr{H}_2^0$ are (algebraic) modules over the ring $L^0(\mathfrak{m})$, it makes sense to consider their tensor product $\mathscr{H}_1^0 \otimes_{\mathrm{Alg}} \mathscr{H}_2^0$, which is the space of formal finite sums of objects of the form $v \otimes w$, with $(v, w) \mapsto v \otimes w$ being $L^0(\mathfrak{m})$-bilinear. We endow it with a pointwise scalar product in the following way: first we declare

$$\langle v \otimes v', w \otimes w' \rangle := \langle v, v' \rangle \langle w, w' \rangle \in L^0(\mathfrak{m}) \quad \text{for every } v, v' \in \mathscr{H}_1^0 \text{ and } w, w' \in \mathscr{H}_2^0,$$

then we can uniquely extend it to an $L^0(\mathfrak{m})$-bilinear operator $\langle \cdot, \cdot \rangle : \left[\mathscr{H}_1^0 \otimes_{\mathrm{Alg}} \mathscr{H}_2^0\right]^2 \to L^0(\mathfrak{m})$. It turns out that such operator is a pointwise scalar product, as we are now going to prove.

Lemma 3.2.17 *Let $\mathscr{H}^0$ be the L^0-completion of a normed module $\mathscr{H}$. Let $v_1, \ldots, v_n \in \mathscr{H}^0$ be given. Then there exist $\mathsf{e}_1, \ldots, \mathsf{e}_n \in \mathscr{H}^0$ with the following properties:*

i) $\langle \mathsf{e}_i, \mathsf{e}_j \rangle = 0$ *holds* $\mathfrak{m}$*-a.e. for every* $i \neq j$.
ii) $|\mathsf{e}_i| = \chi_{\{|\mathsf{e}_i|>0\}}$ *holds* $\mathfrak{m}$*-a.e. for every* $i = 1, \ldots, n$.
iii) *For all* $i = 1, \ldots, n$ *there exist* $(a_{ij})_{j=1}^n \subseteq L^0(\mathfrak{m})$ *such that* $v_i = \sum_{j=1}^n a_{ij}\, \mathsf{e}_j$.

Proof We explicitly build the desired $\mathsf{e}_1, \ldots, \mathsf{e}_n$ by means of a 'Gram-Schmidt orthogonalisation' procedure: we recursively define the e_i's as $\mathsf{e}_1 := \chi_{\{|v_1|>0\}}\, v_1/|v_1|$ and

$$w_k := v_k - \sum_{i=1}^{k-1} \langle v_k, \mathsf{e}_i \rangle\, \mathsf{e}_i, \qquad \mathsf{e}_k := \chi_{\{|w_k|>0\}} \frac{w_k}{|w_k|} \qquad \text{for every } k = 2, \ldots, n.$$

It can be readily checked that $\mathsf{e}_1, \ldots, \mathsf{e}_n$ satisfy the required properties. □

Remark 3.2.18 Let $(\mathsf{e}_i)_{i=1}^n \subseteq \mathscr{H}^0$ satisfy items i), ii) of Lemma 3.2.17. Let $v \in \mathscr{H}^0$ be an element of the form $v = \sum_{i=1}^n a_i\, \mathsf{e}_i$, for some $(a_i)_{i=1}^n \subseteq L^0(\mathfrak{m})$. Then it is easy to check that there is a unique choice of $(b_i)_{i=1}^n \subseteq L^0(\mathfrak{m})$ such that:

a) $v = \sum_{i=1}^n b_i\, \mathsf{e}_i$.
b) $b_i = 0$ holds $\mathfrak{m}$-a.e. on $\{\mathsf{e}_i = 0\}$ for all $i = 1, \ldots, n$.

Moreover, we have that $|v|^2 = \sum_{i=1}^n |b_i|^2$ is satisfied $\mathfrak{m}$-a.e. on X. ■

Lemma 3.2.19 *Let $A \in \mathscr{H}_1^0 \otimes_{\mathrm{Alg}} \mathscr{H}_2^0$ be given. Then $\langle A, A \rangle \geq 0$ holds $\mathfrak{m}$-a.e. on X. Moreover, we have that $\langle A, A \rangle = 0$ holds $\mathfrak{m}$-a.e. on some Borel set $E \subseteq \mathrm{X}$ if and only if $\chi_E\, A = 0$.*

Proof Say $A = \sum_{i=1}^n v_i \otimes w_i$. Associate $\mathsf{e}_1, \ldots, \mathsf{e}_n \in \mathscr{H}_1^0$ and $\mathsf{f}_1, \ldots, \mathsf{f}_n \in \mathscr{H}_2^0$ to $v_1, \ldots, v_n$ and $w_1, \ldots, w_n$, respectively, as in Lemma 3.2.17. Let $b_{ij}, c_{ik} \in L^0(\mathfrak{m})$ be as in Remark 3.2.18, with $v_i = \sum_{j=1}^n b_{ij}\, \mathsf{e}_j$ and $w_i = \sum_{k=1}^n c_{ik}\, \mathsf{f}_k$ for all

$i = 1, \ldots, n$. If $a_{jk} := \sum_{i=1}^n b_{ij}\, c_{ik}$ then

$$\langle A, A \rangle = \sum_{j,k=1}^{n} |a_{jk}|^2\, |\mathsf{e}_j|^2\, |\mathsf{f}_k|^2 \quad \text{holds } \mathfrak{m}\text{-a.e. on X},$$

whence the statement easily follows. □

Accordingly, it makes sense to define the *pointwise Hilbert-Schmidt norm* as

$$|A|_{\mathsf{HS}} := \sqrt{\langle A, A \rangle} \in L^0(\mathfrak{m})^+ \quad \text{for every } A \in \mathscr{H}_1^0 \otimes_{\mathrm{Alg}} \mathscr{H}_2^0.$$

It immediately stems from Lemma 3.2.19 that $|A|_{\mathsf{HS}} = 0$ holds $\mathfrak{m}$-a.e. on a Borel set $E \subseteq \mathrm{X}$ if and only if $\chi_E A = 0$.

Definition 3.2.20 (Tensor Product of Hilbert Modules) We define $\mathscr{H}_1 \otimes \mathscr{H}_2$ as the completion of the space

$$\left\{ A \in \mathscr{H}_1^0 \otimes_{\mathrm{Alg}} \mathscr{H}_2^0 \;:\; |A|_{\mathsf{HS}} \in L^2(\mathfrak{m}) \right\}$$

with respect to the norm $A \mapsto \sqrt{\int |A|_{\mathsf{HS}}^2 \, \mathrm{d}\mathfrak{m}}$. It turns out that $\mathscr{H}_1 \otimes \mathscr{H}_2$ is a Hilbert module. Moreover, we denote by $\mathscr{H}_1^0 \otimes \mathscr{H}_2^0$ the L^0-completion of $\mathscr{H}_1 \otimes \mathscr{H}_2$.

It can be readily checked that $(\mathscr{H}_1^0 \otimes \mathscr{H}_2^0) \otimes \mathscr{H}_3^0$ and $\mathscr{H}_1^0 \otimes (\mathscr{H}_2^0 \otimes \mathscr{H}_3^0)$ are isomorphic, in other words the operation $\otimes$ is associative. Then for any $k \in \mathbb{N}$ it makes sense to define

$$(\mathscr{H}^0)^{\otimes k} := \underbrace{\mathscr{H}^0 \otimes \ldots \otimes \mathscr{H}^0}_{k\text{-times}}$$

for every $L^0(\mathfrak{m})$-normed Hilbert module $\mathscr{H}^0$.

Lemma 3.2.21 *Let $D_1 \subseteq \mathscr{H}_1$ and $D_2 \subseteq \mathscr{H}_2$ be dense subsets such that $|v|, |w| \in L^\infty(\mathfrak{m})$ for every $v \in D_1$ and $w \in D_2$. Then the set*

$$\tilde{D} := \left\{ \sum_{i=1}^{n} v_i \otimes w_i \;:\; v_i \in D_1,\; w_i \in D_2 \right\}$$

is dense in $\mathscr{H}_1 \otimes \mathscr{H}_2$. In particular, $\mathscr{H}_1 \otimes \mathscr{H}_2$ is separable as soon as $\mathscr{H}_1, \mathscr{H}_2$ are separable.

Proof To prove the first part of the statement, it is clearly sufficient to show that

$$v \otimes w \text{ is in the closure of } \tilde{D} \text{ for all } v \in \mathscr{H}_1,\; w \in \mathscr{H}_2 \text{ with } v \otimes w \in \mathscr{H}_1 \otimes \mathscr{H}_2. \tag{3.30}$$

First of all, the closure of $\tilde{D}$ contains $\{v \otimes w \,:\, v \in \mathscr{H}_1,\ w \in D_2\}$: chosen any $(v_n)_n \subseteq D_1$ converging to v, we have that $|v_n \otimes w - v \otimes w|_{\mathsf{HS}} = \big|(v_n - v) \otimes w\big|_{\mathsf{HS}} = |v_n - v||w| \to 0$ in $L^2(\mathfrak{m})$. In a symmetric way, one can prove that the closure of $\tilde{D}$ contains also $\{v \otimes w \,:\, v \in D_1,\ w \in \mathscr{H}_2\}$. Therefore $\{v \otimes w \,:\, v \in \mathscr{H}_1,\ w \in \mathscr{H}_2,\ |w| \in L^\infty(\mathfrak{m})\}$ is contained in the closure of $\tilde{D}$: given any $v \in \mathscr{H}_1$, $w \in \mathscr{H}_2$ with $|w| \in L^\infty(\mathfrak{m})$ and a sequence $(v_n)_n \subseteq D_1$ with $v_n \to v$, we have

$$|v_n \otimes w - v \otimes w|_{\mathsf{HS}} \leq |v_n - v||w| \to 0 \quad \text{in } L^2(\mathfrak{m}).$$

Finally, take any $v \in \mathscr{H}_1$, $w \in \mathscr{H}_2$ such that $v \otimes w \in \mathscr{H}_1 \otimes \mathscr{H}_2$ and define $w_n := \chi_{\{|w| \leq n\}} w \in \mathscr{H}_2$ for all $n \in \mathbb{N}$. Given that $|v \otimes w_n - v \otimes w|_{\mathsf{HS}} = |v||w_n - w| = \chi_{\{|w| > n\}}|v||w|$ holds $\mathfrak{m}$-a.e. on X for any $n \in \mathbb{N}$, by applying the dominated convergence theorem we conclude that $v \otimes w_n \to v \otimes w$. Therefore the claim (3.30) is proved, thus showing the first part of the statement.

The last part of the statement follows by noticing that any separable Hilbert module admits a countable dense subset made of bounded elements. □

Remark 3.2.22 Given any Hilbert module $\mathscr{H}$, we obtain the *transposition* operator

$$\mathsf{t} \colon \mathscr{H} \otimes \mathscr{H} \to \mathscr{H} \otimes \mathscr{H}$$

by first declaring that $\mathsf{t}(v \otimes w) := w \otimes v \in \mathscr{H}_2^0 \otimes_{\mathrm{Alg}} \mathscr{H}_1^0$ for all $v \in \mathscr{H}_1^0$, $w \in \mathscr{H}_2^0$ and then extending it by linearity and continuity (notice that it preserves the pointwise norm). It turns out that t is an isometric $L^\infty(\mathfrak{m})$-linear map. Since it is also an involution, i.e. $\mathsf{t} \circ \mathsf{t} = \mathrm{id}_{\mathscr{H} \otimes \mathscr{H}}$, we also see that it is an isomorphism of modules. We shall say that $A \in \mathscr{H} \otimes \mathscr{H}$ is *symmetric* provided $A^{\mathsf{t}} := \mathsf{t}(A) = A$. ■

Given any $L^0(\mathfrak{m})$-normed Hilbert module $\mathscr{H}^0$ and some number $k \in \mathbb{N}$, we define the *exterior power* $\Lambda^k \mathscr{H}^0$ as follows: we set $\Lambda^0 \mathscr{H}^0 := L^0(\mathfrak{m})$ and $\Lambda^1 \mathscr{H}^0 := \mathscr{H}^0$, while for $k \geq 2$

$$\Lambda^k \mathscr{H}^0 := (\mathscr{H}^0)^{\otimes k} / \mathsf{V}_k, \qquad \begin{array}{l}\text{where we call } \mathsf{V}_k \text{ the closed subspace generated by} \\ \text{the elements } v_1 \otimes \ldots \otimes v_k, \text{ with } v_1, \ldots, v_k \in \mathscr{H}^0 \\ \text{and } v_i = v_j \text{ for some } i \neq j.\end{array} \tag{3.31}$$

The equivalence class of an element $v_1 \otimes \ldots \otimes v_k$ is denoted by $v_1 \wedge \ldots \wedge v_k$. The pointwise scalar product between any two such elements is given by

$$\langle v_1 \wedge \ldots \wedge v_k, w_1 \wedge \ldots \wedge w_k \rangle(x) = \det\big(\langle v_i, w_j \rangle(x)\big)_{i,j} \qquad \text{for } \mathfrak{m}\text{-a.e. } x \in \mathrm{X}, \tag{3.32}$$

up to a factor $k!$.

3.2.3 Pullback of Normed Modules

We now introduce the notion of 'pullback module'. In order to explain the ideas underlying its construction, we first see in an example in the classical case of smooth manifolds how such notion pops out and why it is relevant.

Let $\varphi : M \to N$ be a smooth map between two smooth manifolds M and N. Given a point $x \in M$ and a tangent vector $v \in T_x M$, we have that $\mathrm{d}\varphi_x(v) \in T_{\varphi(x)} N$ is the unique element for which $\mathrm{d}\varphi_x(v)(f) = \mathrm{d}(f \circ \varphi)_x(v)$ holds for any smooth function f on N. However, in our framework vector fields are not pointwise defined, so we are rather interested in giving a meaning to the object $\mathrm{d}\varphi(X)$, where X is a vector field on M. Unless φ is a diffeomorphism, we cannot hope to define $\mathrm{d}\varphi(X)$ as a vector field on N. What we need is the notion of 'pullback bundle': informally speaking, given a bundle E over N, we define $\varphi^* E$ as that bundle over M such that the fiber at a point $x \in M$ is exactly the fiber of E at $\varphi(x)$. Hence the object $\mathrm{d}\varphi(X)$ can be defined as the section of $\varphi^* TN$ satisfying $\mathrm{d}\varphi(X)(x) = \mathrm{d}\varphi_x\big(X(x)\big)$ for every $x \in M$.

Definition 3.2.23 (Maps of Bounded Compression) Let $(\mathrm{X}, \mathsf{d}_\mathrm{X}, \mathfrak{m}_\mathrm{X})$ and $(\mathrm{Y}, \mathsf{d}_\mathrm{Y}, \mathfrak{m}_\mathrm{Y})$ be metric measure spaces. Then a map $\varphi : \mathrm{Y} \to \mathrm{X}$ is said to be *of bounded compression* provided it is Borel and there exists a constant $C > 0$ such that $\varphi_* \mathfrak{m}_\mathrm{Y} \leq C \mathfrak{m}_\mathrm{X}$. The least such constant $C > 0$ will be denoted by $\mathrm{Comp}(\varphi)$ and called *compression constant* of φ.

We introduce the notion of 'pullback module':

Theorem 3.2.24 (Pullback Module) *Let* $(\mathrm{X}, \mathsf{d}_\mathrm{X}, \mathfrak{m}_\mathrm{X})$ *and* $(\mathrm{Y}, \mathsf{d}_\mathrm{Y}, \mathfrak{m}_\mathrm{Y})$ *be metric measure spaces. Let* $\mathscr{M}$ *be an* $L^2(\mathfrak{m}_\mathrm{X})$*-normed module. Let* $\varphi : \mathrm{Y} \to \mathrm{X}$ *be a map of bounded compression. Then there exists a unique couple* $(\varphi^* \mathscr{M}, \varphi^*)$*, where* $\varphi^* \mathscr{M}$ *is an* $L^2(\mathfrak{m}_\mathrm{Y})$*-normed module and* $\varphi^* : \mathscr{M} \to \varphi^* \mathscr{M}$ *is a linear continuous operator, such that*

i) $|\varphi^* v| = |v| \circ \varphi$ *holds* $\mathfrak{m}_\mathrm{Y}$*-a.e. for every* $v \in \mathscr{M}$,
ii) *the set* $\{\varphi^* v : v \in \mathscr{M}\}$ *generates* $\varphi^* \mathscr{M}$ *as a module.*

Uniqueness is up to unique isomorphism: given another couple $(\widetilde{\varphi^* \mathscr{M}}, \widetilde{\varphi^*})$ *with the same properties, there is a unique module isomorphism* $\Phi : \varphi^* \mathscr{M} \to \widetilde{\varphi^* \mathscr{M}}$ *such that* $\Phi \circ \varphi^* = \widetilde{\varphi^*}$.

Proof The proof goes as follows:
UNIQUENESS. We define the space $V \subseteq \varphi^* \mathscr{M}$ of simple elements as

$$V := \left\{ \sum_{i=1}^{n} \chi_{A_i} \varphi^* v_i \;\middle|\; (A_i)_i \text{ Borel partition of } \mathrm{Y},\ (v_i)_i \subseteq \mathscr{M} \right\}.$$

We are obliged to define $\Phi\big(\sum_i \chi_{A_i}\,\varphi^* v_i\big) := \sum_i \chi_{A_i}\,\widetilde{\varphi^* v_i}$ for any $\sum_i \chi_{A_i}\,\varphi^* v_i \in V$. Since

$$\Big|\sum_i \chi_{A_i}\,\widetilde{\varphi^* v_i}\Big| = \sum_i \chi_{A_i}\,|\widetilde{\varphi^* v_i}| = \sum_i \chi_{A_i}\,|v_i| \circ \varphi = \sum_i \chi_{A_i}\,|\varphi^* v_i|$$
$$= \Big|\sum_i \chi_{A_i}\,\varphi^* v_i\Big| \quad \mathfrak{m}\text{.a.e.},$$

we see that such Φ is well-defined. Moreover, it is also linear and continuous, whence it can be uniquely extended to a map $\Phi : \varphi^*\mathscr{M} \to \widetilde{\varphi^*\mathscr{M}}$. It can be readily proven that Φ is a module isomorphism satisfying $\Phi \circ \varphi^* = \widetilde{\varphi}^*$, thus showing uniqueness.

EXISTENCE. We define the 'pre-pullback module' Ppb as

$$\mathsf{Ppb} := \big\{(A_i, v_i)_{i=1}^n \;\big|\; n \in \mathbb{N},\ (A_i)_{i=1}^n \text{ Borel partition of Y},\ (v_i)_{i=1}^n \subseteq \mathscr{M}\big\}.$$

We consider the following equivalence relation on Ppb: we declare $(A_i, v_i)_i \sim (B_j, w_j)_j$ provided $|v_i - w_j| \circ \varphi = 0$ holds $\mathfrak{m}_{\mathrm{Y}}$-a.e. on $A_i \cap B_j$ for every i, j. We shall denote by $[A_i, v_i]_i$ the equivalence class of $(A_i, v_i)_i$. Hence we introduce some operations on $\mathsf{Ppb}/\sim$:

$$[A_i, v_i]_i + [B_j, w_j]_j := [A_i \cap B_j, v_i + w_j]_{i,j},$$
$$\lambda\,[A_i, v_i]_i := [A_i, \lambda\, v_i]_i,$$
$$\Big(\sum_j \alpha_j\,\chi_{B_j}\Big) \cdot [A_i, v_i]_i := [A_i \cap B_j, \alpha_j\, v_i]_{i,j},$$
$$\big|[A_i, v_i]_i\big| := \sum_i \chi_{A_i}\,|v_i| \circ \varphi \in L^2(\mathfrak{m}_{\mathrm{Y}}),$$
$$\big\|[A_i, v_i]_i\big\| := \Big(\int \big|[A_i, v_i]_i\big|^2\,\mathrm{d}\mathfrak{m}_{\mathrm{Y}}\Big)^{1/2}.$$

One can prove that $\big(\mathsf{Ppb}/\sim, \|\cdot\|\big)$ is a normed space, then we define $\varphi^*\mathscr{M}$ as its completion and we call $\varphi^* : \mathscr{M} \to \varphi^*\mathscr{M}$ the map sending any $v \in \mathscr{M}$ to $[\mathrm{Y}, v]$. It can be seen that the above operations can be uniquely extended by continuity to $\varphi^*\mathscr{M}$, thus endowing it with the structure of an $L^2(\mathfrak{m}_{\mathrm{Y}})$-normed module, and that $(\varphi^*\mathscr{M}, \varphi^*)$ satisfies the required properties. This concludes the proof of the statement. □

Example 3.2.25 Consider $\mathscr{M} := L^2(\mathfrak{m}_{\mathrm{X}})$. Then $\varphi^*\mathscr{M} = L^2(\mathfrak{m}_{\mathrm{Y}})$ and $\varphi^* f = f \circ \varphi$ holds for every $f \in L^2(\mathfrak{m}_{\mathrm{X}})$. ■

Example 3.2.26 Suppose that we have $\mathrm{Y} = \mathrm{X} \times \mathrm{Z}$, for some metric measure space $(\mathrm{Z}, \mathsf{d}_\mathrm{Z}, \mathfrak{m}_\mathrm{Z})$ such that $\mathfrak{m}_\mathrm{Z}(\mathrm{Z}) < +\infty$. Let us define $\mathsf{d}_\mathrm{Y}\big((x_1, z_1), (x_2, z_2)\big)^2 := \mathsf{d}_\mathrm{X}(x_1, x_2)^2 + \mathsf{d}_\mathrm{Z}(z_1, z_2)^2$ for every pair $(x_1, z_1), (x_2, z_2) \in \mathrm{X} \times \mathrm{Z}$ and $\mathfrak{m}_\mathrm{Y} := \mathfrak{m}_\mathrm{X} \otimes \mathfrak{m}_\mathrm{Z}$. Denote by $\varphi : \mathrm{Y} \to \mathrm{X}$ the canonical projection, which has bounded compression as $\varphi_* \mathfrak{m}_\mathrm{Y} = \mathfrak{m}_\mathrm{Z}(\mathrm{Z})\, \mathfrak{m}_\mathrm{X}$.

Now fix an $L^2(\mathfrak{m}_\mathrm{X})$-normed module $\mathscr{M}$ and consider the space $L^2(\mathrm{Z}, \mathscr{M})$, which can be naturally endowed with the structure of an $L^2(\mathfrak{m}_\mathrm{Y})$-normed module. For any $f \in L^\infty(\mathfrak{m}_\mathrm{Y})$ and $V_\cdot \in L^2(\mathrm{Z}, \mathscr{M})$, we have that $f \cdot V_\cdot \in L^2(\mathrm{Z}, \mathscr{M})$ is defined as $z \mapsto f(\cdot, z) V_z \in \mathscr{M}$. Given any element $V_\cdot$ of $L^2(\mathrm{Z}, \mathscr{M})$, say $z \mapsto V_z$, we have that the pointwise norm $|V_\cdot|$ is ($\mathfrak{m}_\mathrm{Y}$-a.e.) given by the function $(x, z) \mapsto |V_z|(x)$. Moreover, consider the operator $\hat{\cdot} : \mathscr{M} \to L^2(\mathrm{Z}, \mathscr{M})$ sending any $v \in \mathscr{M}$ to the function $\hat{v} : \mathrm{Z} \to \mathscr{M}$ that is identically equal to v. We claim that

$$\big(\varphi^* \mathscr{M}, \varphi^*\big) \sim \big(L^2(\mathrm{Z}, \mathscr{M}), \hat{\cdot}\big). \tag{3.33}$$

To prove property i) of Theorem 3.2.24 observe that

$$|\hat{v}_\cdot|(x, z) = |V_z|(x) = |v|(x) = \big(|v| \circ \varphi\big)(x, z) \quad \text{for } \mathfrak{m}_\mathrm{Y}\text{-a.e. } (x, z),$$

while ii) follows from density of the simple functions in $L^2(\mathrm{Z}, \mathscr{M})$. ■

Remark 3.2.27 Suppose that $\mathfrak{m}_\mathrm{X}$ is a Dirac delta. Hence any Banach space $\mathbb{B}$ can be viewed as an $L^2(\mathfrak{m}_\mathrm{X})$-normed module (since $L^\infty(\mathfrak{m}_\mathrm{X}) \sim \mathbb{R}$). Then it holds that

$$(\varphi^* \mathbb{B}, \varphi^*) \sim \big(L^2(\mathrm{Z}, \mathbb{B}), \hat{\cdot}\big) \tag{3.34}$$

as a consequence of the previous example. ■

Example 3.2.28 Fix an $L^2(\mathfrak{m}_\mathrm{X})$-normed module $\mathscr{M}$. Suppose that the space Y is a subset of X with $\mathfrak{m}_\mathrm{X}(\mathrm{Y}) > 0$. Call $\varphi : \mathrm{Y} \to \mathrm{X}$ the inclusion map, which has bounded compression provided Y is equipped with the measure $\mathfrak{m}_\mathrm{Y} := \mathfrak{m}_\mathrm{X}|_\mathrm{Y}$. Consider the quotient $L^2(\mathfrak{m}_\mathrm{Y})$-normed module $\mathscr{M}/\sim$, where $v \sim w$ if and only if $|v - w| = 0$ holds $\mathfrak{m}_\mathrm{X}$-a.e. on Y. Then

$$(\varphi^* \mathscr{M}, \varphi^*) \sim \big(\mathscr{M}/\sim, \pi\big), \tag{3.35}$$

where $\pi : \mathscr{M} \to \mathscr{M}/\sim$ is the canonical projection. ■

Proposition 3.2.29 *Let* $(\mathrm{X}, \mathsf{d}_\mathrm{X}, \mathfrak{m}_\mathrm{X})$, $(\mathrm{Y}, \mathsf{d}_\mathrm{Y}, \mathfrak{m}_\mathrm{Y})$ *be metric measure spaces. Let* $\varphi : \mathrm{Y} \to \mathrm{X}$ *be a map of bounded compression and* $\mathscr{M}$ *an* $L^2(\mathfrak{m}_\mathrm{X})$*-normed module. Consider a generating linear subspace* V *of* $\mathscr{M}$*. Let* $\mathscr{N}$ *be an* $L^2(\mathfrak{m}_\mathrm{Y})$*-normed module and* $T : V \to \mathscr{N}$ *a linear map satisfying the inequality*

$$\big|T(v)\big| \leq C\, |v| \circ \varphi \quad \mathfrak{m}_\mathrm{Y}\text{-a.e.} \quad \text{for every } v \in V, \tag{3.36}$$

for some constant $C > 0$. *Then there is a unique linear continuous extension* $\hat{T} : \mathscr{M} \to \mathscr{N}$ *of* T *such that* $|\hat{T}(v)| \leq C\,|v| \circ \varphi$ *holds* $\mathfrak{m}_Y$*-a.e. for every* $v \in \mathscr{M}$.

Proof First of all, we claim that any extension $\hat{T}$ as in the statement must satisfy

$$\hat{T}(\chi_A v) = \chi_A \circ \varphi\, T(v) \quad \text{for every } v \in V \text{ and } A \subseteq \mathrm{X} \text{ Borel.} \tag{3.37}$$

To prove the claim, observe that

$$\hat{T}(\chi_A v) + \hat{T}(\chi_{A^c} v) = T(v) = \chi_A \circ \varphi\, T(v) + \chi_{A^c} \circ \varphi\, T(v). \tag{3.38}$$

Moreover, we have that $\chi_A \circ \varphi\, |\hat{T}(\chi_{A^c} v)| \leq C\, \chi_A \circ \varphi\, |\chi_{A^c} v| \circ \varphi = 0$, i.e. $\chi_A \circ \varphi\, \hat{T}(\chi_{A^c} v) = 0$. Similarly, one has that $\chi_{A^c} \circ \varphi\, \hat{T}(\chi_A v) = 0$. Hence by multiplying both sides of (3.38) by the function $\chi_A \circ \varphi$ we get $\chi_A \circ \varphi\, \hat{T}(\chi_A v) = \chi_A \circ \varphi\, T(v)$ and accordingly

$$\hat{T}(\chi_A v) = \chi_A \circ \varphi\, \hat{T}(\chi_A v) + \chi_{A^c} \circ \varphi\, \hat{T}(\chi_A v) = \chi_A \circ \varphi\, \hat{T}(\chi_A v) = \chi_A \circ \varphi\, T(v),$$

thus proving the validity of (3.37).

In light of (3.37), we necessarily have to define $\hat{T}\big(\sum_i \chi_{A_i} v_i\big) := \sum_i \chi_{A_i} \circ \varphi\, T(v_i)$ for any finite Borel partition $(A_i)_i$ of X and for any $(v_i)_i \subseteq V$. Well-posedness of such definition stems from the $\mathfrak{m}_Y$-a.e. inequality

$$\Big|\sum_i \chi_{A_i} \circ \varphi\, T(v_i)\Big| = \sum_i \chi_{\varphi^{-1}(A_i)}\, |T(v_i)| \leq C \sum_i \big(\chi_{A_i} |v_i|\big) \circ \varphi = C\, \Big|\sum_i \chi_{A_i} v_i\Big| \circ \varphi,$$

which also grants (linearity and) continuity of $\hat{T}$. Therefore the operator $\hat{T}$ admits a unique extension $\hat{T} : \mathscr{M} \to \mathscr{N}$ with the required properties. □

Remark 3.2.30 The operator $\hat{T}$ in Proposition 3.2.29 also satisfies

$$\hat{T}(f\, v) = f \circ \varphi\, \hat{T}(v) \quad \text{for every } f \in L^\infty(\mathfrak{m}_X) \text{ and } v \in \mathscr{M}. \tag{3.39}$$

Such property can be easily obtained by means of an approximation argument. ■

The ideas contained in the proof of Proposition 3.2.29 can be adapted to show the following result, whose proof will be omitted.

Proposition 3.2.31 *Let* $(\mathrm{X}, \mathsf{d}, \mathfrak{m})$ *be a metric measure space. Let* $\mathscr{M}_1$, $\mathscr{M}_2$ *be* $L^2(\mathfrak{m})$*-normed modules and* $T : \mathscr{M}_1 \to \mathscr{M}_2$ *a linear map such that*

$$|T(v)| \leq C\,|v| \quad \mathfrak{m}\text{-a.e.} \quad \text{for every } v \in \mathscr{M}_1, \tag{3.40}$$

for some constant $C > 0$. *Then* T *is* $L^\infty(\mathfrak{m})$*-linear and continuous.*

Exercise 3.2.32 Let $T : L^2(\mathfrak{m}) \to L^2(\mathfrak{m})$ be an $L^\infty(\mathfrak{m})$-linear and continuous operator. Prove that there exists a unique $g \in L^\infty(\mathfrak{m})$ such that $T(f) = gf$ for every $f \in L^2(\mathfrak{m})$. ■

Theorem 3.2.33 (Universal Property) *Let* $(\mathrm{X}, \mathsf{d}_\mathrm{X}, \mathfrak{m}_\mathrm{X})$, $(\mathrm{Y}, \mathsf{d}_\mathrm{Y}, \mathfrak{m}_\mathrm{Y})$ *be two metric measure spaces. Let* $\varphi : \mathrm{Y} \to \mathrm{X}$ *be a map of bounded compression. Consider an* $L^2(\mathfrak{m}_\mathrm{X})$*-normed module* $\mathscr{M}$, *an* $L^2(\mathfrak{m}_\mathrm{Y})$*-normed module* $\mathscr{N}$ *and a linear map* $T : \mathscr{M} \to \mathscr{N}$. *Suppose that there exists a constant* $C > 0$ *such that*

$$\big|T(v)\big| \leq C\,|v| \circ \varphi \quad \mathfrak{m}_\mathrm{Y}\text{-a.e.} \qquad \text{for every } v \in \mathscr{M}. \tag{3.41}$$

Then there exists a unique $L^\infty(\mathfrak{m}_\mathrm{Y})$*-linear continuous operator* $\hat{T} : \varphi^*\mathscr{M} \to \mathscr{N}$, *called* lifting *of* T, *such that* $\big|\hat{T}(w)\big| \leq C\,|w|$ *holds* $\mathfrak{m}_\mathrm{Y}$*-a.e. for any* $w \in \varphi^*\mathscr{M}$ *and such that*

$$\begin{array}{ccc} \mathscr{M} & \xrightarrow{\ \varphi^*\ } & \varphi^*\mathscr{M} \\ & {}_{T}\searrow & \downarrow{\scriptstyle \hat{T}} \\ & & \mathscr{N} \end{array} \tag{3.42}$$

is a commutative diagram.

Proof Call $V := \{\varphi^* v : v \in \mathscr{M}\}$, then V is a generating linear subspace of $\varphi^*\mathscr{M}$. We define the map $S : V \to \mathscr{N}$ as $S(\varphi^* v) := T(v)$ for every $v \in \mathscr{M}$. The $\mathfrak{m}_\mathrm{Y}$-a.e. inequality

$$\big|T(v)\big| \leq C\,|v| \circ \varphi = C\,|\varphi^* v|$$

grants that S is well-defined. Hence Proposition 3.2.31 guarantees that S admits a unique extension $\hat{T} : \varphi^*\mathscr{M} \to \mathscr{N}$ with the required properties. □

Theorem 3.2.34 (Functoriality) *Let* $(\mathrm{X}, \mathsf{d}_\mathrm{X}, \mathfrak{m}_\mathrm{X})$, $(\mathrm{Y}, \mathsf{d}_\mathrm{Y}, \mathfrak{m}_\mathrm{Y})$ *and* $(\mathrm{Z}, \mathsf{d}_\mathrm{Z}, \mathfrak{m}_\mathrm{Z})$ *be metric measure spaces. Let* $\varphi : \mathrm{Y} \to \mathrm{X}$ *and* $\psi : \mathrm{Z} \to \mathrm{Y}$ *be maps of bounded compression. Fix an* $L^2(\mathfrak{m}_\mathrm{X})$*-normed module* $\mathscr{M}$. *Then the map* $\varphi \circ \psi$ *has bounded compression and*

$$\big(\psi^*(\varphi^*\mathscr{M}), \psi^* \circ \varphi^*\big) \sim \big((\varphi \circ \psi)^*\mathscr{M}, (\varphi \circ \psi)^*\big). \tag{3.43}$$

Proof It is trivial to check that $\varphi \circ \psi$ has bounded compression. It only remains to show that

$$\big|\psi^*(\varphi^* v)\big| = |v| \circ \varphi \circ \psi \qquad \mathfrak{m}_\mathrm{Z}\text{-a.e.} \qquad \text{for every } v \in \mathscr{M},$$

$$\big\{\psi^*(\varphi^* v) : v \in \mathscr{M}\big\} \qquad \text{generates } \psi^*(\varphi^*\mathscr{M}) \text{ as a module.}$$

To prove the former, just notice that $\left|\psi^*(\varphi^*v)\right| = |\varphi^*v| \circ \psi = |v| \circ \varphi \circ \psi$. For the latter, notice that the set V of all finite sums of the form $\sum_i \chi_{A_i} \varphi^* v_i$, with $(A_i)_i$ Borel partition of Y and $(v_i)_i \subseteq \mathscr{M}$, is a dense vector subspace of $\varphi^*\mathscr{M}$. Hence the set of all finite sums of the form $\sum_j \chi_{B_j} \psi^* w_j$, with $(B_j)_j$ Borel partition of Z and $(w_j)_j \subseteq V$, is dense in $\psi^*(\varphi^*\mathscr{M})$, thus proving that $\left\{\psi^*(\varphi^*v) : v \in \mathscr{M}\right\}$ generates $\psi^*(\varphi^*\mathscr{M})$. □

Remark 3.2.35 Suppose that the map $\varphi : \mathrm{Y} \to \mathrm{X}$ is invertible and that both φ, φ^{-1} have bounded compression. Then Theorem 3.2.34 grants that $(\varphi^{-1})^*(\varphi^*\mathscr{M}) \sim \mathscr{M}$, thus in particular one has that $\varphi^* : \mathscr{M} \to \varphi^*\mathscr{M}$ is bijective. Hence, morally speaking, $\mathscr{M}$ and $\varphi^*\mathscr{M}$ are the same module, up to identifying the spaces $L^\infty(\mathfrak{m}_{\mathrm{X}})$ and $L^\infty(\mathfrak{m}_{\mathrm{Y}})$ via the invertible map $f \mapsto f \circ \varphi$. ■

We now investigate the relation between $(\varphi^*\mathscr{M})^*$ and $\varphi^*\mathscr{M}^*$. Under suitable assumptions, it will turn out that the operations of taking the dual and passing to the pullback commute.

Proposition 3.2.36 *Let* $(\mathrm{X}, \mathsf{d}_{\mathrm{X}}, \mathfrak{m}_{\mathrm{X}})$, $(\mathrm{Y}, \mathsf{d}_{\mathrm{Y}}, \mathfrak{m}_{\mathrm{Y}})$ *be metric measure spaces and* $\varphi : \mathrm{Y} \to \mathrm{X}$ *a map of bounded compression. Then there exists a unique* $L^\infty(\mathfrak{m}_{\mathrm{Y}})$*-bilinear and continuous map* $B : \varphi^*\mathscr{M} \times \varphi^*\mathscr{M}^* \to L^1(\mathfrak{m}_{\mathrm{Y}})$ *such that* $B(\varphi^*v, \varphi^*L) = L(v) \circ \varphi$ *is satisfied* $\mathfrak{m}_{\mathrm{Y}}$*-a.e. for every* $v \in \mathscr{M}$ *and* $L \in \mathscr{M}^*$.

Proof We are forced to declare $B\big(\sum_i \chi_{E_i} \varphi^* v_i, \sum_j \chi_{F_j} \varphi^* L_j\big) := \sum_{i,j} \chi_{E_i \cap F_j} L_j(v_i) \circ \varphi$. Since

$$
\begin{aligned}
\left|\sum_{i,j} \chi_{E_i\cap F_j} L_j(v_i)\circ\varphi\right| &= \sum_{i,j} \chi_{E_i\cap F_j} \left|L_j(v_i)\right| \circ\varphi \le \sum_{i,j} \chi_{E_i\cap F_j} |L_j|\circ\varphi\,|v_i|\circ\varphi \\
&= \Big(\sum_i \chi_{E_i}\,|v_i|\circ\varphi\Big)\Big(\sum_j \chi_{F_j}\,|L_j|\circ\varphi\Big) \\
&= \left|\sum_i \chi_{E_i}\,\varphi^* v_i\right| \left|\sum_j \chi_{F_j}\,\varphi^* L_j\right|,
\end{aligned}
$$

we see that B is (well-defined and) continuous, whence it can be uniquely extended to an operator $B : \varphi^*\mathscr{M} \times \varphi^*\mathscr{M}^* \to L^1(\mathfrak{m}_{\mathrm{Y}})$ satisfying all of the required properties. □

Proposition 3.2.37 *Under the assumptions of Proposition 3.2.36, the map*

$$
I : \varphi^*\mathscr{M}^* \longrightarrow (\varphi^*\mathscr{M})^*, \quad W \longmapsto B(\cdot, W) \tag{3.44}
$$

is well-defined, $L^\infty(\mathfrak{m}_{\mathrm{Y}})$*-linear continuous and preserving the pointwise norm, i.e. the* $\mathfrak{m}_{\mathrm{Y}}$*-a.e. equality* $\left|I(W)\right| = |W|$ *holds for every* $W \in \varphi^*\mathscr{M}^*$.

Proof The map $I(W) : \varphi^*\mathscr{M} \to L^1(\mathfrak{m}_Y)$ is $L^\infty(\mathfrak{m}_Y)$-linear continuous by Proposition 3.2.36, in other words $I(W) \in (\varphi^*\mathscr{M})^*$, which shows that I is well-posed. Moreover, notice that

$$\big|I(W)\big| = \operatorname*{ess\,sup}_{\substack{V\in\varphi^*\mathscr{M},\\ |V|\leq 1\ \mathfrak{m}_Y\text{-a.e.}}} \big|B(V,W)\big| \leq \operatorname*{ess\,sup}_{\substack{V\in\varphi^*\mathscr{M},\\ |V|\leq 1\ \mathfrak{m}_Y\text{-a.e.}}} |V||W| \leq |W| \quad \mathfrak{m}_Y\text{-a.e.},$$

whence I can be easily proven to be $L^\infty(\mathfrak{m}_Y)$-linear and continuous. Finally, to conclude it suffices to prove that also $\big|I(W)\big| \geq |W|$ holds $\mathfrak{m}_Y$-a.e. in Y. By density, it is actually enough to obtain it for W of the form $\sum_{j=1}^n \chi_{F_j}\,\varphi^* L_j$. Then observe that

$$\big|I(W)\big| \geq \operatorname*{ess\,sup}_{\substack{v_1,\ldots,v_n\in\mathscr{M},\\ |v_1|,\ldots,|v_n|\leq 1\ \mathfrak{m}_X\text{-a.e.}}} I(W)\bigg(\sum_{j=1}^n \chi_{F_j}\,\varphi^* v_j\bigg) = \sum_{j=1}^n \chi_{F_j} \operatorname*{ess\,sup}_{\substack{v_j\in\mathscr{M},\\ |v_j|\leq 1\ \mathfrak{m}_X\text{-a.e.}}} L_j(v_j)\circ\varphi$$

$$= \sum_{j=1}^n \chi_{F_j}\,|L_j|\circ\varphi = \sum_{j=1}^n \chi_{F_j}\,|\varphi^* L_j| = |W|$$

holds $\mathfrak{m}_Y$-a.e. in Y. Therefore the statement is achieved. □

Remark 3.2.38 In particular, Proposition 3.2.37 shows that the map I is an isometric embedding of $\varphi^*\mathscr{M}^*$ into $(\varphi^*\mathscr{M})^*$. However—as we are going to show in the next example—the operator I needs not be surjective. ■

Example 3.2.39 Suppose that $\mathrm{X} := \{\bar{x}\}$ and $\mathfrak{m}_X := \delta_{\bar{x}}$. Moreover, let $\mathrm{Y} := [0,1]$ be endowed with the Lebesgue measure and denote by φ the unique map from Y to X, which is clearly of bounded compression. Given that $L^\infty(\mathfrak{m}_X) \sim \mathbb{R}$, we can view any Banach space $\mathbb{B}$ as an $L^2(\mathfrak{m}_X)$-normed module, so that Remark 3.2.27 yields

$$(\varphi^*\mathbb{B})^* \sim \big(L^2([0,1],\mathbb{B})\big)',$$
$$\varphi^*\mathbb{B}^* \sim L^2([0,1],\mathbb{B}').$$

In general, $L^2([0,1],\mathbb{B}')$ is only embedded into $\big(L^2([0,1],\mathbb{B})\big)'$, via the map that sends any element $\ell_\cdot \in L^2([0,1],\mathbb{B}')$ to $L^2([0,1],\mathbb{B}) \ni v_\cdot \mapsto \int_0^1 \ell_t(v_t)\,\mathrm{d}t$, which clearly belongs to the space $\big(L^2([0,1],\mathbb{B})\big)'$. Now consider e.g. the case in which $\mathbb{B} := L^1(0,1)$. Let us define the map $T : L^2\big([0,1],L^1(0,1)\big) \to \mathbb{R}$ as

$$T(f) := \int_0^1\!\!\int_0^1 f_t(x)\,g_t(x)\,\mathrm{d}x\,\mathrm{d}t \qquad \text{for every } f \in L^2\big([0,1],L^1(0,1)\big),$$

where $g_t := \chi_{[0,t]}$. Hence T does not come from any element of $L^2\big([0,1],L^\infty(0,1)\big)$: it should come from the map $t \mapsto g_t \in L^\infty(0,1)$, which is not Borel (and not essentially separably valued). This shows that the space $L^2([0,1], L^\infty(0,1))$ and the dual of $L^2([0,1], L^1(0,1))$ are different. ■

Lemma 3.2.40 *Let* $(\mathrm{X}, \mathsf{d}_\mathrm{X}, \mathfrak{m}_\mathrm{X})$, $(\mathrm{Y}, \mathsf{d}_\mathrm{Y}, \mathfrak{m}_\mathrm{Y})$ *be metric measure spaces and* $\varphi : \mathrm{Y} \to \mathrm{X}$ *a map of bounded compression. Let* $\mathscr{H}$ *be a Hilbert module on* X. *Then* $\varphi^*\mathscr{H}$ *is a Hilbert module.*

Proof Notice that

$$\begin{aligned}2\big(|\varphi^*v|^2 + |\varphi^*w|^2\big) &= 2\big(|v|^2 + |w|^2\big) \circ \varphi = |v+w|^2 \circ \varphi + |v-w|^2 \circ \varphi \\ &= |\varphi^*v + \varphi^*w|^2 + |\varphi^*v - \varphi^*w|^2\end{aligned}$$

is satisfied $\mathfrak{m}_\mathrm{Y}$-a.e. for any $v, w \in \mathscr{H}$. Then the pointwise parallelogram identity can be shown to hold for elements of the form $\sum_i \chi_{E_i} \varphi^* v_i$, thus accordingly for all elements of $\varphi^*\mathscr{H}$ by an approximation argument. This proves that $\varphi^*\mathscr{H}$ is a Hilbert module, as required. □

Proposition 3.2.41 *Let* $(\mathrm{X}, \mathsf{d}_\mathrm{X}, \mathfrak{m}_\mathrm{X})$, $(\mathrm{Y}, \mathsf{d}_\mathrm{Y}, \mathfrak{m}_\mathrm{Y})$ *be metric measure spaces and* $\varphi : \mathrm{Y} \to \mathrm{X}$ *a map of bounded compression. Let* $\mathscr{H}$ *be a Hilbert module on* X. *Then*

$$\varphi^*\mathscr{H}^* \sim (\varphi^*\mathscr{H})^*. \tag{3.45}$$

Proof Consider the map $I : \varphi^*\mathscr{H}^* \to (\varphi^*\mathscr{H})^*$ of Proposition 3.2.37. We aim to prove that I is surjective. Denote by $\mathscr{R} : \mathscr{H} \to \mathscr{H}^*$ and $\widehat{\mathscr{R}} : \varphi^*\mathscr{H} \to (\varphi^*\mathscr{H})^*$ the Riesz isomorphisms, as in Theorem 3.2.14. Note that $\varphi^* \circ \mathscr{R} : \mathscr{H} \to \varphi^*\mathscr{H}^*$ satisfies $\big|(\varphi^* \circ \mathscr{R})(v)\big| = |v| \circ \varphi$ $\mathfrak{m}_\mathrm{Y}$-a.e. for any $v \in \mathscr{H}$, whence Theorem 3.2.33 grants that there exists a unique $L^\infty(\mathfrak{m}_\mathrm{Y})$-linear continuous operator $\widehat{\varphi^* \circ \mathscr{R}} : \varphi^*\mathscr{H} \to \varphi^*\mathscr{H}^*$ such that $\widehat{\varphi^* \circ \mathscr{R}}(\varphi^* v) = (\varphi^* \circ \mathscr{R})(v)$ holds for every $v \in \mathscr{H}$. Now let us define $J := \widehat{\varphi^* \circ \mathscr{R}} \circ \widehat{\mathscr{R}}^{-1} : (\varphi^*\mathscr{H})^* \to \varphi^*\mathscr{H}^*$. We claim that

$$I \circ J = \mathrm{id}_{(\varphi^*\mathscr{H})^*}. \tag{3.46}$$

Given that $I \circ J$ is $L^\infty(\mathfrak{m}_\mathrm{Y})$-linear continuous by construction, it suffices to check that $I \circ J$ is the identity on the subspace $\big\{\widehat{\mathscr{R}}(\varphi^* v) \,:\, v \in \mathscr{H}\big\}$, which generates $(\varphi^*\mathscr{H})^*$ as a module. Observe that for any $v, w \in \mathscr{H}$ it holds that

$$\begin{aligned}\widehat{\mathscr{R}}(\varphi^*v)(\varphi^*w) &= \langle \varphi^*v, \varphi^*w\rangle = \langle v, w\rangle \circ \varphi, \\ (I \circ J)\big(\widehat{\mathscr{R}}(\varphi^*v)\big)(\varphi^*w) &= I\big(\widehat{\varphi^* \circ \mathscr{R}}(\varphi^*v)\big)(\varphi^*w) \\ &= I\big((\varphi^* \circ \mathscr{R})(v)\big)(\varphi^*w) = \big(\mathscr{R}(v)(w)\big) \circ \varphi \\ &= \langle v, w\rangle \circ \varphi,\end{aligned}$$

whence (3.46) follows. This grants that I is surjective, thus concluding the proof. □

Bibliographical Remarks

Almost all the material contained in this chapter has been introduced by N. Gigli in [17]. The notion of $L^2(\mathfrak{m})$-normed $L^\infty(\mathfrak{m})$-module is a variant of a similar concept that was investigated by N. Weaver [33, 34], who was in turn inspired by the papers [25, 26] of J.-L. Sauvageot.

Furthermore, the above presentation of the notion of $L^0(\mathfrak{m})$-normed $L^0(\mathfrak{m})$-module follows closely the axiomatisation that can be found in the lecture notes [19], wherefrom even Proposition 3.2.41 is taken.

Chapter 4
First-Order Calculus on Metric Measure Spaces

In this chapter we develop a first-order differential structure on general metric measure spaces. First of all, the key notion of *cotangent module* is obtained by combining the Sobolev calculus (discussed in Chap. 2) with the theory of normed modules (described in Chap. 3). The elements of the cotangent module $L^2(T^*\mathrm{X})$, which are defined and studied in Sect. 4.1, provide a convenient abstraction of the concept of '1-form on a Riemannian manifold'.

By duality one can introduce the so-called *tangent module*, which is denoted by $L^2(T\mathrm{X})$. Another strictly related notion is that of *divergence* operator. Both these objects are treated in Sect. 4.2. The fundamental class of *infinitesimally Hilbertian* metric measure spaces, namely those metric measure spaces whose associated tangent/cotangent modules are Hilbert modules, is studied in detail in Sect. 4.3.

Finally, Sect. 4.4 is devoted to the 'transformations' of metric measure spaces, called *maps of bounded deformation*. Any such map is associated with a natural notion of *differential*, which is a linear and continuous operator between suitable normed modules.

4.1 Cotangent Module

4.1.1 Definition and Basic Properties

In the next result we introduce the important notion of cotangent module, which will play a crucial role in the following discussion. It also motivates our interest toward the theory of $L^2(\mathfrak{m})$-normed $L^\infty(\mathfrak{m})$-modules developed in Chap. 3.

Theorem 4.1.1 (Cotangent Module) *Let* $(\mathrm{X}, \mathsf{d}, \mathfrak{m})$ *be a fixed metric measure space. Then there exists a unique couple* $\big(L^2(T^*\mathrm{X}), \mathrm{d}\big)$, *where* $L^2(T^*\mathrm{X})$ *is an*

N. Gigli, E. Pasqualetto, *Lectures on Nonsmooth Differential Geometry*,
SISSA Springer Series 2, https://doi.org/10.1007/978-3-030-38613-9_4

$L^2(\mathfrak{m})$-normed $L^\infty(\mathfrak{m})$-module and $\mathrm{d} : \mathrm{S}^2(\mathrm{X}) \to L^2(T^*\mathrm{X})$ *is a linear operator, such that:*

i) $|\mathrm{d}f| = |Df|$ *holds* $\mathfrak{m}$*-a.e. for every* $f \in \mathrm{S}^2(\mathrm{X})$.
ii) $L^2(T^*\mathrm{X})$ *is generated by* $\{\mathrm{d}f \,:\, f \in \mathrm{S}^2(\mathrm{X})\}$.

Uniqueness is intended up to unique isomorphism: if another couple $(\widetilde{\mathscr{M}}, \widetilde{\mathrm{d}})$ *satisfies the same properties, then there is a unique module isomorphism* $\Phi : L^2(T^*\mathrm{X}) \to \widetilde{\mathscr{M}}$ *such that* $\Phi \circ \mathrm{d} = \widetilde{\mathrm{d}}$.

We shall refer to $L^2(T^*\mathrm{X})$ *as* cotangent module *and to* d *as* differential.

Proof The proof goes as follows:
UNIQUENESS. Fix any couple $(\widetilde{\mathscr{M}}, \widetilde{\mathrm{d}})$ that satisfies both conditions i) and ii). We claim that for every $f, g \in \mathrm{S}^2(\mathrm{X})$ and $E \subseteq \mathrm{X}$ Borel it holds that

$$\mathrm{d}f = \mathrm{d}g \quad \mathfrak{m}\text{-a.e. on } E \quad \Longleftrightarrow \quad \widetilde{\mathrm{d}}f = \widetilde{\mathrm{d}}g \quad \mathfrak{m}\text{-a.e. on } E. \tag{4.1}$$

Indeed, $\mathrm{d}f = \mathrm{d}g$ $\mathfrak{m}$-a.e. on E if and only if $|\mathrm{d}(f-g)| = |D(f-g)| = |\widetilde{\mathrm{d}}(f-g)|$ $\mathfrak{m}$-a.e. on E if and only if $\widetilde{\mathrm{d}}f = \widetilde{\mathrm{d}}g$ $\mathfrak{m}$-a.e. on E. Now let us define

$$V := \left\{ \sum_{i=1}^n \chi_{E_i} \mathrm{d}f_i \;\middle|\; n \in \mathbb{N},\ (E_i)_{i=1}^n \text{ Borel partition of X},\ (f_i)_{i=1}^n \subseteq \mathrm{S}^2(\mathrm{X}) \right\},$$

$$\widetilde{V} := \left\{ \sum_{i=1}^n \chi_{E_i} \widetilde{\mathrm{d}}f_i \;\middle|\; n \in \mathbb{N},\ (E_i)_{i=1}^n \text{ Borel partition of X},\ (f_i)_{i=1}^n \subseteq \mathrm{S}^2(\mathrm{X}) \right\},$$

which are vector subspaces of $L^2(T^*\mathrm{X})$ and $\widetilde{\mathscr{M}}$, respectively. Note that any module isomorphism $\Phi : L^2(T^*\mathrm{X}) \to \widetilde{\mathscr{M}}$ satisfying $\Phi \circ \mathrm{d} = \widetilde{\mathrm{d}}$ must necessarily restrict to the map $\Phi : V \to \widetilde{V}$ given by

$$\Phi\left(\sum_{i=1}^n \chi_{E_i} \mathrm{d}f_i \right) := \sum_{i=1}^n \chi_{E_i} \widetilde{\mathrm{d}}f_i \quad \text{for every } \sum_{i=1}^n \chi_{E_i} \mathrm{d}f_i \in V. \tag{4.2}$$

Well-posedness of (4.2) stems from (4.1). Moreover, the $\mathfrak{m}$-a.e. equalities

$$\left| \sum_{i=1}^n \chi_{E_i} \widetilde{\mathrm{d}}f_i \right| = \sum_{i=1}^n \chi_{E_i} |\widetilde{\mathrm{d}}f_i| = \sum_{i=1}^n \chi_{E_i} |Df_i| = \sum_{i=1}^n \chi_{E_i} |\mathrm{d}f_i| = \left| \sum_{i=1}^n \chi_{E_i} \mathrm{d}f_i \right|$$

grant that Φ preserves the pointwise norm, whence also the norm. Since V is dense in $L^2(T^*\mathrm{X})$ by property ii) for $(L^2(T^*\mathrm{X}), \mathrm{d})$, the linear continuous map $\Phi : V \to \widetilde{\mathscr{M}}$ can be uniquely extended to an operator $\Phi : L^2(T^*\mathrm{X}) \to \widetilde{\mathscr{M}}$, which is linear continuous and preserves the pointwise norm by Remark 3.1.16. In particular, it is an isometry, whence it is injective and it has closed image. Given that $\Phi(V) = \widetilde{V}$ is dense in $\widetilde{\mathscr{M}}$ by property ii) for $(\widetilde{\mathscr{M}}, \widetilde{\mathrm{d}})$, we deduce that Φ is also surjective. In order to conclude, it only remains to show that Φ is $L^\infty(\mathfrak{m})$-linear. To do so, first notice that $\Phi(\chi_E\, v) = \chi_E\, \Phi(v)$ is satisfied for every $E \subseteq \mathrm{X}$ Borel and $v \in V$. Since Φ and the multiplication by L^∞-functions are continuous, the same property holds for every $v \in L^2(T^*\mathrm{X})$, whence $\Phi(f\, v) = f\, \Phi(v)$ for all $f : \mathrm{X} \to \mathbb{R}$ simple and $v \in L^2(T^*\mathrm{X})$ by linearity of Φ. Finally, the same is true also for every $f \in L^\infty(\mathfrak{m})$ by density of the simple functions in $L^\infty(\mathfrak{m})$. This completes the proof of the uniqueness part of the statement.

EXISTENCE. Let us define the *pre-cotangent module* as the set

$$\mathsf{Pcm} := \Big\{ \{(E_i, f_i)\}_{i=1}^n \;\Big|\; n \in \mathbb{N},\ (E_i)_{i=1}^n \text{ Borel partition of } \mathrm{X},\ (f_i)_{i=1}^n \subseteq \mathrm{S}^2(\mathrm{X}) \Big\}.$$

For simplicity, we shall write $(E_i, f_i)_i$ instead of $\{(E_i, f_i)\}_{i=1}^n$. We introduce an equivalence relation on Pcm: we say $(E_i, f_i)_i \sim (F_j, g_j)_j$ if and only if $|D(f_i - g_j)| = 0$ $\mathfrak{m}$-a.e. in $E_i \cap F_j$ for every i, j. Let us denote by $[E_i, f_i]_i \in \mathsf{Pcm}/\sim$ the equivalence class of $(E_i, f_i)_i \in \mathsf{Pcm}$.

We now define some operations on the quotient $\mathsf{Pcm}/\sim$, which are well-defined by locality of minimal weak upper gradients (recall Theorem 2.1.28):

$$\begin{aligned}
[E_i, f_i]_i + [F_j, g_j]_j &:= [E_i \cap F_j, f_i + g_j]_{i,j},\\
\alpha\, [E_i, f_i]_i &:= [E_i, \alpha\, f_i]_i,\\
\Big(\sum_j \alpha_j\, \chi_{F_j}\Big) \cdot [E_i, f_i]_i &:= [E_i \cap F_j, \alpha_j\, f_i]_{i,j},\\
\big|[E_i, f_i]_i\big| &:= \sum_i \chi_{E_i} |Df_i| \quad \mathfrak{m}\text{-a.e. in } \mathrm{X},\\
\big\|[E_i, f_i]_i\big\| &:= \big\|\,|[E_i, f_i]_i|\,\big\|_{L^2(\mathfrak{m})} = \Big(\sum_i \int_{E_i} |Df_i|^2 \,\mathrm{d}\mathfrak{m}\Big)^{1/2}.
\end{aligned} \tag{4.3}$$

The first two operations in (4.3) give $\mathsf{Pcm}/\sim$ a vector space structure, the third one is the multiplication by simple functions $\cdot : \mathsf{Sf}(\mathfrak{m}) \times (\mathsf{Pcm}/\sim) \to (\mathsf{Pcm}/\sim)$ (where $\mathsf{Sf}(\mathfrak{m})$ denotes the space of all simple functions on X modulo $\mathfrak{m}$-a.e. equality), the fourth one is the pointwise norm $|\cdot| : (\mathsf{Pcm}/\sim) \to L^2(\mathfrak{m})$ and the fifth one is a norm on $\mathsf{Pcm}/\sim$.

We only prove that $\|\cdot\|$ is actually a norm on $\mathsf{Pcm}/\sim$: if $\big\|[E_i, f_i]_i\big\| = 0$ then $|Df_i| = 0$ holds $\mathfrak{m}$-a.e. on E_i for every i, so that $(E_i, f_i)_i \sim (\mathrm{X}, 0)$. Moreover, it directly follows from the definitions in (4.3) that $\big\|\alpha\, [E_i, f_i]_i\big\| = |\alpha|\, \big\|[E_i, f_i]_i\big\|$.

Finally, one has

$$\begin{aligned}\big\|[E_i, f_i]_i + [F_j, g_j]_j\big\| &= \big\|[E_i \cap F_j, f_i + g_j]_{i,j}\big\| = \Big\|\sum_{i,j} \chi_{E_i\cap F_j}\,|D(f_i+g_j)|\Big\|_{L^2(\mathfrak{m})}\\ &\le \Big\|\sum_{i,j}\chi_{E_i\cap F_j}\,|Df_i|\Big\|_{L^2(\mathfrak{m})} + \Big\|\sum_{i,j}\chi_{E_i\cap F_j}\,|Dg_j|\Big\|_{L^2(\mathfrak{m})}\\ &= \Big\|\sum_{i}\chi_{E_i}\,|Df_i|\Big\|_{L^2(\mathfrak{m})} + \Big\|\sum_{j}\chi_{F_j}\,|Dg_j|\Big\|_{L^2(\mathfrak{m})}\\ &= \big\|[E_i, f_i]_i\big\| + \big\|[F_j, g_j]_j\big\|,\end{aligned}$$

which is the triangle inequality for $\|\cdot\|$. Hence $\|\cdot\|$ is a norm on $\mathsf{Pcm}/\sim$.

Let us denote by $\big(L^2(T^*\mathrm{X}), \|\cdot\|_{L^2(T^*\mathrm{X})}\big)$ the completion of $\big(\mathsf{Pcm}/\sim, \|\cdot\|\big)$. One has that the operations $|\cdot| : (\mathsf{Pcm}/\sim) \to L^2(\mathfrak{m})$ and $\cdot : \mathsf{Sf}(\mathfrak{m})\times(\mathsf{Pcm}/\sim) \to (\mathsf{Pcm}/\sim)$, which can be readily proved to be continuous, uniquely extend to suitable

$$\begin{aligned}|\cdot| &: L^2(T^*\mathrm{X}) \to L^2(\mathfrak{m}),\\ \cdot &: L^\infty(\mathfrak{m}) \times L^2(T^*\mathrm{X}) \to L^2(T^*\mathrm{X}),\end{aligned}$$

which endow $L^2(T^*\mathrm{X})$ with the structure of an $L^2(\mathfrak{m})$-normed $L^\infty(\mathfrak{m})$-module.

Finally, let us define the differential operator $\mathrm{d} : \mathrm{S}^2(\mathrm{X}) \to L^2(T^*\mathrm{X})$ as $\mathrm{d}f := [\mathrm{X}, f]$ for every $f \in \mathrm{S}^2(\mathrm{X})$, where we think of $\mathsf{Pcm}/\sim$ as a subset of $L^2(T^*\mathrm{X})$. Note that

$$\begin{aligned}\mathrm{d}(\alpha f + \beta g) &= [\mathrm{X}, \alpha f + \beta g] = \alpha\,[\mathrm{X}, f] + \beta\,[\mathrm{X}, g]\\ &= \alpha\,\mathrm{d}f + \beta\,\mathrm{d}g \quad \forall f, g \in \mathrm{S}^2(\mathrm{X}),\ \alpha, \beta \in \mathbb{R},\end{aligned}$$

proving that d is a linear map. Also $|\mathrm{d}f| = \big|[\mathrm{X}, f]\big| = |Df|$ holds $\mathfrak{m}$-a.e. for any $f \in \mathrm{S}^2(\mathrm{X})$, which shows the validity of i). To conclude, observe that the family of all finite sums of the form $\sum_{i=1}^n \chi_{E_i}\,\mathrm{d}f_i$, with $(E_i)_{i=1}^n$ Borel partition of X and $(f_i)_{i=1}^n \subseteq \mathrm{S}^2(\mathrm{X})$, coincides with the space $\mathsf{Pcm}/\sim$, thus it is dense in $L^2(T^*\mathrm{X})$ by the very definition of $L^2(T^*\mathrm{X})$, proving ii) and accordingly the statement. □

Theorem 4.1.2 (Closure of the Differential) *Let $(f_n)_n \subseteq \mathrm{S}^2(\mathrm{X})$ be a given sequence that pointwise converges $\mathfrak{m}$-a.e. to some limit function f. Suppose that $\mathrm{d}f_n \rightharpoonup \omega$ weakly in $L^2(T^*\mathrm{X})$ for some $\omega \in L^2(T^*\mathrm{X})$. Then $f \in \mathrm{S}^2(\mathrm{X})$ and $\mathrm{d}f = \omega$.*

Moreover, the same conclusion holds if $(f_n)_n \subseteq W^{1,2}(\mathrm{X})$ satisfies $f_n \rightharpoonup f$ and $\mathrm{d}f_n \rightharpoonup \omega$ weakly in $L^2(\mathfrak{m})$ and $L^2(T^*\mathrm{X})$, respectively.

Proof By Mazur's lemma (recall Theorem A.2) we can assume without loss of generality that we have $\mathrm{d}f_n \to \omega$ in the strong topology of $L^2(T^*\mathrm{X})$. In particular,

$|Df_n| = |\mathrm{d}f_n| \to |\omega|$ strongly in $L^2(\mathfrak{m})$ as $n \to \infty$, whence we have that $f \in \mathrm{S}^2(\mathrm{X})$ by Proposition 2.1.13. Moreover, it holds that

$$\begin{aligned}\overline{\lim_{n\to\infty}} \|\mathrm{d}f - \mathrm{d}f_n\|_{L^2(T^*\mathrm{X})} &\leq \overline{\lim_{n\to\infty}} \lim_{k\to\infty} \|\mathrm{d}(f_k - f_n)\|_{L^2(T^*\mathrm{X})} \\ &= \overline{\lim_{n\to\infty}} \lim_{k\to\infty} \big\||\mathrm{d}(f_k - f_n)|\big\|_{L^2(\mathfrak{m})} = 0,\end{aligned}$$

so that $\mathrm{d}f = \omega$ as required. Finally, the last statement follows from the first one by applying twice Mazur's lemma and by recalling that any strongly converging sequence in $L^2(\mathfrak{m})$ has a subsequence that is $\mathfrak{m}$-a.e. convergent to the same limit. □

Remark 4.1.3 We point out that the map

$$\begin{aligned}W^{1,2}(\mathrm{X}) &\longrightarrow L^2(\mathfrak{m}) \times L^2(T^*\mathrm{X}),\\ f &\longmapsto (f, \mathrm{d}f),\end{aligned} \tag{4.4}$$

is a linear isometry, as soon as the target space $L^2(\mathfrak{m}) \times L^2(T^*\mathrm{X})$ is endowed with the product norm $\|(f,\omega)\|^2 := \|f\|^2_{L^2(\mathfrak{m})} + \|\omega\|^2_{L^2(T^*\mathrm{X})}$. ■

4.1.2 Calculus Rules and Their Consequences

Theorem 4.1.4 (Calculus Rules for the Differential) *The following properties hold:*

A) LOCALITY. *Let $f, g \in \mathrm{S}^2(\mathrm{X})$ be given. Then $\mathrm{d}f = \mathrm{d}g$ holds $\mathfrak{m}$-a.e. in $\{f = g\}$.*
B) CHAIN RULE. *Let $f \in \mathrm{S}^2(\mathrm{X})$ be given. Then:*

  B1) *If a Borel set $N \subseteq \mathbb{R}$ is $\mathcal{L}^1$-negligible, then $\mathrm{d}f = 0$ holds $\mathfrak{m}$-a.e. in $f^{-1}(N)$.*
  B2) *If $I \subseteq \mathbb{R}$ is an interval satisfying $(f_*\mathfrak{m})(\mathbb{R} \setminus I) = 0$ and $\varphi : I \to \mathbb{R}$ is a Lipschitz function, then $\varphi \circ f \in \mathrm{S}^2(\mathrm{X})$ and $\mathrm{d}(\varphi \circ f) = \varphi' \circ f\, \mathrm{d}f$. The expression $\varphi' \circ f\, \mathrm{d}f$ is a well-defined element of $L^2(T^*\mathrm{X})$ by B1).*

C) LEIBNIZ RULE. *Let $f, g \in \mathrm{S}^2(\mathrm{X}) \cap L^\infty(\mathfrak{m})$ be given. Then $fg \in \mathrm{S}^2(\mathrm{X}) \cap L^\infty(\mathfrak{m})$ and it holds that $\mathrm{d}(fg) = f\,\mathrm{d}g + g\,\mathrm{d}f$.*

Proof The proof goes as follows:

A) Note that $|\mathrm{d}f - \mathrm{d}g| = |D(f-g)| = 0$ holds $\mathfrak{m}$-a.e. in $\{f - g = 0\}$ by Theorem 2.1.28, whence we have that $\mathrm{d}f = \mathrm{d}g$ holds $\mathfrak{m}$-a.e. in $\{f = g\}$, as required.
B1) We have that $|\mathrm{d}f| = |Df| = 0$ holds $\mathfrak{m}$-a.e. on $f^{-1}(N)$ by Theorem 2.1.28, so that $\mathrm{d}f = 0$ holds $\mathfrak{m}$-a.e. on $f^{-1}(N)$.

B2) The Lipschitz function $\varphi : I \to \mathbb{R}$ can be extended to a Lipschitz function $\overline{\varphi} : \mathbb{R} \to \mathbb{R}$ and the precise choice of such extension is irrelevant for the statement to hold, because the set $f^{-1}(\mathbb{R} \setminus I)$ has null $\mathfrak{m}$-measure. Then assume without loss of generality that $I = \mathbb{R}$. We know that $\varphi \circ f \in \mathrm{S}^2(\mathrm{X})$ by Theorem 2.1.28.

If φ is a linear function, then the chain rule just reduces to the linearity of the differential. If φ is an affine function, say that $\varphi(t) = at + b$, then $\mathrm{d}(\varphi \circ f) = \mathrm{d}(af + b) = a\,\mathrm{d}f = \varphi' \circ f\,\mathrm{d}f$. Now suppose that φ is a piecewise affine function. Say that $(I_n)_n$ is a sequence of intervals whose union covers the whole real line $\mathbb{R}$ and that $(\psi_n)_n$ is a sequence of affine functions such that $\varphi_{|I_n} = \psi_n$ holds for every $n \in \mathbb{N}$. Since φ' and ψ_n' coincide $\mathcal{L}^1$-a.e. in the interior of I_n, we have that $\mathrm{d}(\varphi \circ f) = \mathrm{d}(\psi_n \circ f) = \psi_n' \circ f\,\mathrm{d}f = \varphi' \circ f\,\mathrm{d}f$ holds $\mathfrak{m}$-a.e. on $f^{-1}(I_n)$ for all n, so that $\mathrm{d}(\varphi \circ f) = \varphi' \circ f\,\mathrm{d}f$ is verified $\mathfrak{m}$-a.e. on $\bigcup_n f^{-1}(I_n) = \mathrm{X}$, as required.

To prove the case of a general Lipschitz function $\varphi : \mathbb{R} \to \mathbb{R}$, we want to approximate φ with a sequence of functions φ_n satisfying the following properties:

$$
\begin{gathered}
(\varphi_n)_n \subseteq \mathrm{LIP}(\mathbb{R}) \text{ are piecewise affine functions with } \sup_{n\in\mathbb{N}} \mathrm{Lip}(\varphi_n) \leq \mathrm{Lip}(\varphi),\\
\varphi_n(t) \to \varphi(t) \text{ for every } t \in \mathbb{R} \text{ and } \varphi_n'(t) \to \varphi'(t) \text{ for } \mathcal{L}^1\text{-a.e. } t \in \mathbb{R}.
\end{gathered}
\tag{4.5}
$$

First of all, denote by φ_n the function that coincides with φ at $\{i/n : i \in \mathbb{Z}\}$ and is affine in the interval $\big[i/n, (i+1)/n\big]$ for every $i \in \mathbb{Z}$. One can readily prove that $\mathrm{Lip}(\varphi_n) \leq \mathrm{Lip}(\varphi)$ for all n. Given any $i \in \mathbb{Z}$, we deduce from the fact that the identity $\varphi_n'(t) = ⨍_{i/n}^{(i+1)/n} \varphi'\,\mathrm{d}\mathcal{L}^1$ holds for all $t \in \big[i/n, (i+1)/n\big]$ and from an application of Jensen's inequality that

$$
\begin{aligned}
\int_{i/n}^{(i+1)/n} |\varphi_n'|^2\,\mathrm{d}\mathcal{L}^1 &= \frac{1}{n}\left| ⨍_{i/n}^{(i+1)/n} \varphi'\,\mathrm{d}\mathcal{L}^1\right|^2 \leq \frac{1}{n} ⨍_{i/n}^{(i+1)/n} |\varphi'|^2\,\mathrm{d}\mathcal{L}^1\\
&= \int_{i/n}^{(i+1)/n} |\varphi'|^2\,\mathrm{d}\mathcal{L}^1.
\end{aligned}
\tag{4.6}
$$

Now fix $m \in \mathbb{N}$. It can be readily checked that $\varphi_n \to \varphi$ strongly in $L^2(-m,m)$, while (4.6) grants that $\int_{-m}^m |\varphi_n'|^2\,\mathrm{d}\mathcal{L}^1 \leq \int_{-m}^m |\varphi'|^2\,\mathrm{d}\mathcal{L}^1$ for every n, whence there is a subsequence $(n_k)_k$ such that $\varphi_{n_k}' \rightharpoonup g$ weakly in $L^2(-m,m)$ for some $g \in L^2(-m,m)$. This forces $g = \varphi'_{|(-m,m)}$, so that the original sequence $(\varphi_n')_n$ satisfies $\varphi_n' \rightharpoonup \varphi'$ weakly in $L^2(-m,m)$. Moreover, it holds that $\int_{-m}^m |\varphi'|^2\,\mathrm{d}\mathcal{L}^1 \leq \varliminf_n \int_{-m}^m |\varphi_n'|^2\,\mathrm{d}\mathcal{L}^1 \leq \int_{-m}^m |\varphi'|^2\,\mathrm{d}\mathcal{L}^1$, thus necessarily $\varphi_n' \to \varphi'$ strongly in $L^2(-m,m)$. In particular, there exists a subsequence $(n_k)_k$ such that $\varphi_{n_k}'(t) \to \varphi'(t)$ for a.e. $t \in (-m,m)$. Up to performing a diagonalisation argument, we can therefore build a sequence $(\varphi_n)_n$ that satisfies (4.5), as required.

Now notice that $\int |\varphi_n' - \varphi'|^2 \circ f \, |\mathrm{d}f|^2 \, \mathrm{dm} \to 0$ by (4.5), by B1) and by an application of the dominated convergence theorem, in other words $\varphi_n' \circ f \, \mathrm{d}f \to \varphi' \circ f \, \mathrm{d}f$ in the strong topology of $L^2(T^*\mathrm{X})$. Since (4.5) also grants that $\varphi_n \circ f \to \varphi \circ f$ pointwise $\mathfrak{m}$-a.e. in X and since we have $\mathrm{d}(\varphi_n \circ f) = \varphi_n' \circ f \, \mathrm{d}f$ by the previous part of the proof, we deduce from Theorem 4.1.2 that $\mathrm{d}(\varphi_n \circ f) \to \varphi' \circ f \, \mathrm{d}f$ in $L^2(T^*\mathrm{X})$, thus accordingly $\mathrm{d}(\varphi \circ f) = \varphi' \circ f \, \mathrm{d}f$.

C) We already know that $fg \in \mathrm{S}^2(\mathrm{X}) \cap L^\infty(\mathfrak{m})$ by Theorem 2.1.28. In the case in which $f, g \geq 1$, we deduce from property B2) that

$$\frac{\mathrm{d}(fg)}{fg} = \mathrm{d}\log(fg) = \mathrm{d}\big(\log(f)+\log(g)\big) = \mathrm{d}\log(f)+\mathrm{d}\log(g) = \frac{\mathrm{d}f}{f}+\frac{\mathrm{d}g}{g},$$

whence we get $\mathrm{d}(fg) = f \, \mathrm{d}g + g \, \mathrm{d}f$ by multiplying both sides by fg.

In the general case $f, g \in L^\infty(\mathfrak{m})$, choose a constant $C > 0$ so big that $f + C, g + C \geq 1$. By the previous case, we know that

$$\begin{aligned} \mathrm{d}\big((f+C)(g+C)\big) &= (f+C)\,\mathrm{d}(g+C) + (g+C)\,\mathrm{d}(f+C) \\ &= (f+C)\,\mathrm{d}g + (g+C)\,\mathrm{d}f \\ &= f\,\mathrm{d}g + g\,\mathrm{d}f + C\,\mathrm{d}(f+g), \end{aligned} \tag{4.7}$$

while a direct computation yields

$$\mathrm{d}\big((f+C)(g+C)\big) = \mathrm{d}\big(fg+C(f+g)+C^2\big) = \mathrm{d}(fg)+C\,\mathrm{d}(f+g). \tag{4.8}$$

By subtracting (4.8) from (4.7), we finally get that $\mathrm{d}(fg) = f \, \mathrm{d}g + g \, \mathrm{d}f$, as required. Hence the statement is achieved. □

Proposition 4.1.5 *The set* $\{\mathrm{d}f : f \in W^{1,2}(\mathrm{X})\}$ *generates the tangent module* $L^2(T^*\mathrm{X})$.

Proof Denote by $\mathscr{M}$ the module generated by $\{\mathrm{d}f : f \in W^{1,2}(\mathrm{X})\}$. It clearly suffices to prove that $\mathrm{d}f \in \mathscr{M}$ whenever $f \in \mathrm{S}^2(\mathrm{X})$. Fix any $\bar{x} \in \mathrm{X}$. For any $n, m \in \mathbb{N}$, let us call

$$\begin{aligned} f_n &:= (f \wedge n) \vee (-n) \in L^\infty(\mathfrak{m}), \\ \eta_m &:= \big(1 - \mathsf{d}(\cdot, B_m(\bar{x}))\big)^+, \\ f_{nm} &:= \eta_m \, f_n \in L^2(\mathfrak{m}). \end{aligned}$$

Since the function f_n can be written as $\varphi_n \circ f$, where φ_n is the 1-Lipschitz function defined by $\varphi_n(t) := (t \wedge n) \vee (-n)$, we have that $f_n \in \mathrm{S}^2(\mathrm{X})$ by property B2) of Theorem 4.1.4, thus accordingly $f_{nm} \in W^{1,2}(\mathrm{X})$ by property C) of Theorem 4.1.4.

More precisely, it holds that

$$\mathrm{d}f_n = \varphi_n' \circ f \, \mathrm{d}f = \chi_{\{|f| \leq n\}} \, \mathrm{d}f,$$
$$\chi_{B_m(\bar{x})} \, \mathrm{d}f_{nm} = \chi_{B_m(\bar{x})} \big(\eta_m \, \mathrm{d}f_n + f_n \, \mathrm{d}\eta_m\big) = \chi_{B_m(\bar{x})} \, \mathrm{d}f_n,$$

so that $\mathrm{d}f = \mathrm{d}f_{nm}$ holds $\mathfrak{m}$-a.e. in $A_{nm} := f^{-1}\big([-n, n]\big) \cap B_m(\bar{x})$. Given that $\mathfrak{m}(\mathrm{X} \setminus A_{nm}) \searrow 0$ as $n, m \to \infty$, we deduce from the dominated convergence theorem that $\chi_{A_{nm}} \, \mathrm{d}f_{nm} \to \mathrm{d}f$ in the strong topology of $L^2(T^*\mathrm{X})$ as $n, m \to \infty$. Since each $\chi_{A_{nm}} \, \mathrm{d}f_{nm}$ belongs to $\mathscr{M}$, we conclude that $\mathrm{d}f \in \mathscr{M}$ as well. This proves the statement. □

The ensuing result consists of an equivalent definition of cotangent module/differential:

Proposition 4.1.6 *There exists a unique (up to unique isomorphism) couple* $(\mathscr{M}, \tilde{\mathrm{d}})$, *where the space* $\mathscr{M}$ *is a module and* $\tilde{\mathrm{d}} : W^{1,2}(\mathrm{X}) \to \mathscr{M}$ *is a linear map, such that* $|\tilde{\mathrm{d}}f| = |Df|$ *holds* $\mathfrak{m}$*-a.e. for every* $f \in W^{1,2}(\mathrm{X})$ *and* $\mathscr{M}$ *is generated by* $\big\{\mathrm{d}f \,:\, f \in W^{1,2}(\mathrm{X})\big\}$. *Moreover, given any such couple there exists a unique module isomorphism* $\Psi : \mathscr{M} \to L^2(T^*\mathrm{X})$ *such that*

$$\begin{array}{ccc} W^{1,2}(\mathrm{X}) & \xrightarrow{\ \tilde{\mathrm{d}}\ } & \mathscr{M} \\ \big\downarrow & & \big\downarrow \Psi \\ \mathrm{S}^2(\mathrm{X}) & \xrightarrow[\ \mathrm{d}\ ]{} & L^2(T^*\mathrm{X}) \end{array} \tag{4.9}$$

is a commutative diagram.

Proof The proof goes as follows:

EXISTENCE. One can repeat verbatim the proof of the existence part of Theorem 4.1.1. Alternatively, let us call $\mathscr{M}$ the submodule of $L^2(T^*\mathrm{X})$ that is generated by $\big\{\mathrm{d}f \,:\, f \in W^{1,2}(\mathrm{X})\big\}$ and define $\tilde{\mathrm{d}} := \mathrm{d}_{|W^{1,2}(\mathrm{X})}$. It can be easily seen that $(\mathscr{M}, \tilde{\mathrm{d}})$ satisfies the required properties.

UNIQUENESS. In order to get uniqueness, it is clearly enough to prove the last part of the statement. By the very same arguments that had been used in the proof of the uniqueness part of Theorem 4.1.1, one can see that the requirement that Ψ is an $L^\infty(\mathfrak{m})$-linear operator satisfying $\Psi(\tilde{\mathrm{d}}f) = \mathrm{d}f$ for any $f \in W^{1,2}(\mathrm{X})$ forces a unique choice of $\Psi : \mathscr{M} \to L^2(T^*\mathrm{X})$. The surjectivity of Ψ stems from Proposition 4.1.5. □

The abstract theory of cotangent modules presented above is consistent with the classical one, as shown by the following result:

Proposition 4.1.7 *Fix* $d \in \mathbb{N} \setminus \{0\}$. *Let* $L^2\big(\mathbb{R}^d, (\mathbb{R}^d)^*, \mathcal{L}^d\big)$ *denote the space of all the* $L^2(\mathcal{L}^d)$ *1-forms in* $\mathbb{R}^d$. *Let* $\bar{\mathrm{d}} : W^{1,2}(\mathbb{R}^d) \to L^2\big(\mathbb{R}^d, (\mathbb{R}^d)^*, \mathcal{L}^d\big)$ *be the map assigning to each Sobolev function* $f \in W^{1,2}(\mathbb{R}^d)$ *its distributional differential.*

Then

$$\left(L^2\left(\mathbb{R}^d, (\mathbb{R}^d)^*, \mathcal{L}^d\right), \bar{\mathrm{d}}\right) \sim \left(L^2(T^*\mathbb{R}^d), \mathrm{d}\right), \tag{4.10}$$

in the sense that there exists a unique module isomorphism $\Phi : L^2(T^*\mathbb{R}^d) \to L^2\left(\mathbb{R}^d, (\mathbb{R}^d)^*, \mathcal{L}^d\right)$ *such that* $\Phi \circ \mathrm{d} = \bar{\mathrm{d}}$.

Proof We know by Theorem 2.1.37 that $|\bar{\mathrm{d}} f| = |Df|$ holds $\mathcal{L}^d$-a.e. for every $f \in W^{1,2}(\mathbb{R}^d)$. Moreover, for any bounded Borel subset B of $\mathbb{R}^d$ and any $\omega \in (\mathbb{R}^d)^*$, there exists (by a cut-off argument) a function $f \in W^{1,2}(\mathbb{R}^d)$ such that $\bar{\mathrm{d}} f = \omega$ holds $\mathcal{L}^d$-a.e. in B. Hence the normed module $L^2\left(\mathbb{R}^d, (\mathbb{R}^d)^*, \mathcal{L}^d\right)$ is generated by the elements of the form $\chi_B\, \omega$. We thus conclude by applying Proposition 4.1.6. □

We conclude the section with an alternative approach: it is possible to define a notion of cotangent module whose elements do not satisfy any integrability requirement.

Proposition 4.1.8 *There exists a unique (up to unique isomorphism) couple* $(\mathscr{M}^0, \mathrm{d}^0)$*, where the space* $\mathscr{M}^0$ *is an* L^0*-normed module and* $\mathrm{d}^0 : \mathrm{S}^2_{loc}(\mathrm{X}) \to \mathscr{M}^0$ *is a linear map, such that the equality* $|\mathrm{d}^0 f| = |Df|$ *holds* $\mathfrak{m}$*-a.e. for every* $f \in \mathrm{S}^2_{loc}(\mathrm{X})$ *and such that* L^0*-linear combinations of elements in* $\{\mathrm{d}^0 f : f \in \mathrm{S}^2_{loc}(\mathrm{X})\}$ *are dense in* $\mathscr{M}^0$*. Given any such couple, there exists a unique map* $\iota : L^2(T^*\mathrm{X}) \to \mathscr{M}^0$*—which is* L^∞*-linear, continuous and preserving the pointwise norm—such that*

$$\begin{array}{ccc} W^{1,2}(\mathrm{X}) & \xrightarrow{\ \mathrm{d}\ } & L^2(T^*\mathrm{X}) \\ \big\downarrow & & \big\downarrow \iota \\ \mathrm{S}^2_{loc}(\mathrm{X}) & \xrightarrow[\ \mathrm{d}^0\ ]{} & \mathscr{M}^0 \end{array} \tag{4.11}$$

is a commutative diagram. Moreover, the image of $L^2(T^*\mathrm{X})$ *in* $\mathscr{M}^0$ *via* ι *is dense.*

Proof Uniqueness follows along the same lines of Theorem 4.1.1. For existence, we consider the L^0-completion $(\mathscr{M}^0, \iota)$ of $L^2(T^*\mathrm{X})$. For any $f \in \mathrm{S}^2_{loc}(\mathrm{X})$ there is a partition $(E_n)_n$ of X and functions $f_n \in W^{1,2}(\mathrm{X})$ such that $f = f_n$ $\mathfrak{m}$-a.e. on E_n for every $n \in \mathbb{N}$. It is clear that the series $\sum_n \chi_{E_n}\, \iota(\mathrm{d} f_n)$ converges in $\mathscr{M}^0$ and the locality of the differential grants that its limit, which we shall call $\mathrm{d}^0 f$, does not depend on the particular choice of $(E_n)_n, (f_n)_n$.

Then the identity $|\mathrm{d}^0 f| = |Df|$ follows from the construction and the analogous property of the differential. Also, we know that L^∞-linear combinations of $\{\mathrm{d} f : f \in W^{1,2}(\mathrm{X})\}$ are dense in $L^2(T^*\mathrm{X})$ and that $\iota\left(L^2(T^*\mathrm{X})\right)$ is dense in $\mathscr{M}^0$. Thus L^∞-linear combinations of elements in $\{\iota(\mathrm{d} f) = \mathrm{d}^0 f : f \in W^{1,2}(\mathrm{X})\}$ are dense in $\mathscr{M}^0$. This construction also shows the existence and uniqueness of ι as in (4.11). □

4.2 Tangent Module

4.2.1 Definition and Basic Properties

Definition 4.2.1 (Tangent Module) We define the *tangent module* $L^2(T\mathrm{X})$ as the module dual of $L^2(T^*\mathrm{X})$. Its elements are called *vector fields*.

We can introduce the notion of vector field in an alternative way, which is not based upon the theory of normed modules. Namely, we can define a suitable notion of derivation:

Definition 4.2.2 (L^2-Derivations) A linear map $L : \mathrm{S}^2(\mathrm{X}) \to L^1(\mathfrak{m})$ is an L^2-*derivation* provided there exists $\ell \in L^2(\mathfrak{m})$ such that

$$\big|L(f)\big| \leq \ell\,|Df| \quad \mathfrak{m}\text{-a.e.} \quad \text{for every } f \in \mathrm{S}^2(\mathrm{X}). \tag{4.12}$$

The relation between vector fields and derivations is described by the following result:

Proposition 4.2.3 *Given any $v \in L^2(T\mathrm{X})$, the map $\mathrm{S}^2(\mathrm{X}) \ni f \mapsto \mathrm{d}f(v)$ is a derivation.*

Conversely, for any derivation $L : \mathrm{S}^2(\mathrm{X}) \to L^1(\mathfrak{m})$ there exists a unique $v \in L^2(T\mathrm{X})$ such that $L(f) = \mathrm{d}f(v)$, and $|v|$ is the least function ℓ (in the $\mathfrak{m}$-a.e. sense) for which (4.12) *holds.*

Proof Given any $v \in L^2(T\mathrm{X})$, let us define $L := v \circ \mathrm{d}$. Since $\big|L(f)\big| = \big|\mathrm{d}f(v)\big| \leq |Df||v|$ holds $\mathfrak{m}$-a.e., we have that L is the required derivation.

On the other hand, fix a derivation L and set $V := \big\{\mathrm{d}f \,:\, f \in \mathrm{S}^2(\mathrm{X})\big\}$. By arguing as in the proof of Proposition 3.2.9 one can see that for any $f_1, f_2 \in \mathrm{S}^2(\mathrm{X})$ we have

$$\mathrm{d}f_1 = \mathrm{d}f_2 \quad \mathfrak{m}\text{-a.e. on X} \quad \Longrightarrow \quad L(f_1) = L(f_2) \quad \mathfrak{m}\text{-a.e. on X}. \tag{4.13}$$

Then the map $T : V \to L^1(\mathfrak{m})$, given by $T(\mathrm{d}f) := L(f)$, is well-defined. Moreover, one has that $\big|T(\mathrm{d}f)\big| \leq \ell\,|Df|$ for each $f \in \mathrm{S}^2(\mathrm{X})$, whence Proposition 3.2.9 grants the existence of a unique vector field $v \in L^2(T\mathrm{X})$ such that $\omega(v) = T(\omega)$ for all $\omega \in V$. In other words, we have that $\mathrm{d}f(v) = L(f)$ for every $f \in \mathrm{S}^2(\mathrm{X})$. Now observe that $\big|L(f)\big| = \big|\mathrm{d}f(v)\big| \leq |v||Df|$ holds $\mathfrak{m}$-a.e. for every $f \in \mathrm{S}^2(\mathrm{X})$, which shows that $|v|$ satisfies (4.12). Finally, let us take any function $\ell \in L^2(\mathfrak{m})$ for which (4.12) holds. It can be readily checked that the $\mathfrak{m}$-a.e. equality

$$|v| = \operatorname*{ess\,sup}_{\omega \in L^2(T^*\mathrm{X})} \chi_{\{|\omega|>0\}} \frac{\big|\omega(v)\big|}{|\omega|} = \operatorname*{ess\,sup}_{f \in \mathrm{S}^2(\mathrm{X})} \chi_{\{|Df|>0\}} \frac{\big|\mathrm{d}f(v)\big|}{|Df|} \overset{(4.12)}{\leq} \ell$$

is verified, thus proving the required minimality of $|v|$. This completes the proof. □

Corollary 4.2.4 *Let $L : \mathrm{S}^2(\mathrm{X}) \to L^1(\mathfrak{m})$ be a derivation. Then*

$$L(fg) = f\,L(g) + g\,L(f) \quad \textit{for every } f, g \in \mathrm{S}^2(\mathrm{X}) \cap L^\infty(\mathfrak{m}). \tag{4.14}$$

Proof Direct consequence of Proposition 4.2.3 and of the Leibniz rule for the differential (see item C) of Theorem 4.1.4). □

4.2.2 Divergence Operator and Gradients

The adjoint $\mathrm{d}^* : L^2(T\mathrm{X}) \to L^2(\mathfrak{m})$ of the unbounded operator $\mathrm{d} : L^2(\mathfrak{m}) \to L^2(T^*\mathrm{X})$ is (up to a sign) what we call 'divergence operator'. More explicitly:

Definition 4.2.5 (Divergence) We call $D(\mathrm{div})$ the space of all vector fields $v \in L^2(T\mathrm{X})$ for which there exists $h \in L^2(\mathfrak{m})$ satisfying

$$-\int f\,h\,\mathrm{d}\mathfrak{m} = \int \mathrm{d}f(v)\,\mathrm{d}\mathfrak{m} \quad \text{for every } f \in W^{1,2}(\mathrm{X}). \tag{4.15}$$

The function h, which is unique by density of $W^{1,2}(\mathrm{X})$ in $L^2(\mathfrak{m})$, will be unambiguously denoted by $\mathrm{div}(v)$. Moreover, $D(\mathrm{div})$ is a vector subspace of $L^2(T\mathrm{X})$ and $\mathrm{div} : D(\mathrm{div}) \to L^2(\mathfrak{m})$ is a linear operator.

We show some properties of the divergence operator:

Proposition 4.2.6 *Let $v, w \in D(\mathrm{div})$ be given. Suppose that $v = w$ holds $\mathfrak{m}$-a.e. on some open set $\Omega \subseteq \mathrm{X}$. Then* $\mathrm{div}(v) = \mathrm{div}(w)$ *is satisfied $\mathfrak{m}$-a.e. on Ω.*

Proof By linearity of the divergence, it clearly suffices to prove that $\mathrm{div}(v) = 0$ $\mathfrak{m}$-a.e. on Ω whenever $v = 0$ $\mathfrak{m}$-a.e. on Ω. In order to prove it, notice that a simple cut-off argument gives

$$A := \{f \in W^{1,2}(\mathrm{X}) \;:\; f = 0 \text{ on } \Omega^c\} \text{ is dense in } B := \{g \in L^2(\mathfrak{m}) \;:\; g = 0 \text{ on } \Omega^c\}. \tag{4.16}$$

Moreover, $-\int f\,\mathrm{div}(v)\,\mathrm{d}\mathfrak{m} = \int \mathrm{d}f(v)\,\mathrm{d}\mathfrak{m} = 0$ holds for every $f \in A$, whence property (4.16) ensures that $\int g\,\mathrm{div}(v)\,\mathrm{d}\mathfrak{m} = 0$ for all $g \in B$, i.e. $\mathrm{div}(v)$ vanishes $\mathfrak{m}$-a.e. on Ω. □

Proposition 4.2.7 *Let $v \in D(\mathrm{div})$ be given. Let $f : \mathrm{X} \to \mathbb{R}$ be a bounded Lipschitz function. Then $f v \in D(\mathrm{div})$ and*

$$\mathrm{div}(f v) = \mathrm{d}f(v) + f\,\mathrm{div}(v) \quad \textit{holds } \mathfrak{m}\textit{-a.e. in } \mathrm{X}. \tag{4.17}$$

Proof Observe that the right hand side in (4.17) belongs to $L^2(\mathfrak{m})$. Then pick $g \in W^{1,2}(\mathrm{X})$. By the Leibniz rule for the differential, we have that

$$-\int g\big(\mathrm{d}f(v) + f\,\mathrm{div}(v)\big)\,\mathrm{d}\mathfrak{m} = -\int g\,\mathrm{d}f(v) + fg\,\mathrm{div}(v)\,\mathrm{d}\mathfrak{m} = \int \mathrm{d}(fg)(v) - g\,\mathrm{d}f(v)\,\mathrm{d}\mathfrak{m}$$
$$= \int f\,\mathrm{d}g(v)\,\mathrm{d}\mathfrak{m}.$$

Therefore the statement is achieved. □

To be precise, in the proof of the previous result we made use of this variant of the Leibniz rule for the differential:

Proposition 4.2.8 *Let* $f \in W^{1,2}(\mathrm{X})$ *and* $g \in \mathrm{LIP}(\mathrm{X}) \cap L^\infty(\mathfrak{m})$ *be given. Then* $fg \in W^{1,2}(\mathrm{X})$ *and* $\mathrm{d}(fg) = f\,\mathrm{d}g + g\,\mathrm{d}f$.

Proof Fix $\bar{x} \in \mathrm{X}$ and for any $m \in \mathbb{N}$ pick a 1-Lipschitz function $\chi_m : \mathrm{X} \to [0, 1]$ with bounded support such that $\chi_m = 1$ on $B_m(\bar{x})$. Then define $f_n := (f \wedge n) \vee (-n)$ and $g_m := \chi_m\, g$ for every $n, m \in \mathbb{N}$. Hence $f_n\, g_m \in W^{1,2}(\mathrm{X}) \cap L^\infty(\mathfrak{m})$ and $\mathrm{d}(f_n\, g_m) = f_n\, \mathrm{d}g_m + g_m\, \mathrm{d}f_n$. Given that $\big|\mathrm{d}(f_n\, g_m)\big| \leq \big(\|g\|_{L^\infty(\mathfrak{m})} + \mathrm{Lip}(g)\big)|f| + \|g\|_{L^\infty(\mathfrak{m})}|\mathrm{d}f| \in L^2(\mathfrak{m})$ holds $\mathfrak{m}$-a.e. for every choice of $n, m \in \mathbb{N}$ and $f_n\, g_m \to fg$ pointwise $\mathfrak{m}$-a.e. as $n, m \to \infty$, we deduce that $fg \in \mathrm{S}^2(\mathrm{X})$ by the closure of the differential. Now observe that for any $n \in \mathbb{N}$ we have

$$\chi_{B_m(\bar{x})}\, \mathrm{d}(f_n\, g) = \chi_{B_m(\bar{x})}\, \mathrm{d}(f_n\, g_m) = \chi_{B_m(\bar{x})}\big(f_n\, \mathrm{d}g + g\, \mathrm{d}f_n\big) \quad \text{for every } m \in \mathbb{N},$$

whence $\mathrm{d}(f_n\, g) = f_n\, \mathrm{d}g + g\, \mathrm{d}f_n$ is satisfied for every $n \in \mathbb{N}$. Given that $f_n\, g \to fg$ in $L^2(\mathfrak{m})$ and $f_n\, \mathrm{d}g + g\, \mathrm{d}f_n \to f\, \mathrm{d}g + g\, \mathrm{d}f$ in $L^2(T^*\mathrm{X})$, we conclude that $\mathrm{d}(fg) = f\, \mathrm{d}g + g\, \mathrm{d}f$ by the closure of d. □

We now introduce a special class of vector fields: that of *gradients* of Sobolev functions.

Definition 4.2.9 Let $f \in \mathrm{S}^2(\mathrm{X})$. Then we call $\mathsf{Grad}(f)$ the set of all $v \in L^2(T\mathrm{X})$ such that

$$\mathrm{d}f(v) = |\mathrm{d}f|^2 = |v|^2 \quad \text{holds } \mathfrak{m}\text{-a.e. in X}. \tag{4.18}$$

Remark 4.2.10 As observed in Remark 3.2.6, it holds that $\mathsf{Grad}(f) \neq \emptyset$ for every $f \in \mathrm{S}^2(\mathrm{X})$. However, it can happen that $\mathsf{Grad}(f)$ is not a singleton. Furthermore, even if each $\mathsf{Grad}(f)$ is a singleton, its unique element does not necessarily depend linearly on f. ■

Given any Banach space $\mathbb{B}$, we can define the multi-valued map $\mathsf{Dual} : \mathbb{B} \twoheadrightarrow \mathbb{B}'$ as

$$\mathbb{B} \ni v \longmapsto \Big\{L \in \mathbb{B}' \,:\, L(v) = \|L\|^2_{\mathbb{B}'} = \|v\|^2_{\mathbb{B}}\Big\}. \tag{4.19}$$

The Hahn-Banach theorem grants that $\mathsf{Dual}(v) \neq \emptyset$ for every $v \in \mathbb{B}$.

Exercise 4.2.11 Prove that Dual is single-valued and linear if and only if $\mathbb{B}$ is a Hilbert space. In this case, Dual is the Riesz isomorphism. ■

Coming back to the gradients, we point out that

$$\mathrm{Int}_{L^2(T^*\mathrm{X})}\big(\mathsf{Grad}(f)\big) = \mathsf{Dual}(\mathrm{d}f) \quad \text{for every } f \in \mathrm{S}^2(\mathrm{X}), \tag{4.20}$$

where the map Dual is associated to $\mathbb{B} := L^2(T^*\mathrm{X})$.

Example 4.2.12 Consider the space $(\mathbb{R}^2, \|\cdot\|_\infty)$, where $\|(x,y)\|_\infty = \max\{|x|,|y|\}$. Define the map $f : \mathbb{R}^2 \to \mathbb{R}$ as $f(x,y) := x$. Then $\mathsf{Grad}(f) = \{(x,y) \in \mathbb{R}^2 : x = 1,\ |y| \le 1\}$. ■

Exercise 4.2.13 Prove that the multi-valued map Dual on $(\mathbb{R}^n, \|\cdot\|)$ is single-valued at any point if and only if the norm $\|\cdot\|$ is differentiable. ■

Remark 4.2.14 The inequality $\mathrm{d}f(v) \le \frac{1}{2}|\mathrm{d}f|^2 + \frac{1}{2}|v|^2$ holds $\mathfrak{m}$-a.e. in X for every $f \in \mathrm{S}^2(\mathrm{X})$ and $v \in L^2(T\mathrm{X})$ (by Young inequality). It can be readily proved that the opposite inequality is satisfied $\mathfrak{m}$-a.e. if and only if $v \in \mathsf{Grad}(f)$. ■

Theorem 4.2.15 *The following properties hold:*

A) LOCALITY. *Let $f, g \in \mathrm{S}^2(\mathrm{X})$. Suppose that $f = g$ holds $\mathfrak{m}$-a.e. on some Borel set $E \subseteq \mathrm{X}$. Then for any $v \in \mathsf{Grad}(f)$ there exists $w \in \mathsf{Grad}(g)$ such that $v = w$ $\mathfrak{m}$-a.e. on E.*

B) CHAIN RULE. *Let $f \in \mathrm{S}^2(\mathrm{X})$ and $v \in \mathsf{Grad}(f)$ be given. Then:*

B1) *If a Borel set $N \subseteq \mathbb{R}$ is $\mathcal{L}^1$-negligible, then $v = 0$ holds $\mathfrak{m}$-a.e. on $f^{-1}(N)$.*

B2) *If $\varphi : \mathbb{R} \to \mathbb{R}$ is Lipschitz then $\varphi' \circ f\, v \in \mathsf{Grad}(\varphi \circ f)$, where $\varphi' \circ f$ is arbitrarily defined on $f^{-1}\{$non-differentiability points of $\varphi\}$.*

Proof To prove A), choose any $\widetilde{w} \in \mathsf{Grad}(g)$ and define $w := \chi_E\, v + \chi_{E^c}\, \widetilde{w}$. Then $w \in \mathsf{Grad}(g)$ and $v = w$ holds $\mathfrak{m}$-a.e. on E by locality of the differential, as required.

Property B1) directly follows from the analogous one for differentials (see Theorem 4.1.4), while to show B2) notice that

$$\begin{aligned}
\mathrm{d}(\varphi \circ f)(\varphi' \circ f\, v) &= \varphi' \circ f\, \mathrm{d}(\varphi \circ f)(v) = |\varphi' \circ f|^2\, \mathrm{d}f(v) \\
&= |\varphi' \circ f|^2\, |\mathrm{d}f|^2 = |\varphi' \circ f|^2\, |v|^2 \\
&= \big|\mathrm{d}(\varphi \circ f)\big|^2
\end{aligned}$$

is verified $\mathfrak{m}$-a.e. on X. □

Given any two Sobolev functions $f, g \in \mathrm{S}^2(\mathrm{X})$, let us define

$$H_{f,g}(\varepsilon) := \frac{1}{2}\big|D(g+\varepsilon f)\big|^2 \in L^1(\mathfrak{m}) \quad \text{for every } \varepsilon \in \mathbb{R}. \tag{4.21}$$

Then the map $H_{f,g} : \mathbb{R} \to L^1(\mathfrak{m})$ can be easily proven to be *convex*, meaning that

$$H\big((1-\lambda)\varepsilon_0 + \lambda\varepsilon_1\big) \le (1-\lambda)H(\varepsilon_0) + \lambda H(\varepsilon_1) \quad \mathfrak{m}\text{-a.e.} \quad \text{for all } \varepsilon_0, \varepsilon_1 \in \mathbb{R} \text{ and } \lambda \in [0,1]. \tag{4.22}$$

Therefore the monotonicity of the difference quotients of $H_{f,g}$ grants that

$$\exists\, L^1(\mathfrak{m})\text{-}\lim_{\varepsilon \searrow 0} \frac{H_{f,g}(\varepsilon) - H_{f,g}(0)}{\varepsilon} = \operatorname*{ess\,inf}_{\varepsilon>0} \frac{H_{f,g}(\varepsilon) - H_{f,g}(0)}{\varepsilon} \tag{4.23}$$

and an analogous statement holds for $\varepsilon \nearrow 0$.

Remark 4.2.16 The object in (4.23) could be morally denoted by $\mathrm{d}f(\nabla g)$, for the reasons we are now going to explain. Given a Banach space $\mathbb{B}$, we have that the map Dual defined in (4.19) is (formally) the differential of $\|\cdot\|_{\mathbb{B}}^2/2$. Since $T_v\mathbb{B} \approx \mathbb{B}$ and $T_{\|v\|_{\mathbb{B}}^2/2}\mathbb{R} \approx \mathbb{R}$ for any vector $v \in \mathbb{B}$, we can actually view $\mathrm{d}\big(\|\cdot\|_{\mathbb{B}}^2/2\big)(v) : T_v\mathbb{B} \to T_{\|v\|_{\mathbb{B}}^2/2}\mathbb{R}$ as an element of $\mathbb{B}'$. In our case, if we let $\mathbb{B} = L^2(T^*\mathrm{X})$ then we have that

$$\lim_{\varepsilon\to 0} \frac{\|\mathrm{d}g + \varepsilon\,\mathrm{d}f\|_{\mathbb{B}}^2 - \|\mathrm{d}g\|_{\mathbb{B}}^2}{2\,\varepsilon} = \mathrm{d}\left(\frac{\|\cdot\|_{\mathbb{B}}^2}{2}\right)(\mathrm{d}g)(\mathrm{d}f) = \mathsf{Dual}(\mathrm{d}g)(\mathrm{d}f) = \mathrm{d}f(\nabla g),$$

which leads to our interpretation. ■

Proposition 4.2.17 *Let $f, g \in \mathrm{S}^2(\mathrm{X})$. Then the following properties hold:*

i) *For any $v \in \mathsf{Grad}(g)$ we have that* $\operatorname{ess\,inf}_{\varepsilon>0} \frac{H_{f,g}(\varepsilon)-H_{f,g}(0)}{\varepsilon} \ge \mathrm{d}f(v)$ *holds $\mathfrak{m}$-a.e. in* X.
ii) *There is $v_{f,+} \in \mathsf{Grad}(g)$ such that* $\operatorname{ess\,inf}_{\varepsilon>0} \frac{H_{f,g}(\varepsilon)-H_{f,g}(0)}{\varepsilon} = \mathrm{d}f(v_{f,+})$ *$\mathfrak{m}$-a.e. in* X.
i') *For any $v \in \mathsf{Grad}(g)$ we have that* $\operatorname{ess\,sup}_{\varepsilon<0} \frac{H_{f,g}(\varepsilon)-H_{f,g}(0)}{\varepsilon} \le \mathrm{d}f(v)$ *holds $\mathfrak{m}$-a.e. in* X.
ii') *There is $v_{f,-} \in \mathsf{Grad}(g)$ such that* $\operatorname{ess\,sup}_{\varepsilon<0} \frac{H_{f,g}(\varepsilon)-H_{f,g}(0)}{\varepsilon} = \mathrm{d}f(v_{f,-})$ *$\mathfrak{m}$-a.e. in* X.

Proof The proof goes as follows:

i), i′) Take $v \in \mathsf{Grad}(g)$. By Remark 4.2.14 we have that

$$\mathrm{d}g(v) \ge \frac{1}{2}\,|\mathrm{d}g|^2 + \frac{1}{2}\,|v|^2 \quad \text{holds } \mathfrak{m}\text{-a.e. in X}. \tag{4.24}$$

Moreover, an application of Young's inequality yields

$$\mathrm{d}(g+\varepsilon f)(v) \leq \frac{1}{2}\left|\mathrm{d}(g+\varepsilon f)\right|^2 + \frac{1}{2}|v|^2 \quad \mathfrak{m}\text{-a.e. in X.} \tag{4.25}$$

By subtracting (4.24) from (4.25) we thus obtain

$$\varepsilon\,\mathrm{d}f(v) \leq \frac{\left|\mathrm{d}(g+\varepsilon f)\right|^2 - |\mathrm{d}g|^2}{2} \quad \mathfrak{m}\text{-a.e. in X.} \tag{4.26}$$

Dividing both sides of (4.26) by $\varepsilon > 0$ (resp. $\varepsilon < 0$) and letting $\varepsilon \to 0$, we get i) (resp. i$'$)).

ii), ii$'$) We shall only prove ii), since the proof of ii$'$) is analogous. For any $\varepsilon \in (0,1)$, let us pick some $v_\varepsilon \in \mathsf{Grad}(g+\varepsilon f)$. Notice that

$$\begin{aligned}\|v_\varepsilon\|_{L^2(T\mathrm{X})} &= \left\|\mathrm{d}(g+\varepsilon f)\right\|_{L^2(T^*\mathrm{X})} \\ &\leq \|\mathrm{d}g\|_{L^2(T^*\mathrm{X})} + \|\mathrm{d}f\|_{L^2(T^*\mathrm{X})} \quad \text{for every } \varepsilon \in (0,1),\end{aligned}$$

whence the intersection among all $0 < \varepsilon' < 1$ of the weak*-closure of $\{v_\varepsilon : \varepsilon \in (0,\varepsilon')\}$ is non-empty by Banach-Alaoglu theorem. Then call $v_{f,+}$ one of its elements. By expanding the formula $\mathrm{d}(g+\varepsilon f)(v_\varepsilon) \geq \frac{1}{2}\left|\mathrm{d}(g+\varepsilon f)\right|^2 + \frac{1}{2}|v_\varepsilon|^2$, which holds $\mathfrak{m}$-a.e. for every $\varepsilon \in (0,1)$, we see that

$$\frac{1}{2}|v_\varepsilon|^2 + \frac{1}{2}|\mathrm{d}g|^2 - \mathrm{d}g(v_\varepsilon) \leq G_\varepsilon \quad \text{holds } \mathfrak{m}\text{-a.e. in X,} \tag{4.27}$$

for a suitable $G_\varepsilon \in L^1(\mathfrak{m})$ that $L^1(\mathfrak{m})$-converges to 0 as $\varepsilon \searrow 0$. Observe that for any $E \subseteq \mathrm{X}$ Borel we have that

$$F_E : L^2(T\mathrm{X}) \to \mathbb{R}, \quad v \longmapsto \int_E \frac{1}{2}|v|^2 + \frac{1}{2}|\mathrm{d}g|^2 - \mathrm{d}g(v)\,\mathrm{d}\mathfrak{m} \tag{4.28}$$

is a weakly*-lower semicontinuous operator. Hence (4.27) grants that $F_E(v_{f,+}) \leq 0$ for every Borel set $E \subseteq \mathrm{X}$, or equivalently $\frac{1}{2}|v_{f,+}|^2 + \frac{1}{2}|\mathrm{d}g|^2 - \mathrm{d}g(v_{f,+}) \leq 0$ $\mathfrak{m}$-a.e. in X. Therefore Remark 4.2.14 gives $v_{f,+} \in \mathsf{Grad}(g)$. Finally, observe that it $\mathfrak{m}$-a.e. holds that

$$\begin{aligned}\mathrm{d}(g+\varepsilon f)(v_\varepsilon) &= \frac{1}{2}\left|D(g+\varepsilon f)\right|^2 + \frac{1}{2}|v_\varepsilon|^2, \\ \mathrm{d}g(v_\varepsilon) &\leq \frac{1}{2}|Dg|^2 + \frac{1}{2}|v_\varepsilon|^2.\end{aligned}$$

(The first line is due to the fact that $v_\varepsilon \in \mathsf{Grad}(g+\varepsilon f)$, while the second one follows from Young's inequality, as seen above.) By subtracting the second line from the first one, we obtain the $\mathfrak{m}$-a.e. inequality

$$\begin{aligned} \mathrm{d}f(v_\varepsilon) &\geq \frac{H_{f,g}(\varepsilon) - H_{f,g}(0)}{\varepsilon} \\ &\geq \operatorname*{ess\,inf}_{\varepsilon'>0} \frac{H_{f,g}(\varepsilon') - H_{f,g}(0)}{\varepsilon'} =: \Theta \quad \text{for every } \varepsilon \in (0,1). \end{aligned} \tag{4.29}$$

Recall that $L^2(T\mathrm{X}) \ni v \mapsto \int \omega(v)\,\mathrm{d}\mathfrak{m}$ is weakly*-continuous for any $\omega \in L^2(T^*\mathrm{X})$. By applying this fact with $\omega := \chi_E\,\mathrm{d}f$, where $E \subseteq \mathrm{X}$ is any Borel set, we deduce from (4.29) that

$$\int_E \mathrm{d}f(v_{f,+})\,\mathrm{d}\mathfrak{m} \geq \int_E \Theta\,\mathrm{d}\mathfrak{m} \quad \text{for every } E \subseteq \mathrm{X} \text{ Borel.}$$

This grants that $\mathrm{d}f(v_{f,+}) \geq \Theta$ holds $\mathfrak{m}$-a.e. in X, which together with i) imply ii). □

Exercise 4.2.18 Prove that the norm of a finite-dimensional Banach space is differentiable if and only if its dual norm is strictly convex. ■

4.3 Infinitesimal Hilbertianity

Proposition 4.3.1 *Let* $(\mathrm{X}, \mathsf{d}, \mathfrak{m})$ *be a metric measure space. Then the following conditions are equivalent:*

i) *For every* $f, g \in \mathrm{S}^2(\mathrm{X})$ *it holds that*

$$\operatorname*{ess\,inf}_{\varepsilon>0} \frac{H_{f,g}(\varepsilon) - H_{f,g}(0)}{\varepsilon} = \operatorname*{ess\,sup}_{\varepsilon<0} \frac{H_{f,g}(\varepsilon) - H_{f,g}(0)}{\varepsilon}. \tag{4.30}$$

ii) *For every* $g \in \mathrm{S}^2(\mathrm{X})$ *the set* $\mathsf{Grad}(g)$ *is a singleton.*

Proof The proof goes as follows:

ii)⟹i) It trivially follows from items ii) and ii′) of Proposition 4.2.17.

i)⟹ii) Our aim is to show that if $v, w \in \mathsf{Grad}(g)$ then $v = w$. We claim that it is enough to prove that

$$\mathrm{d}f(v) = \mathrm{d}f(w) \quad \text{for every } f \in \mathrm{S}^2(\mathrm{X}). \tag{4.31}$$

Indeed, if (4.31) holds true then the operator $\mathrm{d}f \mapsto \mathrm{d}f(v - w)$ from the generating linear subspace $V := \{\mathrm{d}f \,:\, f \in \mathrm{S}^2(\mathrm{X})\}$ of $L^2(T^*\mathrm{X})$ to

$L^1(\mathfrak{m})$ is identically null, whence accordingly we have that $v - w = 0$ by Proposition 3.2.9. This shows that it suffices to prove (4.31).

Take any $f \in \mathrm{S}^2(\mathrm{X})$. Suppose that (4.31) fails, then (possibly interchanging v and w) there exists a Borel set $E \subseteq \mathrm{X}$ with $\mathfrak{m}(E) > 0$ such that $\mathrm{d}f(v) < \mathrm{d}f(w)$ holds $\mathfrak{m}$-a.e. in E. Therefore we have that

$$\begin{aligned} \operatorname*{ess\,sup}_{\varepsilon<0} \frac{H_{f,g}(\varepsilon) - H_{f,g}(0)}{\varepsilon} &\leq \mathrm{d}f(v) < \mathrm{d}f(w) \\ &\leq \operatorname*{ess\,inf}_{\varepsilon>0} \frac{H_{f,g}(\varepsilon) - H_{f,g}(0)}{\varepsilon} \quad \mathfrak{m}\text{-a.e. in } E, \end{aligned}$$

which contradicts (4.30). This shows (4.31), as required. □

Definition 4.3.2 (Infinitesimal Strict Convexity) We say that $(\mathrm{X}, \mathsf{d}, \mathfrak{m})$ is *infinitesimally strictly convex* provided the two conditions of Proposition 4.3.1 hold true. For any $g \in \mathrm{S}^2(\mathrm{X})$, we shall denote by ∇g the only element of $\mathsf{Grad}(g)$.

Theorem 4.3.3 *The following conditions are equivalent:*

i) $W^{1,2}(\mathrm{X})$ *is a Hilbert space.*
ii) $2\big(|\mathrm{d}f|^2 + |\mathrm{d}g|^2\big) = \big|\mathrm{d}(f+g)\big|^2 + \big|\mathrm{d}(f-g)\big|^2$ *holds* $\mathfrak{m}$*-a.e. for every* $f, g \in W^{1,2}(\mathrm{X})$.
iii) $(\mathrm{X}, \mathsf{d}, \mathfrak{m})$ *is infinitesimally strictly convex and* $\mathrm{d}f(\nabla g) = \mathrm{d}g(\nabla f)$ *holds* $\mathfrak{m}$*-a.e. in* X *for every* $f, g \in W^{1,2}(\mathrm{X})$.
iv) $L^2(T^*\mathrm{X})$ *and* $L^2(T\mathrm{X})$ *are Hilbert modules.*
v) $(\mathrm{X}, \mathsf{d}, \mathfrak{m})$ *is infinitesimally strictly convex and* $\nabla(f+g) = \nabla f + \nabla g$ *holds* $\mathfrak{m}$*-a.e. in* X *for every* $f, g \in W^{1,2}(\mathrm{X})$.
vi) $(\mathrm{X}, \mathsf{d}, \mathfrak{m})$ *is infinitesimally strictly convex and* $\nabla(fg) = f\,\nabla g + g\,\nabla f$ *holds* $\mathfrak{m}$*-a.e. in* X *for every* $f, g \in W^{1,2}(\mathrm{X}) \cap L^\infty(\mathfrak{m})$.

Proof The proof goes as follows:

i)⟹ii) First of all, observe that $W^{1,2}(\mathrm{X})$ is a Hilbert space if and only if

$$W^{1,2}(\mathrm{X}) \ni f \longmapsto \mathsf{E}(f) := \frac{1}{2}\int |\mathrm{d}f|^2 \,\mathrm{d}\mathfrak{m} \quad \text{satisfies the parallelogram rule.} \tag{4.32}$$

Now suppose that i) holds, then $\mathsf{E}(f+\varepsilon g) + \mathsf{E}(f-\varepsilon g) = 2\,\mathsf{E}(f) + 2\,\varepsilon^2\,\mathsf{E}(g)$ for all $f, g \in W^{1,2}(\mathrm{X})$ and $\varepsilon \neq 0$, or equivalently

$$\frac{\mathsf{E}(f+\varepsilon g) - \mathsf{E}(f)}{\varepsilon} - \frac{\mathsf{E}(f-\varepsilon g) - \mathsf{E}(f)}{\varepsilon} = 2\,\varepsilon\,\mathsf{E}(g). \tag{4.33}$$

Hence (4.33) and Proposition 4.2.17 grant that

$$\int \operatorname*{ess\,sup}_{v\in\mathsf{Grad}(f)} \mathrm{d}g(v)\,\mathrm{dm} = \lim_{\varepsilon\searrow 0}\frac{\mathsf{E}(f+\varepsilon\, g)-\mathsf{E}(f)}{\varepsilon} = \lim_{\varepsilon\nearrow 0}\frac{\mathsf{E}(f+\varepsilon\, g)-\mathsf{E}(f)}{\varepsilon}$$
$$= \int \operatorname*{ess\,inf}_{v\in\mathsf{Grad}(f)} \mathrm{d}g(v)\,\mathrm{dm},$$

thus accordingly $\operatorname{ess\,inf}_{v\in\mathsf{Grad}(f)}\mathrm{d}g(v) = \operatorname{ess\,sup}_{v\in\mathsf{Grad}(f)}\mathrm{d}g(v)$ holds $\mathfrak{m}$-a.e. in X. This guarantees that $\mathsf{Grad}(f)$ is a singleton for every $f \in W^{1,2}(\mathrm{X})$, i.e. $(\mathrm{X},\mathsf{d},\mathfrak{m})$ is infinitesimally strictly convex. We now claim that

$$\int \mathrm{d}f(\nabla g)\,\mathrm{dm} = \int \mathrm{d}g(\nabla f)\,\mathrm{dm} \quad \text{for every } f,g \in W^{1,2}(\mathrm{X}). \tag{4.34}$$

Given $f, g \in W^{1,2}(\mathrm{X})$, denote by $\mathcal{Q} : \mathbb{R}^2 \to \mathbb{R}$ the function $(t,s) \mapsto \mathsf{E}(t\,f + s\,g)$. Since $\mathcal{Q}$ is a quadratic polynomial, in particular smooth, we have $\frac{\mathrm{d}}{\mathrm{d}t}\big|_{t=0}\frac{\mathrm{d}}{\mathrm{d}s}\big|_{s=0}\mathcal{Q}(t,s) = \frac{\mathrm{d}}{\mathrm{d}s}\big|_{s=0}\frac{\mathrm{d}}{\mathrm{d}t}\big|_{t=0}\mathcal{Q}(t,s)$. The left-hand side of the previous equation can be rewritten as

$$\frac{\mathrm{d}}{\mathrm{d}t}\bigg|_{t=0}\left(\lim_{h\to 0}\frac{\mathsf{E}(t\,f+h\,g)-\mathsf{E}(t\,f)}{h}\right) = \frac{\mathrm{d}}{\mathrm{d}t}\bigg|_{t=0}\left(\int \mathrm{d}g\big(\nabla(t\,f)\big)\,\mathrm{dm}\right)$$
$$= \frac{\mathrm{d}}{\mathrm{d}t}\bigg|_{t=0}\left(t\int \mathrm{d}g(\nabla f)\,\mathrm{dm}\right)$$
$$= \int \mathrm{d}g(\nabla f)\,\mathrm{dm}$$

and analogously the right-hand side equals $\int \mathrm{d}f(\nabla g)\,\mathrm{dm}$, proving (4.34).

Fix any function $h \in \mathrm{LIP}(\mathrm{X}) \cap L^\infty(\mathfrak{m})$. We want to prove that

$$W^{1,2}(\mathrm{X})\cap L^\infty(\mathfrak{m}) \ni f \longmapsto \int h\,|\mathrm{d}f|^2\,\mathrm{dm} \quad \text{satisfies the parallelogram rule.} \tag{4.35}$$

To this aim, notice that the Leibniz rule and the chain rule for differentials yield

$$\int h\,|\mathrm{d}f|^2\,\mathrm{dm} = \int h\,\mathrm{d}f(\nabla f)\,\mathrm{dm} = \int \mathrm{d}(fh)(\nabla f) - f\,\mathrm{d}h(\nabla f)\,\mathrm{dm}$$
$$= \int \mathrm{d}(fh)(\nabla f) - \mathrm{d}h\big(\nabla(f^2/2)\big)\,\mathrm{dm}$$
$$\overset{(4.34)}{=} \int \mathrm{d}(fh)(\nabla f) - \mathrm{d}(f^2/2)(\nabla h)\,\mathrm{dm}.$$

Both the addenda $\int \mathrm{d}(fh)(\nabla f)\,\mathrm{dm}$ and $-\int \mathrm{d}(f^2/2)(\nabla h)\,\mathrm{dm}$ are quadratic forms, the former because $(f,g) \mapsto \int \mathrm{d}(fh)(\nabla g)\,\mathrm{dm} = \int \mathrm{d}g\big(\nabla(fh)\big)\,\mathrm{dm}$ is bilinear, whence (4.35). Given that the set $\mathrm{LIP(X)} \cap L^\infty(\mathfrak{m})$ is weakly* dense in $L^\infty(\mathfrak{m})$, we finally deduce from (4.35) that

$$2\int h\,|\mathrm{d}f|^2 + h\,|\mathrm{d}g|^2\,\mathrm{dm} = \int h\,\big|\mathrm{d}(f+g)\big|^2 + h\,\big|\mathrm{d}(f-g)\big|^2\,\mathrm{dm}$$

holds for every $f, g \in W^{1,2}(\mathrm{X})$ and $h \in L^\infty(\mathfrak{m})$. Therefore ii) follows.

ii)$\Longrightarrow$i) By integrating the pointwise parallelogram rule over X, we get the parallelogram rule for $\|\cdot\|_{W^{1,2}(\mathrm{X})}$, so that $W^{1,2}(\mathrm{X})$ is a Hilbert space.

i)$\Longrightarrow$iii) By arguing exactly as in the first implication, we see that $(\mathrm{X}, \mathsf{d}, \mathfrak{m})$ is infinitesimally strictly convex and that (4.35) holds true. By following the argument we used to prove (4.34), we deduce that

$$\int h\,\mathrm{d}f(\nabla g)\,\mathrm{dm} = \int h\,\mathrm{d}g(\nabla f)\,\mathrm{dm} \qquad \begin{array}{l}\text{for every } f, g \in W^{1,2}(\mathrm{X}) \cap L^\infty(\mathfrak{m})\\ \text{and } h \in \mathrm{LIP(X)} \cap L^\infty(\mathfrak{m}).\end{array} \tag{4.36}$$

Given that the set $\mathrm{LIP(X)} \cap L^\infty(\mathfrak{m})$ is weakly* dense in $L^\infty(\mathfrak{m})$, we conclude from (4.36) (by applying a truncation and localisation argument) that $\mathrm{d}f(\nabla g) = \mathrm{d}g(\nabla f)$ holds $\mathfrak{m}$-a.e. for every $f, g \in W^{1,2}(\mathrm{X})$. This shows that iii) is verified.

iii)$\Longrightarrow$i) It suffices to prove that E satisfies the parallelogram rule. Fix $f, g \in W^{1,2}(\mathrm{X})$. Note that the function $[0,1] \ni t \mapsto \mathsf{E}(f + t\,g)$ is Lipschitz and that its derivative is given by

$$\begin{aligned}\frac{\mathrm{d}}{\mathrm{d}t}\,\mathsf{E}(f+t\,g) &= \lim_{h\to 0}\frac{\mathsf{E}\big((f+t\,g)+h\,g\big) - \mathsf{E}(f+t\,g)}{h} = \int \mathrm{d}g\big(\nabla(f+t\,g)\big)\,\mathrm{dm}\\ &= \int \mathrm{d}(f+t\,g)(\nabla g)\,\mathrm{dm} = \int \mathrm{d}f(\nabla g)\,\mathrm{dm} + t\int |\mathrm{d}g|^2\,\mathrm{dm},\end{aligned}$$

whence by integrating on $[0,1]$ we get $\mathsf{E}(f+g) - \mathsf{E}(f) = \int \mathrm{d}f(\nabla g)\,\mathrm{dm} + \int |\mathrm{d}g|^2/2\,\mathrm{dm}$. If we replace g with $-g$, we also obtain that $\mathsf{E}(f-g) - \mathsf{E}(f) = -\int \mathrm{d}f(\nabla g)\,\mathrm{dm} + \int |\mathrm{d}g|^2/2\,\mathrm{dm}$, whence by summing these two equalities we conclude that $\mathsf{E}(f+g) + \mathsf{E}(f-g) = 2\,\mathsf{E}(f) + 2\,\mathsf{E}(g)$.

ii)$\Longrightarrow$iv) Consider two 1-forms ω and η in $L^2(T^*\mathrm{X})$, say $\omega = \sum_i \chi_{E_i}\mathrm{d}f_i$ and $\eta = \sum_j \chi_{F_j}\mathrm{d}g_j$. By locality we see that $|\omega+\eta|^2 + |\omega-\eta|^2 = 2\,|\omega|^2 + 2\,|\eta|^2$ holds $\mathfrak{m}$-a.e. in X, whence by integrating we get $\|\omega+\eta\|^2_{L^2(T^*\mathrm{X})} + \|\omega-\eta\|^2_{L^2(T^*\mathrm{X})} = 2\,\|\omega\|^2_{L^2(T^*\mathrm{X})} + 2\,\|\eta\|^2_{L^2(T^*\mathrm{X})}$. By density of the simple 1-forms in $L^2(T^*\mathrm{X})$, we conclude that $L^2(T^*\mathrm{X})$ (and accordingly also $L^2(T\mathrm{X})$) is a Hilbert module, thus proving iv).

iv)⟹ii) It trivially follows from Proposition 3.2.12.

iv)⟹v) Let $f \in W^{1,2}(\mathrm{X})$ and $v \in \mathsf{Grad}(f)$. By Theorem 3.2.14 applied to $L^2(T\mathrm{X})$ there exists a unique 1-form $\omega \in L^2(T^*\mathrm{X})$ such that $\langle \omega, \eta \rangle = \eta(v)$ for every $\eta \in L^2(T^*\mathrm{X})$. Moreover, it holds that $|\omega|_* = |v| = |\mathrm{d}f|_*$ $\mathfrak{m}$-a.e. in X. Hence by taking $\eta := \mathrm{d}f$ we see that

$$|\omega - \mathrm{d}f|_*^2 = |\omega|_*^2 + |\mathrm{d}f|_*^2 - 2\,\langle \omega, \mathrm{d}f \rangle = 2\,|\mathrm{d}f|_*^2 - 2\,\mathrm{d}f(v) = 0 \quad \mathfrak{m}\text{-a.e.},$$

which grants that $\omega = \mathrm{d}f$. Again by Theorem 3.2.14, we deduce that $(\mathrm{X}, \mathsf{d}, \mathfrak{m})$ is infinitesimally strictly convex and that $f \mapsto \nabla f$ is linear, as required.

v)⟹ii) For any $f, g \in W^{1,2}(\mathrm{X})$, it $\mathfrak{m}$-a.e. holds that

$$\big|\mathrm{d}(f+g)\big|^2 = \mathrm{d}(f+g)\big(\nabla(f+g)\big) = \mathrm{d}f(\nabla f) + \mathrm{d}f(\nabla g) + \mathrm{d}g(\nabla f) + \mathrm{d}g(\nabla g),$$

$$\big|\mathrm{d}(f-g)\big|^2 = \mathrm{d}(f-g)\big(\nabla(f-g)\big) = \mathrm{d}f(\nabla f) - \mathrm{d}f(\nabla g) - \mathrm{d}g(\nabla f) + \mathrm{d}g(\nabla g),$$

hence by summing them we get the $\mathfrak{m}$-a.e. equality $\big|\mathrm{d}(f+g)\big|^2 + \big|\mathrm{d}(f-g)\big|^2 = 2\,|\mathrm{d}f|^2 + 2\,|\mathrm{d}g|^2$, proving the validity of ii).

v)⟺vi) By applying the chain rule for gradients, we see that if $f, g \in W^{1,2}(\mathrm{X}) \cap L^\infty(\mathfrak{m})$ and $f' := \exp(f)$, $g' := \exp(g)$, then we have

$$f'g'\,\nabla(f+g) = f'g'\,\nabla\big(\log(f'g')\big) = \nabla(f'g'),$$

$$f'g'(\nabla f + \nabla g) = f'g'\,\nabla\big(\log(f')\big) + f'g'\,\nabla\big(\log(g')\big) = g'\,\nabla f' + f'\,\nabla g'.$$

Therefore we conclude that v) is equivalent to vi), thus concluding the proof. □

Definition 4.3.4 (Infinitesimal Hilbertianity) We say that $(\mathrm{X}, \mathsf{d}, \mathfrak{m})$ is *infinitesimally Hilbertian* provided the six conditions of Theorem 4.3.3 hold true.

Proposition 4.3.5 *Let* $(\mathrm{X}, \mathsf{d}, \mathfrak{m})$ *be an infinitesimally Hilbertian metric measure space. Then the spaces* $W^{1,2}(\mathrm{X})$, $L^2(T^*\mathrm{X})$ *and* $L^2(T\mathrm{X})$ *are separable.*

Proof The space $W^{1,2}(\mathrm{X})$, being reflexive by hypothesis, is separable by Theorem 2.1.27. Given that the differentials of the functions in $W^{1,2}(\mathrm{X})$ generate the cotangent module, we deduce from Lemma 3.1.17 that even $L^2(T^*\mathrm{X})$ is separable. Finally, Theorem 3.2.14 grants that $L^2(T\mathrm{X})$ is separable as well. □

4.4 Maps of Bounded Deformation

Definition 4.4.1 (Maps of Bounded Deformation) Let $(\mathrm{X}, \mathsf{d}_\mathrm{X}, \mathfrak{m}_\mathrm{X})$ and $(\mathrm{Y}, \mathsf{d}_\mathrm{Y}, \mathfrak{m}_\mathrm{Y})$ be given metric measure spaces. Then a map $\varphi : \mathrm{Y} \to \mathrm{X}$ is said to be *of bounded deformation* provided it is Lipschitz and of bounded compression (recall Definition 3.2.23).

A map of bounded deformation $\varphi : \mathrm{Y} \to \mathrm{X}$ naturally induces a mapping

$$\begin{aligned} \boldsymbol{\varphi} : C\big([0,1],\mathrm{Y}\big) &\longrightarrow C\big([0,1],\mathrm{X}\big), \\ \gamma &\longmapsto \varphi \circ \gamma. \end{aligned} \tag{4.37}$$

It is then easy to prove that

$$\gamma \text{ is an AC curve in Y} \quad \Longrightarrow \quad \begin{array}{l} \boldsymbol{\varphi}(\gamma) \text{ is an AC curve in X and} \\ \big|\dot{\boldsymbol{\varphi}(\gamma)}_t\big| \leq \mathrm{Lip}(\varphi)\,|\dot{\gamma}_t| \text{ for a.e. } t. \end{array} \tag{4.38}$$

Indeed, we have $\mathsf{d}_\mathrm{X}\big(\varphi(\gamma_t),\varphi(\gamma_s)\big) \leq \mathrm{Lip}(\varphi)\,\mathsf{d}_\mathrm{Y}(\gamma_t,\gamma_s) \leq \mathrm{Lip}(\varphi)\int_s^t |\dot{\gamma}_r|\,\mathrm{d}r$ for all $s < t$.

Lemma 4.4.2 *Let $\boldsymbol{\pi}$ be a test plan on Y and $\varphi : \mathrm{Y} \to \mathrm{X}$ a map of bounded deformation. Then $\boldsymbol{\varphi}_*\boldsymbol{\pi}$ is a test plan on X.*

Proof Observe that

$$(\mathrm{e}_t)_*\boldsymbol{\varphi}_*\boldsymbol{\pi} = \varphi_*(\mathrm{e}_t)_*\boldsymbol{\pi} \leq \varphi_*(C\,\mathfrak{m}_\mathrm{Y}) \leq \mathrm{Comp}(\varphi)\,C\,\mathfrak{m}_\mathrm{X} \qquad \text{for every } t \in [0,1],$$

$$\int_0^1\!\!\int |\dot{\gamma}_t|^2\,\mathrm{d}\boldsymbol{\varphi}_*\boldsymbol{\pi}(\gamma)\,\mathrm{d}t = \int_0^1\!\!\int \big|\dot{\boldsymbol{\varphi}(\gamma)}_t\big|^2\,\mathrm{d}\boldsymbol{\pi}(\gamma)\,\mathrm{d}t \leq \mathrm{Lip}(\varphi)^2\int_0^1\!\!\int |\dot{\gamma}_t|^2\,\mathrm{d}\boldsymbol{\pi}(\gamma)\,\mathrm{d}t < +\infty,$$

whence the statement follows. □

By duality with Lemma 4.4.2, we can thus obtain the following result:

Proposition 4.4.3 *Let $\varphi : \mathrm{Y} \to \mathrm{X}$ be a map of bounded deformation and $f \in \mathrm{S}^2(\mathrm{X})$. Then it holds that $f \circ \varphi \in \mathrm{S}^2(\mathrm{Y})$ and*

$$\big|D(f\circ\varphi)\big| \leq \mathrm{Lip}(\varphi)\,|Df|\circ\varphi \quad \textit{holds } \mathfrak{m}_\mathrm{Y}\textit{-a.e. in } \mathrm{Y}. \tag{4.39}$$

Proof Since $|Df|\circ\varphi \in L^2(\mathfrak{m}_\mathrm{Y})$, it only suffices to prove that $\mathrm{Lip}(\varphi)\,|Df|\circ\varphi$ is a weak upper gradient for f. Then fix any test plan $\boldsymbol{\pi}$ on Y. We have that

$$\begin{aligned} \int \big|f\circ\varphi\circ\mathrm{e}_1 - f\circ\varphi\circ\mathrm{e}_0\big|\,\mathrm{d}\boldsymbol{\pi} &= \int |f\circ\mathrm{e}_1 - f\circ\mathrm{e}_0|\,\mathrm{d}\boldsymbol{\varphi}_*\boldsymbol{\pi} \\ &\leq \int_0^1\!\!\int |Df|(\gamma_t)\,|\dot{\gamma}_t|\,\mathrm{d}\boldsymbol{\varphi}_*\boldsymbol{\pi}(\gamma)\,\mathrm{d}t \\ &= \int_0^1\!\!\int |Df|\big(\boldsymbol{\varphi}(\gamma)_t\big)\,\big|\dot{\boldsymbol{\varphi}(\gamma)}_t\big|\,\mathrm{d}\boldsymbol{\pi}(\gamma)\,\mathrm{d}t \\ &\leq \mathrm{Lip}(\varphi)\int_0^1\!\!\int \big(|Df|\circ\varphi\big)(\gamma_t)\,|\dot{\gamma}_t|\,\mathrm{d}\boldsymbol{\pi}(\gamma)\,\mathrm{d}t, \end{aligned}$$

proving that $\mathrm{Lip}(\varphi)\,|Df|\circ\varphi$ is a weak upper gradient, as required. □

Theorem 4.4.4 (Pullback of 1-Forms) *Let* $(\mathrm{X}, \mathsf{d}_{\mathrm{X}}, \mathfrak{m}_{\mathrm{X}})$, $(\mathrm{Y}, \mathsf{d}_{\mathrm{Y}}, \mathfrak{m}_{\mathrm{Y}})$ *be metric measure spaces and* $\varphi : \mathrm{Y} \to \mathrm{X}$ *a map of bounded deformation. Then there exists a unique linear and continuous operator* $\varphi^* : L^2(T^*\mathrm{X}) \to L^2(T^*\mathrm{Y})$ *such that*

$$\begin{aligned} \varphi^* \mathrm{d}f &= \mathrm{d}(f \circ \varphi) \quad \textit{for every } f \in \mathrm{S}^2(\mathrm{X}), \\ \varphi^*(g\,\omega) &= g \circ \varphi\, \varphi^*\omega \quad \textit{for every } g \in L^\infty(\mathfrak{m}_{\mathrm{X}}) \textit{ and } \omega \in L^2(T^*\mathrm{X}). \end{aligned} \tag{4.40}$$

Moreover, it holds that

$$|\varphi^*\omega| \leq \mathrm{Lip}(\varphi)\, |\omega| \circ \varphi \quad \mathfrak{m}_{\mathrm{Y}}\textit{-a.e.} \quad \textit{for every } \omega \in L^2(T^*\mathrm{X}). \tag{4.41}$$

Proof We are obliged to define $\varphi^*\big(\sum_i \chi_{E_i}\, \mathrm{d}f_i\big) := \sum_i \chi_{E_i} \circ \varphi\, \mathrm{d}(f_i \circ \varphi)$. Given that

$$\begin{aligned} \Big|\sum_i \chi_{E_i} \circ \varphi \mathrm{d}(f_i \circ \varphi)\Big| &= \sum_i \chi_{\varphi^{-1}(E_i)} \big|\mathrm{d}(f_i \circ \varphi)\big| \overset{(4.39)}{\leq} \mathrm{Lip}(\varphi) \sum_i \chi_{\varphi^{-1}(E_i)}\, |\mathrm{d}f_i| \circ \varphi \\ &= \mathrm{Lip}(\varphi) \Big|\sum_i \chi_{E_i}\, \mathrm{d}f_i\Big|, \end{aligned}$$

we see that φ^* is well-defined, linear and continuous. Then it can be uniquely extended to an operator $\varphi^* : L^2(T^*\mathrm{X}) \to L^2(T^*\mathrm{Y})$ having all the required properties. □

We have introduced two different notions of pullback for the cotangent module $L^2(T^*\mathrm{X})$. We shall make use of the notation $\varphi^* : L^2(T^*\mathrm{X}) \to L^2(T^*\mathrm{Y})$ for the pullback described in Theorem 4.4.4, while we write $[\varphi^*] : L^2(T^*\mathrm{X}) \to \varphi^* L^2(T^*\mathrm{X})$ for the one of Theorem 3.2.24.

Theorem 4.4.5 (Differential of a Map of Bounded Deformation) *Let us consider two metric measure spaces* $(\mathrm{X}, \mathsf{d}_{\mathrm{X}}, \mathfrak{m}_{\mathrm{X}})$ *and* $(\mathrm{Y}, \mathsf{d}_{\mathrm{Y}}, \mathfrak{m}_{\mathrm{Y}})$. *Suppose* $(\mathrm{X}, \mathsf{d}_{\mathrm{X}}, \mathfrak{m}_{\mathrm{X}})$ *is infinitesimally Hilbertian. Let* $\varphi : \mathrm{Y} \to \mathrm{X}$ *be a map of bounded deformation. Then there exists a unique* $L^\infty(\mathfrak{m}_{\mathrm{Y}})$*-linear continuous map* $\mathrm{d}\varphi : L^2(T\mathrm{Y}) \to \varphi^* L^2(T\mathrm{X})$, *called* differential *of* φ, *such that*

$$[\varphi^*\omega]\big(\mathrm{d}\varphi(v)\big) = \varphi^*\omega(v) \quad \textit{for every } v \in L^2(T\mathrm{Y}) \textit{ and } \omega \in L^2(T^*\mathrm{X}). \tag{4.42}$$

Moreover, it holds that

$$\big|\mathrm{d}\varphi(v)\big| \leq \mathrm{Lip}(\varphi)\, |v| \quad \mathfrak{m}_{\mathrm{Y}}\textit{-a.e.} \quad \textit{for every } v \in L^2(T\mathrm{Y}). \tag{4.43}$$

Proof Denote by V the generating linear subspace $\big\{[\varphi^*\omega] : \omega \in L^2(T^*\mathrm{X})\big\}$ of $\varphi^* L^2(T^*\mathrm{X})$. Fix $v \in L^2(T\mathrm{Y})$ and define $L_v : V \to L^1(\mathfrak{m}_{\mathrm{Y}})$ as $L_v[\varphi^*\omega] :=$

$\varphi^*\omega(v)$. The $\mathfrak{m}_Y$-a.e. inequality

$$\big|\varphi^*\omega(v)\big| \leq |\varphi^*\omega|\,|v| \overset{(4.41)}{\leq} \mathrm{Lip}(\varphi)\,|\omega|\circ\varphi\,|v| = \mathrm{Lip}(\varphi)\,|v|\,\big|[\varphi^*\omega]\big| \tag{4.44}$$

grants that L_v is a well-defined, linear and continuous operator. Hence we know from Propositions 3.2.9 and 3.2.41 that there exists a unique element $\mathrm{d}\varphi(v) \in \big(\varphi^*L^2(T^*\mathrm{X})\big)^* \sim \varphi^*L^2(T\mathrm{X})$ such that $[\varphi^*\omega]\big(\mathrm{d}\varphi(v)\big) = \varphi^*\omega(v)$. Moreover, such element necessarily satisfies

$$\big|\mathrm{d}\varphi(v)\big| \leq \mathrm{Lip}(\varphi)\,|v| \quad \mathfrak{m}_Y\text{-a.e. in Y},$$

again by Proposition 3.2.9. Thus to conclude it only remains to show that the assignment $L^2(T\mathrm{Y}) \ni v \mapsto \mathrm{d}\varphi(v) \in \varphi^*L^2(T\mathrm{X})$ is $L^\infty(\mathfrak{m}_Y)$-linear. This follows from the chain of equalities

$$[\varphi^*\omega]\big(\mathrm{d}\varphi(f\,v)\big) = \varphi^*\omega(f\,v) = f\,\varphi^*\omega(v) = f\,[\varphi^*\omega]\big(\mathrm{d}\varphi(v)\big),$$

which holds $\mathfrak{m}_Y$-a.e. for every choice of $f \in L^\infty(\mathfrak{m}_Y)$ and $v \in L^2(T\mathrm{Y})$. □

In the case in which φ is invertible and its inverse is a map of bounded compression, the differential of φ can be equivalently expressed in the following fashion (based upon what previously discussed in Remark 3.2.35):

Theorem 4.4.6 *Let* $(\mathrm{X}, \mathsf{d}_X, \mathfrak{m}_X)$, $(\mathrm{Y}, \mathsf{d}_Y, \mathfrak{m}_Y)$ *be metric measure spaces and let* $\varphi : \mathrm{Y} \to \mathrm{X}$ *be a map of bounded deformation. Suppose that* φ *is invertible and that* φ^{-1} *has bounded compression. Then there exists a unique linear continuous operator* $\mathrm{d}\varphi : L^2(T\mathrm{Y}) \to L^2(T\mathrm{X})$ *such that*

$$\omega\big(\mathrm{d}\varphi(v)\big) = \big(\varphi^*\omega(v)\big)\circ\varphi^{-1} \quad \mathfrak{m}_X\text{-a.e.} \quad \text{for every } v \in L^2(T\mathrm{Y}) \text{ and } \omega \in L^2(T^*\mathrm{X}). \tag{4.45}$$

Moreover, it holds that

$$\big|\mathrm{d}\varphi(v)\big| \leq \mathrm{Lip}(\varphi)\,|v|\circ\varphi^{-1} \quad \mathfrak{m}_X\text{-a.e.} \quad \text{for every } v \in L^2(T\mathrm{Y}). \tag{4.46}$$

Proof Fix $v \in L^2(T\mathrm{Y})$. Denote by $\mathrm{d}\varphi(v)$ the map $L^2(T^*\mathrm{X}) \ni \omega \mapsto \big(\varphi^*\omega(v)\big)\circ\varphi^{-1} \in L^1(\mathfrak{m}_X)$. Given that $\big|\omega\big(\mathrm{d}\varphi(v)\big)\big| \leq \mathrm{Lip}(\varphi)\,|\omega|\,|v|\circ\varphi^{-1}$, we know that $\mathrm{d}\varphi(v)$ is (linear and) continuous. Moreover, for any $f \in L^\infty(\mathfrak{m}_X)$ it holds

$$\big(\varphi^*(f\,\omega)(v)\big)\circ\varphi^{-1} = \big(f\circ\varphi\,\varphi^*\omega(v)\big)\circ\varphi^{-1} = f\,\big(\varphi^*\omega(v)\big)\circ\varphi^{-1},$$

thus proving the $L^\infty(\mathfrak{m}_X)$-linearity of $\mathrm{d}\varphi(v)$. Hence we have a map $\mathrm{d}\varphi : L^2(T\mathrm{Y}) \to L^2(T\mathrm{X})$, which can be easily seen to satisfy all the required properties. □

In the following result, the function $(\gamma, t) \mapsto |\dot\gamma_t|$ is defined everywhere, as in Remark 1.2.6.

Theorem 4.4.7 (Speed of a Test Plan) *Let* $(\mathrm{X}, \mathsf{d}, \mathfrak{m})$ *be an infinitesimally Hilbertian metric measure space. Let* $\boldsymbol{\pi}$ *be a test plan on* X. *Then for almost every* $t \in [0, 1]$ *there exists an element* $\boldsymbol{\pi}'_t \in \mathrm{e}_t^* L^2(T\mathrm{X})$ *such that*

$$L^1(\boldsymbol{\pi})\text{-}\lim_{h\to 0} \frac{f\circ \mathrm{e}_{t+h} - f\circ \mathrm{e}_t}{h} = [\mathrm{e}_t^*\mathrm{d}f](\boldsymbol{\pi}'_t) \quad \textit{for every } f \in W^{1,2}(\mathrm{X}). \tag{4.47}$$

Moreover, the following hold:

i) *the element of* $\mathrm{e}_t^* L^2(T\mathrm{X})$ *satisfying* (4.47) *is unique,*
ii) *we have that* $|\boldsymbol{\pi}'_t|(\gamma) = |\dot\gamma_t|$ *for* $(\boldsymbol{\pi} \times \mathcal{L}_1)$*-a.e.* (γ, t).

Proof We divide the proof into several steps:

STEP 1. Notice that Proposition 4.3.5 grants that $W^{1,2}(\mathrm{X})$ is separable, thus there exists a countable dense $\mathbb{Q}$-linear subspace D of $W^{1,2}(\mathrm{X})$. By applying Theorem 2.1.21 we see that for any function $f \in D$ it holds that $(f\circ \mathrm{e}_{t+h} - f\circ \mathrm{e}_t)/h$ admits a strong $L^1(\boldsymbol{\pi})$-limit as $h \to 0$ for a.e. t. Moreover, the function $M : [0, 1] \to \mathbb{R}$, $M(t) := \int |\dot\gamma_t|^2 \,\mathrm{d}\boldsymbol{\pi}(\gamma)$ belongs to $L^1(0, 1)$ and the function $(\gamma, t) \mapsto |\dot\gamma_t|$ belongs to $L^2(\boldsymbol{\pi} \times \mathcal{L}_1)$. Hence we can pick a Borel negligible subset $N \subseteq [0, 1]$ such that for every $t \in [0, 1] \setminus N$ the following hold:

- $\mathsf{Der}_t(f) := \lim_{h\to 0} \big(f\circ \mathrm{e}_{t+h} - f\circ \mathrm{e}_t\big)/h \in L^1(\boldsymbol{\pi})$ exists for every $f \in D$,
- the inequality

$$\big|\mathsf{Der}_t(f)\big|(\gamma) \le |Df|(\gamma_t)\,|\dot\gamma_t| \quad \text{for } \boldsymbol{\pi}\text{-a.e. } \gamma \tag{4.48}$$

  is satisfied for every $f \in D$,
- t is a Lebesgue point for M, so that in particular there exists a constant $C_t > 0$ with

$$⨍_t^{t+h} M(s)\,\mathrm{d}s \le C_t \quad \text{for every } h \neq 0 \text{ such that } t+h \in [0, 1], \tag{4.49}$$

- the function $\gamma \mapsto |\dot\gamma_t|$ belongs to $L^2(\boldsymbol{\pi})$.

Since for any $t \in [0, 1] \setminus N$ we have that $\mathsf{Der}_t : D \to L^1(\boldsymbol{\pi})$ is a $\mathbb{Q}$-linear operator satisfying (4.48) for every $f \in D$, it uniquely extends to a linear continuous $\mathsf{Der}_t : W^{1,2}(\mathrm{X}) \to L^1(\boldsymbol{\pi})$ satisfying the inequality (4.48) for all $f \in W^{1,2}(\mathrm{X})$.

STEP 2. Observe that for any $t \in [0, 1] \setminus N$ and $g \in W^{1,2}(\mathrm{X})$ we have that

$$\begin{aligned}\left\| \frac{g \circ \mathrm{e}_{t+h} - g \circ \mathrm{e}_t}{h} \right\|_{L^1(\pi)} &\leq \int\!\!\fint_t^{t+h} |Dg|(\gamma_s)\, |\dot{\gamma}_s| \,\mathrm{d}s\, \mathrm{d}\pi(\gamma) \\ &\leq \left(\int\!\!\fint_t^{t+h} |Dg|^2(\gamma_s)\, \mathrm{d}s\, \mathrm{d}\pi(\gamma) \right)^{1/2} \left(\fint_t^{t+h} M(s)\, \mathrm{d}s \right)^{1/2} \\ &\leq \sqrt{C}\, \big\| |Dg| \big\|_{L^2(\mathfrak{m})} \sqrt{C_t}\end{aligned} \tag{4.50}$$

(where $C := \mathrm{Comp}(\pi)$ stands for the compression constant of π) holds for every $h \neq 0$ such that $t + h \in [0, 1]$. Now fix $t \in [0, 1] \setminus N$ and $f \in W^{1,2}(\mathrm{X})$. Choose any sequence $(f_n)_n \subseteq D$ that converges to f in $W^{1,2}(\mathrm{X})$. Therefore one has that

$$\begin{aligned}&\left\| \frac{f \circ \mathrm{e}_{t+h} - f \circ \mathrm{e}_t}{h} - \mathsf{Der}_t(f) \right\|_{L^1(\pi)} \\ \leq& \sqrt{C\, C_t}\, \big\| |D(f - f_n)| \big\|_{L^2(\mathfrak{m})} + \left\| \frac{f_n \circ \mathrm{e}_{t+h} - f_n \circ \mathrm{e}_t}{h} - \mathsf{Der}_t(f_n) \right\|_{L^1(\pi)} \\ &+ \big\| \mathsf{Der}_t(f_n - f) \big\|_{L^1(\pi)},\end{aligned}$$

so by first letting $h \to 0$ and then $n \to \infty$ we conclude that $\mathsf{Der}_t(f)$ is the strong $L^1(\pi)$-limit of $(f \circ \mathrm{e}_{t+h} - f \circ \mathrm{e}_t)/h$ as $h \to 0$.

STEP 3. Call $V_t := \big\{ [\mathrm{e}_t^* \mathrm{d} f] \,:\, f \in W^{1,2}(\mathrm{X}) \big\}$ for every $t \in [0, 1] \setminus N$. Define $L_t : V_t \to L^1(\pi)$ as $L_t[\mathrm{e}_t^* \mathrm{d} f] := \mathsf{Der}_t(f)$. Given that for any $f \in W^{1,2}(\mathrm{X})$ property (4.48) yields

$$\big| L_t[\mathrm{e}_t^* \mathrm{d} f] \big|(\gamma) \leq \big| [\mathrm{e}_t^* \mathrm{d} f] \big|(\gamma)\, |\dot{\gamma}_t| \quad \text{for } \pi\text{-a.e. } \gamma,$$

we see that the operator L_t (is well-defined, linear, continuous and) can be uniquely extended—by Propositions 3.2.9 and 3.2.41—to an element $\pi'_t \in \mathrm{e}_t^* L^2(T\mathrm{X}) \sim \big(\mathrm{e}_t^* L^2(T^*\mathrm{X}) \big)^*$. Therefore one has $\mathsf{Der}_t(f) = [\mathrm{e}_t^* \mathrm{d} f](\pi'_t)$ for every $f \in W^{1,2}(\mathrm{X})$ and $|\pi'_t|(\gamma) \leq |\dot{\gamma}_t|$ for π-a.e. γ.

STEP 4. Given any $f \in \mathrm{LIP}_{bs}(\mathrm{X})$ and $\gamma : [0, 1] \to \mathrm{X}$ AC, it holds that $f \circ \gamma$ is AC as well and that for π-a.e. γ we have $\big(f(\gamma_{t+h}) - f(\gamma_t) \big)/h \to \frac{\mathrm{d}}{\mathrm{d}t} f(\gamma_t)$ as $h \to 0$ for a.e. t. Then

$$[\mathrm{e}_t^* \mathrm{d} f](\pi'_t)(\gamma) = \frac{\mathrm{d}}{\mathrm{d}t} f(\gamma_t) \quad \text{for } (\pi \times \mathcal{L}_1)\text{-a.e. } (\gamma, t).$$

Since $[\mathrm{e}_t^* \mathrm{d} f](\pi'_t)(\gamma) \leq \big| [\mathrm{e}_t^* \mathrm{d} f] \big|(\gamma)\, |\pi'_t|(\gamma) \leq \mathrm{Lip}(f)\, |\pi'_t|(\gamma)$ holds for π-a.e. γ, we deduce from the previous formula that $\frac{\mathrm{d}}{\mathrm{d}t} f(\gamma_t) \leq \mathrm{Lip}(f)\, |\pi'_t|(\gamma)$ for π-a.e. γ. In order to conclude, it is thus sufficient to provide the existence of a countable family $D' \subseteq \mathrm{LIP}_{bs}(\mathrm{X})$ of 1-Lipschitz functions such that for every AC curve $\gamma : [0,1] \to \mathrm{X}$ it holds

$$|\dot{\gamma}_t| = \sup_{f \in D'} \frac{\mathrm{d}}{\mathrm{d}t} f(\gamma_t) \quad \text{for a.e. } t \in [0,1]. \tag{4.51}$$

To do so, fix a countable dense subset $(x_n)_n$ of X and let us define $f_{n,m} := \big(m - \mathsf{d}(\cdot, x_n)\big)^+$ for every $n, m \in \mathbb{N}$. Then the family $D' := (f_{n,m})_{n,m}$ does the job: given any $x, y \in \mathrm{X}$ it clearly holds that $\mathsf{d}(x,y) = \sup_{n,m}\big[f_{n,m}(x) - f_{n,m}(y)\big]$, whence for all $0 \le s < t \le 1$ we have

$$\begin{aligned} \mathsf{d}(\gamma_t, \gamma_s) = \sup_{n,m}\big[f_{n,m}(\gamma_t) - f_{n,m}(\gamma_s)\big] &= \sup_{n,m} \int_s^t \frac{\mathrm{d}}{\mathrm{d}r} f_{n,m}(\gamma_r)\,\mathrm{d}r \\ &\le \int_s^t \sup_{n,m} \frac{\mathrm{d}}{\mathrm{d}r} f_{n,m}(\gamma_r)\,\mathrm{d}r. \end{aligned}$$

Therefore the statement is achieved. □

Bibliographical Remarks

This chapter is entirely taken from [17, 19].

Chapter 5
Heat Flow on Metric Measure Spaces

In order to develop a second-order differential calculus on spaces with curvature bounds we need to make use of the regularising effects of the heat flow, to which this chapter is dedicated.

In Sect. 5.1 we establish the genera of *gradient flows* on Hilbert spaces. More precisely, we prove existence, uniqueness and several properties of the gradient flow associated to any convex and lower semicontinuous functional defined on a Hilbert space.

In Sect. 5.2 we concentrate our attention on the *heat flow* over metric measure spaces that are infinitesimally Hilbertian. In Sect. 5.2.1 we introduce the *Laplace* operator, while in Sect. 5.2.2 we define the heat flow as the gradient flow in $L^2(\mathfrak{m})$ of the Cheeger energy $f \mapsto \frac{1}{2}\int |Df|^2 \, d\mathfrak{m}$ and we show its basic features.

5.1 Gradient Flows on Hilbert Spaces

5.1.1 Set-Up of the Theory

Let H be a Hilbert space. Let $E : H \to [0, +\infty]$ be a convex lower semicontinuous functional. Given any point $x \in H$ such that $E(x) < \infty$, we define the *subdifferential* of E at x as

$$\partial^- E(x) := \big\{ v \in H \, : \, E(x) + \langle v, y - x \rangle \leq E(y) \text{ for every } y \in H \big\}. \tag{5.1}$$

It trivially holds that $0 \in \partial^- E(x)$ if and only if x is a minimum point of E.

N. Gigli, E. Pasqualetto, *Lectures on Nonsmooth Differential Geometry*, SISSA Springer Series 2, https://doi.org/10.1007/978-3-030-38613-9_5

Exercise 5.1.1 Consider $H := \mathbb{R}$ and $E(x) := |x|$ for every $x \in \mathbb{R}$. Then

$$\partial^- E(x) := \begin{cases} \{1\} & \text{if } x > 0, \\ [-1, 1] & \text{if } x = 0, \\ \{-1\} & \text{if } x < 0. \end{cases} \tag{5.2}$$

■

Proposition 5.1.2 *The following properties hold:*

i) *The multivalued map* $\partial^- E : H \to 2^H$ *is a* monotone operator, *i.e.*

$$\langle x - y, v - w \rangle \geq 0 \quad \textit{for every } x, y \in H,\ v \in \partial^- E(x) \textit{ and } w \in \partial^- E(y). \tag{5.3}$$

ii) *The set* $\{(x, v) \in H \times H : v \in \partial^- E(x)\}$ *is* strongly-weakly closed *in* $H \times H$, *i.e.*

$$\left.\begin{array}{r} x_n \to x \text{ strongly in } H, \\ v_n \rightharpoonup v \text{ weakly in } H, \\ v_n \in \partial^- E(x_n) \text{ for all } n \end{array}\right\} \implies v \in \partial^- E(x). \tag{5.4}$$

Proof The proof goes as follows:

i) From $v \in \partial^- E(x)$ and $w \in \partial^- E(y)$ we deduce that

$$\begin{aligned} E(x) + \langle v, y - x \rangle &\leq E(y), \\ E(y) + \langle w, x - y \rangle &\leq E(x), \end{aligned} \tag{5.5}$$

respectively. By summing the two in (5.5) we obtain $\langle v - w, y - x \rangle \leq 0$, proving (5.3).

ii) Fix two sequences $(x_n)_n, (v_n)_n \subseteq H$ such that $x_n \to x$, $v_n \to v$ and $v_n \in \partial^- E(x_n)$. Hence for any $y \in H$ it holds that

$$E(x) + \langle v, y - x \rangle \leq \varliminf_{n \to \infty} E(x_n) + \lim_{n \to \infty} \langle v_n, y - x_n \rangle \leq E(y),$$

thus showing that $v \in \partial^- E(x)$. This proves the statement.

□

Lemma 5.1.3 *Let H be a Hilbert space. Let* $[0, 1] \ni t \mapsto v_t \in H$ *be an AC curve. Then*

$$\exists \lim_{h \to 0} \frac{v_{t+h} - v_t}{h} =: v'_t \in H \quad \textit{for a.e. } t \in [0, 1]. \tag{5.6}$$

Moreover, the map $t \mapsto v'_t$ belongs to $L^1\big([0,1],H\big)$ and satisfies

$$v_t - v_s = \int_s^t v'_r \, \mathrm{d}r \quad \textit{for every } s,t \in [0,1] \textit{ with } s < t. \tag{5.7}$$

Proof Since v is essentially separably valued (as it is continuous), we assume with no loss of generality that H is separable. Fix an orthonormal basis $(\mathsf{e}_n)_n$ of H. Given any $n \in \mathbb{N}$, we have that $t \mapsto v_t \cdot \mathsf{e}_n \in \mathbb{R}$ is AC and accordingly a.e. differentiable. Hence there exists a Borel negligible set $N \subseteq [0,1]$ such that

$$\exists\, \ell_n(t) := \lim_{h\to 0} \frac{v_{t+h}\cdot \mathsf{e}_n - v_t \cdot \mathsf{e}_n}{h} \in \mathbb{R} \quad \text{for every } n \in \mathbb{N} \text{ and } t \in [0,1]\setminus N.$$

For any $k \in \mathbb{N}$, call $L_k(t) := \sum_{n=0}^k \ell_n(t)\,\mathsf{e}_n \in H$ if $t \in [0,1]\setminus N$ and $L_k(t) := 0 \in H$ if $t \in N$. Clearly the map $L_k : [0,1] \to H$ is strongly Borel. Moreover, for any $k \in \mathbb{N}$ it holds that

$$\begin{aligned}\sum_{n=0}^{\infty} |\ell_n(t)|^2 &= \lim_{k\to\infty} \sum_{n=0}^{k} |\ell_n(t)|^2 = \lim_{k\to\infty}\lim_{h\to 0} \sum_{n=0}^{k} \left| \frac{v_{t+h}-v_t}{h} \cdot \mathsf{e}_n \right|^2 \\ &\leq \lim_{h\to 0} \left\| \frac{v_{t+h}-v_t}{h} \right\|_H^2 = |\dot v_t|^2 < +\infty \qquad \text{for a.e. } t \in [0,1]\setminus N.\end{aligned} \tag{5.8}$$

In particular, for a.e. $t \in [0,1]\setminus N$ there exists $L(t) \in H$ such that $\lim_k \|L_k(t) - L(t)\|_H = 0$. We also deduce from (5.8) that $\|L(t)\|_H \leq |\dot v_t|$ for a.e. $t \in [0,1]$, whence $L : [0,1] \to H$ is Bochner integrable by Proposition 1.3.6. By applying the dominated convergence theorem, we see that $\int_s^t L(r)\,\mathrm{d}r = \lim_k \int_s^t L_k(r)\,\mathrm{d}r$ for every $t,s \in [0,1]$ with $s \leq t$, so that

$$\begin{aligned} v_t - v_s &= \lim_{k\to\infty} \sum_{n=0}^{k} \big[(v_t - v_s)\cdot \mathsf{e}_n\big]\mathsf{e}_n \\ &= \lim_{k\to\infty} \sum_{n=0}^{k} \left(\int_s^t \ell_n(r)\,\mathrm{d}r \right) \mathsf{e}_n \overset{(1.45)}{=} \lim_{k\to\infty} \int_s^t L_k(r)\,\mathrm{d}r \\ &= \int_s^t L(r)\,\mathrm{d}r. \end{aligned}$$

Hence v is a.e. differentiable, with derivative $v' := L$, proving the statement. □

Let us now define

$$D(E) := \{x \in H \ : \ E(x) < +\infty\},$$
$$D(\partial^- E) := \{x \in H \ : \ \partial^- E(x) \neq \emptyset\} \subseteq D(E).$$

The *slope* of E is the functional $|\partial^- E| : H \to [0, +\infty]$ given by

$$|\partial^- E|(x) := \begin{cases} \sup_{y \neq x} \big(E(y) - E(x)\big)^- / |x - y| & \text{if } x \in D(E), \\ +\infty & \text{otherwise.} \end{cases} \tag{5.9}$$

Observe that $|\partial^- E|(x) = 0$ if and only if x is a minimum point of E.

Remark 5.1.4 In general, the slope $|\partial^- E|$ is defined as

$$|\partial^- E|(x) := \begin{cases} \varlimsup_{y \to x} \big(E(y) - E(x)\big)^- / |x - y| & \text{if } x \in D(E), \\ +\infty & \text{otherwise.} \end{cases}$$

In this case, this definition is equivalent to (5.9) thanks to the convexity of E. ■

Remark 5.1.5 We claim that

$$|\partial^- E|(x) \leq |v| \quad \text{for every } v \in \partial^- E(x). \tag{5.10}$$

Indeed, we know that $E(x) + \langle v, y - x \rangle \leq E(y)$ for any $y \in H$, so that $E(x) - E(y) \leq |v|\,|x - y|$ and accordingly $\big(E(x) - E(y)\big)^+ \leq |v|\,|x - y|$ for any $y \in H$, which gives (5.10). ■

Exercise 5.1.6 Let H be a Hilbert space. Given any $x \in H$ and $\tau > 0$, let us define

$$F_{x,\tau}(\cdot) := E(\cdot) + \frac{|\cdot - x|^2}{2\tau}. \tag{5.11}$$

Then it holds that $\partial^- F_{x,\tau}(y) = \partial^- E(y) + \frac{y-x}{\tau}$ for every $y \in H$. ■

Proposition 5.1.7 *Let $x \in H$ and $\tau > 0$. Then there exists a unique minimiser $x_\tau \in H$ of the functional $F_{x,\tau}$ defined in* (5.11)*. Moreover, it holds that $\frac{x_\tau - x}{\tau} \in -\partial^- E(x_\tau)$.*

Proof Since E is convex lower semicontinuous and $|\cdot - x|^2/(2\tau)$ is strictly convex and continuous, we get that the functional $F_{x,\tau}$ is strictly convex and lower semicontinuous. This grants that the sublevels of $F_{x,\tau}$ are convex and strongly closed, so that they are also weakly closed by Hahn-Banach theorem, in other words $F_{x,\tau}$ is weakly lower semicontinuous. Moreover, the sublevels of $|\cdot - x|^2/(2\tau)$ are bounded, whence those of $F_{x,\tau}$ are bounded as well, thus in particular they are weakly compact. Then the Bolzano-Weierstrass theorem yields existence of a minimum point $x_\tau \in H$ of $F_{x,\tau}$, which is unique by strict convexity of $F_{x,\tau}$.

Finally, since x_τ is a minimiser for $F_{x,\tau}$, we know from Exercise 5.1.6 that $0 \in \partial^- F_{x,\tau}(x_\tau) = \partial^- E(x_\tau) + \frac{x_\tau - x}{\tau}$, or equivalently $\frac{x_\tau - x}{\tau} \in -\partial^- E(x_\tau)$, which gives the last statement. □

Corollary 5.1.8 *It holds that $D(\partial^- E)$ is dense in $D(E)$ and that*

$$|\partial^- E|(x_\tau) \leq \frac{|x_\tau - x|}{\tau} \leq |\partial^- E|(x) \quad \textit{for every } x \in H \textit{ and } \tau > 0. \tag{5.12}$$

Proof Given any $x \in D(E)$, we deduce from the very definition of x_τ that

$$\varlimsup_{\tau \searrow 0} |x_\tau - x|^2 \leq \varlimsup_{\tau \searrow 0} 2\,\tau \left(E(x_\tau) + \frac{|x_\tau - x|^2}{2\,\tau} \right) \leq \lim_{\tau \searrow 0} 2\,\tau\, E(x) = 0,$$

whence the first statement follows. Moreover, since $\frac{x - x_\tau}{\tau} \in \partial^- E(x_\tau)$ by Proposition 5.1.7, we infer from (5.10) that $|x_\tau - x|/\tau \geq |\partial^- E|(x_\tau)$. To conclude, define $z_\lambda := (1-\lambda)\,x + \lambda\, x_\tau$ for every $\lambda \in [0,1]$. The minimality of x_τ and the convexity of E give

$$E(x_\tau) + \frac{|x_\tau - x|^2}{2\,\tau} \leq E(z_\lambda) + \frac{|z_\lambda - x|^2}{2\,\tau} \leq (1-\lambda)\, E(x) + \lambda\, E(x_\tau) + \lambda^2 \frac{|x_\tau - x|^2}{2\,\tau}$$

for every $\lambda \in [0,1]$, which can be rewritten as

$$(1-\lambda)\big(E(x) - E(x_\tau)\big) \geq (1-\lambda^2)\, \frac{|x_\tau - x|^2}{2\,\tau} \qquad \text{for every } \lambda \in [0,1],$$

so that $\frac{E(x)-E(x_\tau)}{|x_\tau - x|} \geq (1+\lambda)\, \frac{|x_\tau - x|}{2\,\tau}$ for all $\lambda \in [0,1]$. By letting $\lambda \nearrow 1$ in such inequality, we conclude that $|\partial^- E|(x) \geq \frac{E(x)-E(x_\tau)}{|x_\tau - x|} \geq \frac{|x_\tau - x|}{\tau}$. Hence the statement is achieved. □

Remark 5.1.9 We claim that the functional $|\partial^- E| : H \to [0, +\infty]$ is lower semicontinuous.

In order to prove it, for any $y \in H$ we define $G_y : H \to [0, +\infty]$ as

$$G_y(x) := \begin{cases} \big(E(y) - E(x)\big)^- / |x - y| & \text{if } x \neq y, \\ 0 & \text{if } x = y, \end{cases}$$

with the convention that $\big(E(y) - E(x)\big)^- := +\infty$ when $E(x) = E(y) = +\infty$. It can be readily checked that $|\partial^- E|(x) = \sup_{y \in H} G_y(x)$ for every $x \in H$. Given that each G_y is a lower semicontinuous functional by construction, we conclude that $|\partial^- E|$ is lower semicontinuous as well. ■

Lemma 5.1.10 *It holds that*

$$|\partial^- E|(x) = \min_{v \in \partial^- E(x)} |v| \quad \textit{for every } x \in D(\partial^- E). \tag{5.13}$$

Proof The inequality $\leq$ is granted by (5.10). To prove $\geq$, notice that $|\partial^- E|(x) \geq |x - x_\tau|/\tau$ for all $\tau > 0$ by (5.12). Then there exists a sequence $(\tau_n)_n \searrow 0$ such that $\frac{x - x_{\tau_n}}{\tau_n} \rightharpoonup v$ weakly in H as $n \to \infty$, for some $v \in H$. Since we have that $\frac{x - x_{\tau_n}}{\tau_n} \in \partial^- E(x_{\tau_n})$ for all $n \in \mathbb{N}$, we conclude that $v \in \partial^- E(x)$ by item ii) of Proposition 5.1.2. Given that

$$|v| \leq \varliminf_n |x_{\tau_n} - x|/\tau_n \leq |\partial^- E|(x),$$

we proved (5.13). □

Remark 5.1.11 It is clear that the set $\partial^- E(x)$ is closed and convex for every $x \in H$.
In particular, if x belongs to $D(\partial^- E)$, then the set $\partial^- E(x)$ admits a unique element of minimal norm. ■

5.1.2 Existence and Uniqueness of the Gradient Flow

We are now ready to state and prove—by using the language and the results that have been introduced in the previous subsection—the main result of this chapter, which concerns existence and uniqueness of gradient flows:

Theorem 5.1.12 (Gradient Flow) *Let H be a Hilbert space. Let $E : H \to [0, +\infty]$ be a convex lower semicontinuous functional. Let $x \in \overline{D(E)}$ be fixed. Then there exists a unique continuous curve $[0, +\infty) \ni t \mapsto x_t \in H$ starting from x, called* gradient flow trajectory, *which is locally AC on $(0, +\infty)$ and satisfies $x_t' \in -\partial^- E(x_t)$ for a.e. $t \in [0, +\infty)$. Moreover, the following hold:*

1) (CONTRACTION PROPERTY) *Given two gradient flow trajectories (x_t) and (y_t), we have*

$$|x_t - y_t| \leq |x_0 - y_0| \quad \textit{for every } t \geq 0. \tag{5.14}$$

2) *The maps $t \mapsto x_t$ and $t \mapsto E(x_t)$ are locally Lipschitz on $(0, +\infty)$.*
3) *The functions $t \mapsto E(x_t)$ and $t \mapsto |\partial^- E|(x_t)$ are non increasing on $[0, +\infty)$.*
4) *For any $y \in H$, we have that $E(x_t) + \langle x_t', x_t - y\rangle \leq E(y)$ holds for a.e. $t \in (0, +\infty)$.*
5) *We have that $-\frac{\mathrm{d}}{\mathrm{d}t} E(x_t) = |\dot{x}_t|^2 = |\partial^- E|^2(x_t)$ for a.e. $t \in [0, +\infty)$.*
6) *The following inequalities are satisfied:*

   6a) $E(x_t) \leq E(y) + \frac{|x_0 - y|^2}{2t}$ *for every $y \in H$ and $t \geq 0$.*
   6b) $|\partial^- E|^2(x_t) \leq |\partial^- E|^2(y) + \frac{|x_0 - y|^2}{t^2}$ *for every $y \in H$ and $t \geq 0$.*

7) *For any $t > 0$, we have that the difference quotients $\frac{x_{t+h} - x_t}{h}$ converge to the element of minimal norm of $\partial^- E(x_t)$ as $h \searrow 0$. The same holds for $t = 0$ provided $\partial^- E(x_0) \neq \emptyset$.*

Proof We divide the proof into several steps:

STEP 1. We start by proving existence in the case $x \in D(E)$. Fix $\tau > 0$. We recursively define the sequence $(x^\tau_{(n)})_n \subseteq H$ as $x^\tau_{(0)} := x$ and

$$x^\tau_{(n+1)} := \operatorname*{argmin}_H \left(E(\cdot) + \frac{|\cdot - x^\tau_{(n)}|^2}{2\tau} \right) \qquad \text{for every } n \in \mathbb{N}.$$

Then define (x^τ_t) as the unique curve in H such that $x^\tau_{n\tau} = x^\tau_{(n)}$ for all $n \in \mathbb{N}$ and that is affine on each interval $[n\tau, (n+1)\tau]$. For any $n \in \mathbb{N}$, we clearly have that

$$(x^\tau_t)' = \frac{x^\tau_{(n+1)} - x^\tau_{(n)}}{\tau} \qquad \text{for every } t \in (n\tau, (n+1)\tau). \tag{5.15}$$

Since $E(x^\tau_{(n+1)}) + \left|x^\tau_{(n+1)} - x^\tau_{(n)}\right|^2/(2\,\tau) \le E(x^\tau_{(n)})$ for all $n \in \mathbb{N}$, we infer from (5.15) that

$$\frac{1}{2}\int_0^{+\infty} |\dot{x}^\tau_t|^2 \,\mathrm{d}t = \sum_{n=0}^{\infty} \frac{\left|x^\tau_{(n+1)} - x^\tau_{(n)}\right|^2}{2\,\tau} \le E(x) < +\infty. \tag{5.16}$$

Given $\tau, \eta > 0$ and $k, k' \in \mathbb{N}$ such that $t \in ((k-1)\tau, k\tau] \cap ((k'-1)\eta, k'\eta]$, it holds that

$$\begin{aligned}
\frac{\mathrm{d}}{\mathrm{d}t}\frac{|x^\tau_t - x^\eta_t|^2}{2} &= \underbrace{\big\langle (x^\tau_t)' - (x^\eta_t)', x^\tau_{k\tau} - x^\eta_{k'\eta} \big\rangle}_{\le 0 \text{ by } (5.3)} + \big\langle (x^\tau_t)' - (x^\eta_t)', (x^\tau_t - x^\tau_{k\tau}) - (x^\eta_t - x^\eta_{k'\eta}) \big\rangle \\
&\le \big(|(x^\tau_t)'| + |(x^\eta_t)'| \big)\big(\tau\,|(x^\tau_t)'| + \eta\,|(x^\eta_t)'| \big) \\
&\le |(x^\tau_t)'|^2 \left(\tau + \frac{\tau+\eta}{2} \right) + |(x^\eta_t)'|^2 \left(\eta + \frac{\tau+\eta}{2} \right).
\end{aligned}$$

By integrating over the interval $[0, T]$, we thus deduce from (5.16) that

$$\frac{|x^\tau_T - x^\eta_T|^2}{2} \le 2\,E(x)\,(\tau + \eta) \qquad \text{for every } \tau, \eta > 0. \tag{5.17}$$

This grants that $\sup_{t\ge 0} |x^\tau_t - x^\eta_t| \to 0$ as $\tau, \eta \searrow 0$, so there exists a continuous curve (x_t), with $x_0 = x$, which is the uniform limit of (x^τ_t) as $\tau \searrow 0$.
Notice that $\big\{(x^\tau_\cdot)' \in L^2([0,+\infty), H) \,\big|\, \tau > 0\big\}$ is norm bounded by (5.16), so that there exists $(\tau_n)_n \searrow 0$ such that $(x^{\tau_n}_\cdot)' \rightharpoonup v_\cdot$ weakly in $L^2([0,+\infty), H)$ as $n \to \infty$, for a suitable limit $v_\cdot \in L^2([0,+\infty), H)$. Given any $t > s > 0$, we know that

$$\int_s^t (x^{\tau_n}_r)' \,\mathrm{d}r = x^{\tau_n}_t - x^{\tau_n}_s \xrightarrow{\;n\;} x_t - x_s \qquad \text{in the strong topology of } H.$$

Moreover, for any $w \in H$ it holds that the map $r \mapsto \chi_{[s,t]}(r)\, w$ belongs to $L^2([0,+\infty), H)$, thus the fact that $(x_\cdot^{\tau_n})' \rightharpoonup v_\cdot$ ensures that

$$\left\langle w, \int_s^t (x_r^{\tau_n})'\,\mathrm{d}r \right\rangle \overset{(1.45)}{=} \int_s^t \big\langle w, (x_r^{\tau_n})' \big\rangle\,\mathrm{d}r \overset{n}{\longrightarrow} \int_s^t \langle w, v_r \rangle\,\mathrm{d}r \overset{(1.45)}{=} \left\langle w, \int_s^t v_r\,\mathrm{d}r \right\rangle.$$

Therefore we deduce that $x_t - x_s = \int_s^t v_r\,\mathrm{d}r$ is satisfied for every $t > s > 0$. This ensures that the curve (x_t) is locally AC on $(0,+\infty)$ and its derivative is given by (v_t). Now fix $y \in H$. We claim that

$$\int_{t_0}^{t_1} E(x_t) + \langle x_t', x_t - y \rangle\,\mathrm{d}t \leq (t_1 - t_0)\, E(y) \quad \text{for every } 0 \leq t_0 \leq t_1 < +\infty. \tag{5.18}$$

Recall that $-(x_{(n+1)}^\tau - x_{(n)}^\tau)/\tau \in \partial^- E(x_{(n+1)}^\tau)$ for all $n \in \mathbb{N}$. Moreover, it holds that

$$\int_0^\tau E(x_t^\tau)\,\mathrm{d}t \leq \int_0^\tau \left(1 - \frac{t}{\tau}\right) E(x_0) + \frac{t}{\tau}\, E(x_{(1)}^\tau)\,\mathrm{d}t = \frac{\tau}{2}\, E(x_0) + \frac{\tau}{2}\, E(x_{(1)}^\tau).$$

Therefore we deduce from Proposition 5.1.7 that

$$\begin{aligned}
\int_{t_0}^{t_1} E(x_t) + \langle x_t', x_t - y \rangle\,\mathrm{d}t &\leq \varliminf_{\tau \searrow 0} \int_{t_0}^{t_1} E(x_t^\tau) + \big\langle (x_t^\tau)', x_t^\tau - y \big\rangle\,\mathrm{d}t \\
&\leq \varliminf_{\tau \searrow 0} \int_{t_0}^{t_1} E(x_{[t/\tau+1]\tau}^\tau) + \big\langle (x_t^\tau)', x_{[t/\tau+1]\tau}^\tau - y \big\rangle\,\mathrm{d}t \\
&\leq \varliminf_{\tau \searrow 0} \int_{t_0}^{t_1} E(y)\,\mathrm{d}t = (t_1 - t_0)\, E(y),
\end{aligned}$$

which proves the validity of our claim (5.18). Finally, take $t > 0$ that is both a Lebesgue point for $E(x_\cdot)$ and a differentiability point for $x_\cdot$ (almost every $t > 0$ has this property). Then it follows from (5.18) that the formula in item 4) is verified at such t, proving that (x_t) is a gradient flow starting from x. Hence existence and item 4) are proven for $x \in D(E)$. Note that item 4) is trivially satisfied if $y \in H \setminus D(E)$.

STEP 2. Suppose that (x_t), (y_t) are gradient flows starting from points in $\overline{D(E)}$. Then the function $t \mapsto \frac{|x_t - y_t|^2}{2}$ is continuous on $[0,+\infty)$ and locally AC on $(0,+\infty)$. Item i) of Proposition 5.1.2 yields

$$\frac{\mathrm{d}}{\mathrm{d}t} \frac{|x_t - y_t|^2}{2} = \langle x_t' - y_t', x_t - y_t \rangle \leq 0 \quad \text{for a.e. } t > 0.$$

Hence $|x_t - y_t| \leq |x_0 - y_0|$ for every $t \geq 0$, proving 1) and uniqueness of the gradient flow.

STEP 3. We aim to prove 3). Fix $0 \leq t_0 \leq t_1 < +\infty$. Call (x_t) the gradient flow starting from some point $x \in \overline{D(E)}$, then (y_t) the gradient flow starting from x_{t_0}. By uniqueness, we have that $x_{t_1} = y_{t_1-t_0}$. Furthermore, one has $E(x_{t_1}) = E(y_{t_1-t_0}) \leq E(y_0) = E(x_{t_0})$ by construction. This shows that $t \mapsto E(x_t)$ is a non increasing function. A similar argument based on (5.12) and on Remark 5.1.9 grants that $t \mapsto |\partial^- E|(x_t)$ is non increasing as well. Then item 3) is proven.

STEP 4. We want to prove 6a). Fix $x \in \overline{D(E)}$ and call (x_t) the gradient flow with $x_0 = x$. Let $y \in H$ and $t \geq 0$. By integrating the inequality in 4) on $[0, t]$ and by recalling 3), we get

$$t\,E(x_t) \leq \int_0^t E(x_s)\,\mathrm{d}s \leq t\,E(y) - \frac{|x_t - y|^2}{2} + \frac{|x - y|^2}{2} \leq t\,E(y) + \frac{|x - y|^2}{2},$$

whence 6a) immediately follows.

STEP 5. We aim to prove existence of the gradient flow and item 4) for any $x \in \overline{D(E)}$. Choose a sequence $(x^n)_n \subseteq D(E)$ such that $x^n \to x$. Call (x^n_t) the gradient flow with initial datum x^n. We know from the contraction property 1) that

$$\sup_{t \geq 0} |x^n_t - x^m_t| \leq |x^n - x^m| \to 0 \quad \text{as } n, m \to \infty,$$

so there is a continuous curve (x_t) that is uniform limit of (x^n_t) and such curve starts from x. Given $y \in D(E)$ and $t_0 > 0$, we know from item 6a) that there exists a constant $C(t_0) > 0$ such that

$$E(x^n_{t_0}) \leq E(y) + \frac{|x_n - y|^2}{2\,t_0} \leq C(t_0) \quad \text{for every } n \in \mathbb{N},$$

whence from (5.16) it follows that $\frac{1}{2}\int_{t_0}^{+\infty} |\dot{x}^n_t|^2\,\mathrm{d}t \leq C(t_0)$ holds for every $n \in \mathbb{N}$. In other words, $(x^n_\cdot)$ are uniformly AC on $[t_0, +\infty)$. Hence $(x^n_\cdot)' \rightharpoonup x'_\cdot$ weakly in $L^2([t_0, +\infty), H)$, which is enough to conclude by passing to the limit in the inequality

$$\int_s^t E(x^n_r) + \big\langle (x^n_r)', x^n_r - y \big\rangle\,\mathrm{d}r \leq (t - s)\,E(y) \quad \text{for all } t_0 \leq s < t < +\infty \text{ and } y \in H$$

(that is granted by (5.18)) and arguing as in the last part of STEP 1.

STEP 6. Fix $\varepsilon > 0$. Since the curve (x_t) is locally AC on $(0, \varepsilon)$, there exists $t_0 \in (0, \varepsilon)$ such that x'_{t_0} exists. Moreover, for any $s \geq 0$ it holds that $t \mapsto x_{t+s}$ is the gradient flow starting from x_s. Therefore we have that

$$|\dot{x}_{t_0+s}| = \lim_{t \searrow t_0} \frac{|x_{t+s} - x_{t_0+s}|}{|t - t_0|} \overset{1)}{\leq} \lim_{t \searrow t_0} \frac{|x_t - x_{t_0}|}{|t - t_0|} = |\dot{x}_{t_0}| \quad \text{holds for a.e. } s \geq 0,$$

which grants that the metric speed $|\dot{x}|$ is bounded in $[\varepsilon, \infty)$. This means that (x_t) is Lipschitz on $[\varepsilon, +\infty)$. Now call L_ε its Lipschitz constant. Item 4) ensures that for any $y \in H$ one has

$$E(x_t) - L_\varepsilon\,|x_t - x| \le E(x_t) - |\dot{x}_t|\,|x_t - y| \le E(x_t) - \langle x_t', x_t - y\rangle \le E(y)$$

for a.e. $t \in (\varepsilon, +\infty)$, thus also for every $t > \varepsilon$ by lower semicontinuity of E. By choosing $y = x_s$, we see that the inequality $E(x_t) - E(x_s) \le L_\varepsilon |x_t - x_s|$ holds for all $s, t > \varepsilon$. This shows that $t \mapsto E(x_t)$ is locally Lipschitz, thus concluding the proof of 2).

STEP 7. We now prove item 5). Since $\frac{E(x_t)-E(y)}{|x_t-y|} \le |\dot{x}_t|$ holds for every $y \in H$ and a.e. t by property 4), we deduce that

$$|\partial^- E|(x_t) = \sup_{y \neq x_t} \frac{\big(E(x_t) - E(y)\big)^+}{|x_t - y|} \le |\dot{x}_t| \quad \text{for a.e. } t \ge 0. \tag{5.19}$$

Moreover, observe that for a.e. $t \ge 0$ it holds that

$$\begin{aligned} -\frac{\mathrm{d}}{\mathrm{d}t} E(x_t) &= \lim_{h\to 0} \frac{E(x_t) - E(x_{t+h})}{h} \le |\partial^- E|(x_t) \lim_{h\to 0} \frac{|x_{t+h} - x_t|}{|h|} = |\partial^- E|(x_t)\,|\dot{x}_t| \\ &\le \frac{1}{2}\,|\partial^- E|^2(x_t) + \frac{1}{2}\,|\dot{x}_t|^2. \end{aligned} \tag{5.20}$$

By integrating the inequality in item 4) over the interval $[t, t+h]$, we obtain that

$$\frac{|x_{t+h} - y|^2}{2} - \frac{|x_t - y|^2}{2} + \int_t^{t+h} E(x_s)\,\mathrm{d}s \le h\,E(y) \quad \text{for every } y \in H \text{ and } t, h \ge 0.$$

By using such inequality with $y = x_t$ and the dominated convergence theorem, we get

$$\begin{aligned} \frac{|\dot{x}_t|^2}{2} &= \lim_{h\searrow 0} \frac{|x_{t+h} - x_t|^2}{2\,h^2} \le \lim_{h\searrow 0} ⨍_t^{t+h} \frac{E(x_t) - E(x_s)}{h}\,\mathrm{d}s \\ &= \lim_{h\searrow 0} \int_0^1 \frac{E(x_t) - E(x_{t+hr})}{h\,r}\,r\,\mathrm{d}r = -\frac{\mathrm{d}}{\mathrm{d}t} E(x_t) \int_0^1 r\,\mathrm{d}r \\ &= -\frac{1}{2}\frac{\mathrm{d}}{\mathrm{d}t} E(x_t) \quad \text{for a.e. } t > 0. \end{aligned} \tag{5.21}$$

Finally, we obtain 5) by putting together (5.19), (5.20) and (5.21).

STEP 8. We want to prove 6b). Since the slope $|\partial^- E|$ is lower semicontinuous (cf. Remark 5.1.9), it suffices to prove it for $x_0 \in D(E)$. Notice that the Young's inequality yields

$$t\big(E(y) - E(x_t)\big) \le t\,|\partial^- E|(y)\,|y - x_t| \le \frac{t^2\,|\partial^- E|^2(y)}{2} + \frac{|x_t - y|^2}{2}. \tag{5.22}$$

By using (5.22) and items 3), 4), 5), we see that

$$\begin{aligned}\frac{t^2\,|\partial^- E|^2(x_t)}{2} &\le \int_0^t s\,|\partial^- E|^2(x_s)\,\mathrm{d}s = -\int_0^t s\,\frac{\mathrm{d}}{\mathrm{d}s}\,E(x_s)\,\mathrm{d}s = \int_0^t E(x_s)\,\mathrm{d}s - t\,E(x_t)\\ &\le t\,E(y) + \frac{|x_0 - y|^2}{2} - \frac{|x_t - y|^2}{2} - t\,E(x_t)\\ &\le \frac{t^2\,|\partial^- E|^2(y)}{2} + \frac{|x_0 - y|^2}{2},\end{aligned}$$

which proves 6b).

STEP 9. It only remains to prove 7). It is enough to prove it for $t = 0$ and $|\partial^- E|(x_0) < +\infty$. Observe that $\left|\frac{x_h - x}{h}\right| \le ⨍_0^h |\dot{x}_t|\,\mathrm{d}t \le |\partial^- E|(x_0)$ for all $h > 0$ by 3) and 5). Hence there exists a sequence $(h_n)_n \searrow 0$ such that $\frac{x_{h_n} - x_0}{h_n} \rightharpoonup v \in H$. Clearly $|v| \le |\partial^- E|(x_0)$. By recalling Lemma 5.1.10, we thus see that it just remains to show that $v \in \partial^- E(x_0)$. Notice that

$$⨍_0^{h_n} \langle x_t', x_t - y\rangle\,\mathrm{d}t = \Big\langle ⨍_0^{h_n} x_t'\,\mathrm{d}t, x_0 - y\Big\rangle + ⨍_0^{h_n} \langle x_t', x_t - x_0\rangle\,\mathrm{d}t \overset{n\to\infty}{\longrightarrow} \langle v, x_0 - y\rangle.$$

Therefore we finally conclude that

$$E(x_0) + \langle v, x_0 - y\rangle \le \varliminf_{n\to\infty} ⨍_0^{h_n} E(x_t) + \langle x_t', x_t - y\rangle\,\mathrm{d}t \le E(y),$$

which proves that $v \in \partial^- E(x_0)$, as required.

□

5.2 Heat Flow on Infinitesimally Hilbertian Spaces

5.2.1 Laplace Operator

Given an infinitesimally Hilbertian space $(\mathrm{X}, \mathsf{d}, \mathfrak{m})$ and any two vector fields $v, w \in L^2(T\mathrm{X})$, we shall often use the shorthand notation $v \cdot w$ in place of $\langle v, w\rangle$.

Definition 5.2.1 (Laplacian) Let $(\mathrm{X}, \mathsf{d}, \mathfrak{m})$ be an infinitesimally Hilbertian metric measure space. Then a function $f \in W^{1,2}(\mathrm{X})$ is in $D(\Delta)$ provided there exists

$g \in L^2(\mathfrak{m})$ such that

$$\int g\,h\,\mathrm{d}\mathfrak{m} = -\int \nabla f \cdot \nabla h\,\mathrm{d}\mathfrak{m} \quad \text{for every } h \in W^{1,2}(\mathrm{X}). \tag{5.23}$$

In this case the function g, which is uniquely determined by density of $W^{1,2}(\mathrm{X})$ in $L^2(\mathfrak{m})$, will be denoted by Δf.

Remark 5.2.2 One has $f \in D(\Delta)$ if and only if $\nabla f \in D(\mathrm{div})$. In this case, $\Delta f = \mathrm{div}(\nabla f)$.

In order to prove it, just observe that

$$\int \mathrm{d}h(\nabla f)\,\mathrm{d}\mathfrak{m} = \int \nabla f \cdot \nabla h\,\mathrm{d}\mathfrak{m} \quad \text{holds for every } h \in W^{1,2}(\mathrm{X}).$$

In particular, $D(\Delta)$ is a vector space and the map $\Delta : D(\Delta) \to L^2(\mathfrak{m})$ is linear. ■

Proposition 5.2.3 *Let* $(\mathrm{X}, \mathsf{d}, \mathfrak{m})$ *be infinitesimally Hilbertian. Then the following hold:*

i) Δ *is a closed operator from* $L^2(\mathfrak{m})$ *to itself.*
ii) *If* $f \in \mathrm{LIP}(\mathrm{X}) \cap D(\Delta)$ *and* $\varphi \in C^2(\mathbb{R})$ *satisfies* $\varphi'' \in L^\infty(\mathbb{R})$, *then* $\varphi \circ f \in D(\Delta)$ *and*

$$\Delta(\varphi \circ f) = \varphi' \circ f\,\Delta f + \varphi'' \circ f\,|\nabla f|^2. \tag{5.24}$$

iii) *If* $f, g \in \mathrm{LIP}_b(\mathrm{X}) \cap D(\Delta)$, *then* $fg \in D(\Delta)$ *and*

$$\Delta(fg) = f\,\Delta g + g\,\Delta f + 2\,\nabla f \cdot \nabla g. \tag{5.25}$$

Proof The proof goes as follows:

i) We aim to show that if $f_n \to f$ and $\Delta f_n \to g$ in $L^2(\mathfrak{m})$, then $f \in D(\Delta)$ and $\Delta f = g$. There exists a constant $C > 0$ such that $\|f_n\|_{L^2(\mathfrak{m})}, \|\Delta f_n\|_{L^2(\mathfrak{m})} \leq C$ for any $n \in \mathbb{N}$, so that

$$\int |\nabla f_n|^2\,\mathrm{d}\mathfrak{m} = -\int f_n\,\Delta f_n\,\mathrm{d}\mathfrak{m} \leq C \quad \text{for every } n \in \mathbb{N}.$$

This grants that $(f_n)_n$ is bounded in the reflexive space $W^{1,2}(\mathrm{X})$, whence there exists a subsequence $(n_i)_i$ such that $f_{n_i} \rightharpoonup \widetilde{f}$ weakly in $W^{1,2}(\mathrm{X})$, for some $\widetilde{f} \in W^{1,2}(\mathrm{X})$. We already know that $f_{n_i} \to f$ in $L^2(\mathfrak{m})$, then $\widetilde{f} = f$ and accordingly the original sequence $(f_n)_n$ is weakly converging in $W^{1,2}(\mathrm{X})$ to f. Since the differential operator $\mathrm{d} : W^{1,2}(\mathrm{X}) \to L^2(T^*\mathrm{X})$ is linear continuous, we infer that $\mathrm{d}f_n \rightharpoonup \mathrm{d}f$ weakly in $L^2(T^*\mathrm{X})$. By the Riesz isomorphism, this is equivalent to saying that $\nabla f_n \rightharpoonup \nabla f$ weakly in $L^2(TX)$. Therefore

$$-\int h\,g\,\mathrm{dm} = -\lim_{n\to\infty}\int h\,\Delta f_n\,\mathrm{dm} = \lim_{n\to\infty}\int \nabla f_n\cdot\nabla h\,\mathrm{dm} = \int \nabla f\cdot\nabla h\,\mathrm{dm}$$

is satisfied for every $h \in W^{1,2}(X)$, thus proving that $f \in D(\Delta)$ and $\Delta f = g$.

ii) Note that $\varphi\circ f \in S^2(X)$ and $\nabla(\varphi\circ f) = \varphi'\circ f\,\nabla f$. Since $\nabla f \in D(\mathrm{div})$ by Remark 5.2.2 and $\varphi'\circ f \in \mathrm{LIP}_b(X)$, we deduce from Proposition 4.2.7 that $\nabla(\varphi\circ f) \in D(\mathrm{div})$ and

$$\Delta(\varphi\circ f) = \mathrm{div}\big(\varphi'\circ f\,\nabla f\big) = \mathrm{d}(\varphi'\circ f)(\nabla f) + \varphi'\circ f\,\mathrm{div}(\nabla f) = \varphi''\circ f\,|\nabla f|^2 + \varphi'\circ f\,\Delta f,$$

which proves (5.24).

iii) Note that $fg \in S^2(X)$ and $\nabla(fg) = f\,\nabla g + g\,\nabla f$. By applying again Proposition 4.2.7, we deduce that $\nabla(fg) \in D(\mathrm{div})$ and

$$\begin{aligned}\Delta(fg) &= \mathrm{div}\big(f\,\nabla g + g\,\nabla f\big) = \mathrm{d}f(\nabla g) + f\,\mathrm{div}(\nabla g) + \mathrm{d}g(\nabla f) + g\,\mathrm{div}(\nabla f)\\ &= f\,\Delta g + g\,\Delta f + 2\,\nabla f\cdot\nabla g,\end{aligned}$$

which proves (5.25).

□

Given an infinitesimally Hilbertian space $(X, \mathsf{d}, \mathfrak{m})$, we denote by $E : L^2(\mathfrak{m}) \to [0, +\infty]$ the associated *Cheeger's energy* (recall Definition 2.2.3), which is the convex lower semicontinuous functional

$$E(f) := \begin{cases} \frac{1}{2}\int |\nabla f|^2\,\mathrm{dm} & \text{if } f \in W^{1,2}(X),\\ +\infty & \text{otherwise.}\end{cases} \tag{5.26}$$

We can now provide an alternative characterisation of the Laplace operator.

Proposition 5.2.4 *Let* $(X, \mathsf{d}, \mathfrak{m})$ *be infinitesimally Hilbertian. Then a function* $f \in W^{1,2}(X)$ *belongs to* $D(\Delta)$ *if and only if* $\partial^- E(f) \neq \emptyset$. *In this case, it holds that* $\partial^- E(f) = \{-\Delta f\}$.

Proof First of all, observe that for any $f, g \in W^{1,2}(X)$ we have that

$$\mathbb{R} \ni \varepsilon \mapsto E(f+\varepsilon g) \text{ is convex and } \lim_{\varepsilon\to 0}\frac{E(f+\varepsilon g) - E(f)}{\varepsilon} = \int \nabla f\cdot\nabla g\,\mathrm{dm}, \tag{5.27}$$

as one can readily deduce from the fact that $E(f+\varepsilon g) = \frac{1}{2}\int |\nabla f|^2 + 2\varepsilon\,\nabla f\cdot\nabla g + \varepsilon^2\,|\nabla g|^2\,\mathrm{dm}$.

Let $f \in D(\Delta)$. We want to show that $E(f) - \int g\,\Delta f\,\mathrm{dm} \leq E(f+g)$ for every $g \in W^{1,2}(\mathrm{X})$. In order to prove it, just notice that (5.27) yields

$$E(f+g) - E(f) \geq \lim_{\varepsilon \searrow 0} \frac{E(f+\varepsilon g) - E(f)}{\varepsilon} = \int \nabla f \cdot \nabla g\,\mathrm{dm} = -\int g\,\Delta f\,\mathrm{dm},$$

which grants that $-\Delta f \in \partial^- E(f)$.

Conversely, let $v \in \partial^- E(f)$. Then $\varepsilon \int v\,g\,\mathrm{dm} \leq E(f+\varepsilon g) - E(f)$ holds for every $\varepsilon \in \mathbb{R}$ and $g \in W^{1,2}(\mathrm{X})$. Therefore we have that

$$\int \nabla f \cdot \nabla g\,\mathrm{dm} = \lim_{\varepsilon \searrow 0} \frac{E(f-\varepsilon g) - E(f)}{-\varepsilon} \leq \int v\,g\,\mathrm{dm} \leq \lim_{\varepsilon \searrow 0} \frac{E(f+\varepsilon g) - E(f)}{\varepsilon}$$
$$= \int \nabla f \cdot \nabla g\,\mathrm{dm}$$

for every $g \in W^{1,2}(\mathrm{X})$. This says that $f \in D(\Delta)$ and $\Delta f = -v$. □

5.2.2 Heat Flow and Its Properties

Definition 5.2.5 (Heat Flow) Let $(\mathrm{X}, \mathsf{d}, \mathfrak{m})$ be an infinitesimally Hilbertian metric measure space. Then for any $f \in L^2(\mathfrak{m})$ and $t \geq 0$, we denote by $\mathsf{h}_t f$ the gradient flow of the Cheeger energy E (defined in (5.26)) on $L^2(\mathfrak{m})$, starting from f (at time t). We shall call it *heat flow*. This defines a family $(\mathsf{h}_t)_{t\geq 0}$ of operators $\mathsf{h}_t : L^2(\mathfrak{m}) \to L^2(\mathfrak{m})$.

Proposition 5.2.6 *Let* $(\mathrm{X}, \mathsf{d}, \mathfrak{m})$ *be infinitesimally Hilbertian. Then the following hold:*

i) *The operator* $\mathsf{h}_t : L^2(\mathfrak{m}) \to L^2(\mathfrak{m})$ *is linear for every* $t \geq 0$.
ii) *For every* $f \in L^2(\mathfrak{m})$ *and* $t > 0$, *it holds that* $\mathsf{h}_t f \in D(\Delta)$ *and*

$$\frac{\mathsf{h}_{t+\varepsilon} f - \mathsf{h}_t f}{\varepsilon} \longrightarrow \Delta \mathsf{h}_t f \quad \text{in } L^2(\mathfrak{m}) \quad \text{as } \varepsilon \to 0. \tag{5.28}$$

The same holds also at $t = 0$ *provided* $f \in D(\Delta)$.

Proof The proof goes as follows:

i) It directly follows from Theorem 5.1.12, Proposition 5.2.4 and the linearity of Δ.
ii) Proposition 5.2.4 and Theorem 5.1.12 grant that $\mathsf{h}_t f \in D(\partial^- E) = D(\Delta)$ for every $t > 0$, thus it is sufficient to prove the claim for the case $t = 0$ and $f \in D(\Delta)$. In this case, we have that $\partial^- E(f) = \{-\Delta f\}$ and thus the conclusion follows from 7) of Theorem 5.1.12.

□

Proposition 5.2.7 (Δ and h_t Commute) *Let $f \in D(\Delta)$. Then $\mathsf{h}_t \Delta f = \Delta \mathsf{h}_t f$ for all $t \geq 0$.*

Proof Notice that

$$\Delta \mathsf{h}_t f = \lim_{\varepsilon \searrow 0} \frac{\mathsf{h}_t(\mathsf{h}_\varepsilon f) - \mathsf{h}_t f}{\varepsilon} = \mathsf{h}_t \left(\lim_{\varepsilon \searrow 0} \frac{\mathsf{h}_\varepsilon f - f}{\varepsilon} \right) = \mathsf{h}_t \Delta f,$$

which proves the statement. □

Proposition 5.2.8 (Δ is Symmetric) *Let $f, g \in D(\Delta)$. Then*

$$\int g \, \Delta f \, \mathrm{dm} = \int f \, \Delta g \, \mathrm{dm}. \tag{5.29}$$

Proof Just notice that $\int g \, \Delta f \, \mathrm{dm} = -\int \nabla f \cdot \nabla g \, \mathrm{dm} = \int f \, \Delta g \, \mathrm{dm}$. □

Corollary 5.2.9 (h_t is Self-Adjoint) *Let $f, g \in L^2(\mathrm{m})$ and $t \geq 0$. Then*

$$\int g \, \mathsf{h}_t f \, \mathrm{dm} = \int f \, \mathsf{h}_t g \, \mathrm{dm}. \tag{5.30}$$

Proof Define $F(s) := \int \mathsf{h}_s f \, \mathsf{h}_{t-s} g \, \mathrm{dm}$ for every $s \in [0, t]$. Then the function F is AC and

$$F'(s) = \int \Delta \mathsf{h}_s f \, \mathsf{h}_{t-s} g - \mathsf{h}_s f \, \Delta \mathsf{h}_{t-s} g \, \mathrm{dm} \overset{(5.29)}{=} 0 \quad \text{for a.e. } s \in [0, t],$$

whence accordingly $\int g \, \mathsf{h}_t f \, \mathrm{dm} = F(t) = F(0) = \int f \, \mathsf{h}_t g \, \mathrm{dm}$. □

Proposition 5.2.10 *Let $f \in L^2(\mathrm{m})$. Then we have $f \in D(\Delta)$ if and only if $\frac{\mathsf{h}_t f - f}{t}$ admits a strong limit $g \in L^2(\mathrm{m})$ as $t \searrow 0$. In this case, it holds that $g = \Delta f$.*

Proof We separately prove the two implications:
NECESSITY. Already established in point ii) of Proposition 5.2.6.
SUFFICIENCY. Suppose that $\frac{\mathsf{h}_t f - f}{t} \to g$ in $L^2(\mathrm{m})$ as $t \searrow 0$. We first claim that $f \in W^{1,2}(\mathrm{X})$. To prove it, notice that for every $\varepsilon > 0$ we have—because of our assumption and the self-adjointness of h_ε—that

$$\int \mathsf{h}_\varepsilon f \, g \, \mathrm{dm} = \lim_{t \downarrow 0} \int \mathsf{h}_\varepsilon f \frac{\mathsf{h}_t f - f}{t} \, \mathrm{dm} = \lim_{t \downarrow 0} \int f \frac{\mathsf{h}_t \mathsf{h}_\varepsilon f - \mathsf{h}_\varepsilon f}{t} \, \mathrm{dm}.$$

Hence the fact that $\mathsf{h}_\varepsilon f \in D(\Delta)$, the 'necessity' proved before and Proposition 5.2.7 give

$$\int \mathsf{h}_\varepsilon f \, g \, \mathrm{dm} = \int f \, \Delta \mathsf{h}_\varepsilon f \, \mathrm{dm} = \int \mathsf{h}_{\varepsilon/2} f \, \Delta \mathsf{h}_{\varepsilon/2} f \, \mathrm{dm} = -\int |\nabla \mathsf{h}_{\varepsilon/2} f|^2 \, \mathrm{dm}.$$

Since $f \in L^2(\mathfrak{m})$, the (absolute value of the) leftmost side of this last identity remains bounded as $\varepsilon \searrow 0$, hence the same holds for the rightmost one. Hence the lower semicontinuity of the Cheeger energy E gives

$$E(f) \leq \varliminf_{\varepsilon \downarrow 0} E(\mathsf{h}_\varepsilon f) = \varliminf_{\varepsilon \downarrow 0} \frac{1}{2} \int |\nabla \mathsf{h}_\varepsilon f|^2 \, d\mathfrak{m} < \infty,$$

thus giving our claim $f \in W^{1,2}(\mathrm{X})$. Now observe that the inequality $E(\mathsf{h}_s f) \leq E(f)$, valid for all $s \geq 0$, ensures that $(\mathsf{h}_\varepsilon f)_\varepsilon$ is bounded in $W^{1,2}(\mathrm{X})$ and thus weakly relatively compact. Since $\mathsf{h}_\varepsilon f \to f$ in $L^2(\mathfrak{m})$ as $\varepsilon \searrow 0$, we deduce that $\mathsf{h}_\varepsilon f \rightharpoonup f$ weakly in $W^{1,2}(\mathrm{X})$. Given any Sobolev function $\ell \in W^{1,2}(\mathrm{X})$, we thus have that

$$\begin{aligned}
\int g \, \ell \, d\mathfrak{m} &= \lim_{t \searrow 0} \int \frac{\mathsf{h}_t f - f}{t} \, \ell \, d\mathfrak{m} = \lim_{t \searrow 0} \fint_0^t \int \Delta \mathsf{h}_s f \, \ell \, d\mathfrak{m} \, ds \\
&= -\lim_{t \searrow 0} \fint_0^t \int \nabla \mathsf{h}_s f \cdot \nabla \ell \, d\mathfrak{m} \, ds \\
&= -\int \nabla f \cdot \nabla \ell \, d\mathfrak{m},
\end{aligned}$$

which shows that $f \in D(\Delta)$ and $\Delta f = g$. □

Remark 5.2.11 Given any $f \in L^2(\mathfrak{m})$ and $t > 0$, it holds that

$$E(\mathsf{h}_t f) \leq \frac{\|f\|^2_{L^2(\mathfrak{m})}}{2t} \quad \text{and} \quad \|\Delta \mathsf{h}_t f\|^2_{L^2(\mathfrak{m})} \leq \frac{\|f\|^2_{L^2(\mathfrak{m})}}{t^2}. \tag{5.31}$$

This claim directly follows from item 6) of Theorem 5.1.12. ■

Proposition 5.2.12 *Let $f \in L^2(\mathfrak{m})$ be fixed. Then the following hold:*

i) *The map $(0, +\infty) \ni t \mapsto \mathsf{h}_t f$ belongs to $C^\infty\big((0, +\infty), W^{1,2}(\mathrm{X})\big)$.*
ii) *It holds that $\mathsf{h}_t f \in D(\Delta^{(n)})$ for every $n \in \mathbb{N}$ and $t > 0$.*

Proof The proof goes as follows:

i) Fix $\varepsilon > 0$. First of all, we prove that $t \mapsto \mathsf{h}_t f$ belongs to $C^1\big((\varepsilon, +\infty), W^{1,2}(\mathrm{X})\big)$. Recall that we have $\frac{\mathrm{d}}{\mathrm{d}t} \mathsf{h}_t f = \Delta \mathsf{h}_t f$ for a.e. $t > \varepsilon$ and that $(\varepsilon, +\infty) \ni t \mapsto \Delta \mathsf{h}_t f = \mathsf{h}_{t-\varepsilon} \Delta \mathsf{h}_\varepsilon f \in L^2(\mathfrak{m})$ is continuous. Call $g := \Delta \mathsf{h}_\varepsilon f$. Since even the map

$$(\varepsilon, +\infty) \ni t \longmapsto \int |\nabla \mathsf{h}_{t-\varepsilon} g|^2 \, d\mathfrak{m} = -\int \mathsf{h}_{t-\varepsilon} g \, \Delta \mathsf{h}_{t-\varepsilon} g \, d\mathfrak{m}$$

is continuous, we conclude that $(\varepsilon, +\infty) \ni t \mapsto \frac{\mathrm{d}}{\mathrm{d}t} \mathsf{h}_t f = \mathsf{h}_{t-\varepsilon} g \in W^{1,2}(\mathrm{X})$ is continuous as well. This grants that $(t \mapsto \mathsf{h}_t f) \in C^1\big((\varepsilon, +\infty), W^{1,2}(\mathrm{X})\big)$.

We now argue by induction: assume that for some $n \in \mathbb{N}$ it holds that the map $t \mapsto \mathsf{h}_t g$ belongs to $C^n\big((\varepsilon, +\infty), W^{1,2}(\mathrm{X})\big)$ for every $g \in L^2(\mathfrak{m})$. This means that the map $t \mapsto \frac{\mathrm{d}}{\mathrm{d}t}\mathsf{h}_t f = \Delta \mathsf{h}_t f = \mathsf{h}_{t-\varepsilon}\Delta \mathsf{h}_\varepsilon f$ belongs to the space $C^n\big((\varepsilon, +\infty), W^{1,2}(\mathrm{X})\big)$, thus accordingly $(t \mapsto \mathsf{h}_t f) \in C^{n+1}\big((\varepsilon, +\infty), W^{1,2}(\mathrm{X})\big)$.

ii) By Proposition 5.2.7 it suffices to show that $\Delta \mathsf{h}_t f \in D(\Delta)$ for all $f \in L^2(\mathfrak{m})$ and $t > 0$. This immediately follows from the fact that $\Delta \mathsf{h}_t f = \mathsf{h}_{t/2}\Delta \mathsf{h}_{t/2} f \in D(\Delta)$.

□

Lemma 5.2.13 *Let $u : \mathbb{R} \to [0, +\infty]$ be convex lower semicontinuous and $u(0) = 0$. Define*

$$\mathcal{C} := \Big\{ v \in C^\infty(\mathbb{R}) \;\Big|\; v \geq 0 \text{ is convex, } v(0) = v'(0) = 0,\ v',\, v'' \text{ are bounded} \Big\}.$$

Then there exists a sequence $(u_n)_n \subseteq \mathcal{C}$ such that $u_n(t) \nearrow u(t)$ for all $t \in \mathbb{R}$.

Proof Let us define $\tilde{u}(t) := \sup\big\{ v(t) \,\big|\, v \in \mathcal{C},\ v \leq u \big\} \leq u(t)$ for all $t \in \mathbb{R}$. It can be readily checked that actually $\tilde{u} = u$. Now call $I := \{u < +\infty\}$ and fix any compact interval $K \subseteq I$ such that $\mathrm{dist}(K, \mathbb{R} \setminus I) > 0$. Then there exists a constant $C(K, u) > 0$ such that each $v \in \mathcal{C}$ with $v \leq u$ is $C(K, u)$-Lipschitz in K. Moreover, for a suitable sequence $(v_n)_n \subseteq \mathcal{C}$ we have that $\operatorname{ess\,sup}\big\{ v \in \mathcal{C} : v \leq u \big\} = \sup_n v_n$ holds a.e. in K. These two facts grant that actually the equality $\tilde{u} = \sup_n v_n$ holds everywhere in K. Since $\mathrm{int}(I)$ can be written as countable union of intervals K as above, we deduce that there exists $(w_n)_n \subseteq \mathcal{C}$ such that $\tilde{u} = \sup_n w_n$. Finally, we would like to define $u_n := \max_{i \leq n} w_i$ for all $n \in \mathbb{N}$, but such functions have all the required properties apart from smoothness. Therefore the desired functions u_n can be easily built by recalling the facts that $\max\{w_1, w_2\} = \frac{1}{2}\big(|w_1 - w_2| + w_1 + w_2\big)$ and that for all $t \in \mathbb{R}$ one has $|w_1 - w_2|(t) = \sup_{\varepsilon > 0} \sqrt{|w_1 - w_2|^2(t) + \varepsilon^2} - \varepsilon$. □

Proposition 5.2.14 *Let $f \in L^2(\mathfrak{m})$ be fixed. Then the following properties hold:*

i) WEAK MAXIMUM PRINCIPLE. *Suppose that $f \leq c$ holds $\mathfrak{m}$-a.e. for some constant $c \in \mathbb{R}$. Then $\mathsf{h}_t f \leq c$ holds $\mathfrak{m}$-a.e. for every $t > 0$.*
ii) *Let $u : \mathbb{R} \to [0, +\infty]$ be any convex lower semicontinuous function satisfying $u(0) = 0$. Then the function $[0, +\infty) \ni t \mapsto \int u(\mathsf{h}_t f)\,\mathrm{d}\mathfrak{m}$ is non-increasing.*
iii) *Let $p \in [1, \infty]$ be given. Then $\|\mathsf{h}_t f\|_{L^p(\mathfrak{m})} \leq \|f\|_{L^p(\mathfrak{m})}$ holds for every $t > 0$.*

Proof The proof goes as follows:

i) By recalling the 'minimising movements' technique that we used in STEP 1 of Theorem 5.1.12 to prove existence of the gradient flow, one can easily realise that it is enough to show that for any $\tau > 0$ the minimum f_τ of $g \mapsto E(g) + \|f - g\|^2_{L^2(\mathfrak{m})}/(2\tau)$ is $\mathfrak{m}$-a.e. smaller than or equal to c. We argue by contradiction: if not, then the function $\bar{f} := f_\tau \wedge c$ would satisfy the inequalities $E(\bar{f}) \leq E(f_\tau)$ and $\|f - \bar{f}\|_{L^2(\mathfrak{m})} < \|f - f_\tau\|_{L^2(\mathfrak{m})}$, thus contradicting the minimality of f_τ. Hence the weak maximum principle i) is proved.

ii) First of all, we prove it for $u \in C^\infty(\mathbb{R})$ such that $u(0) = u'(0) = 0$ and u', u'' are bounded. Say $|u'(t)|, |u''(t)| \leq C$ for all $t \in \mathbb{R}$. For any $t \geq s$, we thus have that

$$\begin{aligned} |u(t) - u(s)| &= \left| \int_s^t u'(r)\,\mathrm{d}r \right| = \left| (t-s)\,u'(s) + \int_s^t \big(u'(r) - u'(s)\big)\,\mathrm{d}r \right| \\ &\leq C\,|s|\,(t-s) + \int_s^t \int_s^r u''(r')\,\mathrm{d}r'\,\mathrm{d}r \leq C\big[(t-s)^2 + |s|\,(t-s)\big]. \end{aligned} \tag{5.32}$$

Given that $(0,+\infty) \ni t \mapsto \mathsf{h}_t f \in L^2(\mathfrak{m})$ is locally Lipschitz, we deduce from (5.32) that the function $t \mapsto \int u(\mathsf{h}_t f)\,\mathrm{d}\mathfrak{m}$, which is continuous on $[0,+\infty)$, is locally Lipschitz on $(0,+\infty)$. By passing to the limit as $\varepsilon \searrow 0$ in the equalities

$$\begin{aligned} \int \frac{u(\mathsf{h}_{t+\varepsilon} f) - u(\mathsf{h}_t f)}{\varepsilon}\,\mathrm{d}\mathfrak{m} &= \int \fint_t^{t+\varepsilon} u'(\mathsf{h}_s f)\,\Delta \mathsf{h}_s f\,\mathrm{d}s\,\mathrm{d}\mathfrak{m} \\ &= \int_0^1 \int u'(\mathsf{h}_{t+\varepsilon r} f)\,\Delta \mathsf{h}_{t+\varepsilon r} f\,\mathrm{d}\mathfrak{m}\,\mathrm{d}r, \end{aligned}$$

we see that $\frac{\mathrm{d}}{\mathrm{d}t} \int u(\mathsf{h}_t f)\,\mathrm{d}\mathfrak{m} = \int u'(\mathsf{h}_t f)\,\Delta \mathsf{h}_t f\,\mathrm{d}\mathfrak{m}$ for a.e. $t > 0$. Hence by using the chain rule for the differential and the fact that $u'' \geq 0$ we finally conclude that

$$\begin{aligned} \frac{\mathrm{d}}{\mathrm{d}t} \int u(\mathsf{h}_t f)\,\mathrm{d}\mathfrak{m} &= \int u'(\mathsf{h}_t f)\,\Delta \mathsf{h}_t f\,\mathrm{d}\mathfrak{m} = -\int \nabla u'(\mathsf{h}_t f) \cdot \nabla \mathsf{h}_t f\,\mathrm{d}\mathfrak{m} \\ &= -\int u''(\mathsf{h}_t f)\,|\nabla \mathsf{h}_t f|^2\,\mathrm{d}\mathfrak{m} \leq 0 \quad \text{for a.e. } t > 0, \end{aligned}$$

which ensures that the function $[0,+\infty) \ni t \mapsto \int u(\mathsf{h}_t f)\,\mathrm{d}\mathfrak{m}$ is non-increasing.

Now consider the case of a general function u. Consider an approximating sequence $(u_n)_n$ as in Lemma 5.2.13. By monotone convergence theorem, we thus see that

$$\int u(\mathsf{h}_t f)\,\mathrm{d}\mathfrak{m} = \sup_{n \in \mathbb{N}} \int u_n(\mathsf{h}_t f)\,\mathrm{d}\mathfrak{m} \quad \text{for every } t \geq 0.$$

Hence $t \mapsto \int u(\mathsf{h}_t f)\,\mathrm{d}\mathfrak{m}$ is non-increasing as pointwise supremum of non-increasing functions.

iii) To prove the statement for $p \in [1,\infty)$, just apply ii) with $u := |\cdot|^p$. For the case $p = \infty$, notice that $-\|f\|_{L^\infty(\mathfrak{m})} \leq f \leq \|f\|_{L^\infty(\mathfrak{m})}$ holds $\mathfrak{m}$-a.e., whence

$-\|f\|_{L^\infty(\mathfrak{m})} \le \mathsf{h}_t f \le \|f\|_{L^\infty(\mathfrak{m})}$ holds $\mathfrak{m}$-a.e. for every $t > 0$ by i), so that $\|\mathsf{h}_t f\|_{L^\infty(\mathfrak{m})} \le \|f\|_{L^\infty(\mathfrak{m})}$ for all $t > 0$.

□

Proposition 5.2.15 (Heat Flow in $L^p(\mathfrak{m})$) *Let $p \in [1, \infty)$ be given. Then the heat flow uniquely extends to a family of linear contractions in $L^p(\mathfrak{m})$.*

Proof It follows from Proposition 5.2.14 and the density of $L^2(\mathfrak{m}) \cap L^p(\mathfrak{m})$ in $L^p(\mathfrak{m})$. □

Definition 5.2.16 (Heat Flow in $L^\infty(\mathfrak{m})$) Let $f \in L^\infty(\mathfrak{m})$ be given. Then for every $t > 0$ we define $\mathsf{h}_t f \in L^\infty(\mathfrak{m})$ as the function corresponding to $\big[L^1(\mathfrak{m}) \ni g \mapsto \int f\, \mathsf{h}_t g \, d\mathfrak{m} \in \mathbb{R}\big] \in L^1(\mathfrak{m})'$.

Notice that the previous definition is well-posed because $\big|\int f\, \mathsf{h}_t g \, d\mathfrak{m}\big| \le \|f\|_{L^\infty(\mathfrak{m})} \|g\|_{L^1(\mathfrak{m})}$ is verified by item iii) of Proposition 5.2.14.

Exercise 5.2.17 Given $p \in [1, \infty]$ and $t > 0$, we (provisionally) denote by h_t^p the heat flow in $L^p(\mathfrak{m})$ at time t. Prove that $\mathsf{h}_t^p f = \mathsf{h}_t^q f$ for all $p, q \in [1, \infty]$ and $f \in L^p(\mathfrak{m}) \cap L^q(\mathfrak{m})$. ■

Proposition 5.2.18 *Let $\varphi \in C_c^\infty(0, +\infty)$ and $p \in [1, \infty]$ be given. For any $f \in L^2(\mathfrak{m}) \cap L^p(\mathfrak{m})$, let us define the* mollified heat flow $\mathsf{h}_\varphi f \in L^2(\mathfrak{m}) \cap L^p(\mathfrak{m})$ *as*

$$\mathsf{h}_\varphi f := \int_0^{+\infty} \mathsf{h}_s f \, \varphi(s) \, ds. \tag{5.33}$$

Then $\mathsf{h}_\varphi f \in D(\Delta)$ and $\|\Delta \mathsf{h}_\varphi f\|_{L^p(\mathfrak{m})} \le C(\varphi) \|f\|_{L^p(\mathfrak{m})}$ for some constant $C(\varphi) > 0$.

Proof By applying Theorem 1.3.15, we see that $\mathsf{h}_\varphi f \in D(\Delta)$ and that

$$\Delta \mathsf{h}_\varphi f = \int_0^{+\infty} \Delta \mathsf{h}_s f \, \varphi(s) \, ds = \int_0^{+\infty} \frac{d}{ds} \mathsf{h}_s f \, \varphi(s) \, ds = -\int_0^{+\infty} \mathsf{h}_s f \, \varphi'(s) \, ds,$$

whence accordingly item iii) of Proposition 5.2.14 yields

$$\|\Delta \mathsf{h}_\varphi f\|_{L^p(\mathfrak{m})} \le \int_0^{+\infty} \|\mathsf{h}_s f\|_{L^p(\mathfrak{m})} |\varphi'|(s) \, ds \le \|f\|_{L^p(\mathfrak{m})} \int_0^{+\infty} |\varphi'|(s) \, ds.$$

Therefore the statement is verified with $C(\varphi) := \int_0^{+\infty} |\varphi'|(s) \, ds$. □

A direct consequence of Proposition 5.2.18 is given by the next result:

Corollary 5.2.19 *The family $\big\{f \in L^2(\mathfrak{m}) \cap L^\infty(\mathfrak{m}) \,\big|\, f \ge 0,\ f \in D(\Delta),\ \Delta f \in W^{1,2}(\mathrm{X})\big\}$ is strongly $L^2(\mathfrak{m})$-dense in $\big\{f \in L^2(\mathfrak{m}) \,\big|\, f \ge 0\big\}$.*

Bibliographical Remarks

For the material presented in Sect. 5.1 we recommend the thorough monograph [2] and the references contained therein. On the other hand, the results of Sect. 5.2 constitute the outcome of a reformulation of the achievements that have been carried out in [4].

Chapter 6
Second-Order Calculus on RCD Spaces

In this conclusive chapter we introduce the class of those metric measure spaces that satisfy the *Riemannian curvature-dimension condition*, briefly called RCD spaces, and we develop a thorough second-order differential calculus over these structures.

In Sect. 6.1 we lay the groundwork for the theory of RCD spaces. An $\mathsf{RCD}(K,\infty)$ space, where K is a given real constant, is an infinitesimally Hilbertian metric measure space having Ricci curvature bounded from below by K (in some synthetic sense); the definition of this concept is provided in Sect. 6.1.1. In Sect. 6.1.2 we show that the added regularity of RCD spaces guarantees nicer properties of the heat flow. In Sect. 6.1.3 we introduce a fundamental class of functions on RCD spaces, called *test functions*, which will be used as test objects in order to give meaningful definitions of higher-order Sobolev spaces.

By building on top of the abstract first-order differential calculus that has been investigated in Chap. 4, we are thus able to define, e.g., the notions of *Hessian*, of *covariant derivative* and of *exterior derivative* over any $\mathsf{RCD}(K,\infty)$ space; these goals are achieved in Sects. 6.2, 6.3 and 6.4, respectively. We finally conclude by presenting the *Ricci curvature operator* and its properties in Sect. 6.5. (We point out that some of the proofs in these conclusive sections are just sketched.)

6.1 The Theory of RCD Spaces

6.1.1 Definition of RCD *Space*

Consider any smooth function $f:\mathbb{R}^d\to\mathbb{R}$. An easy computation yields the following formula:

$$\Delta\frac{|\nabla f|^2}{2}=|\mathrm{H}f|_{\mathsf{HS}}^2+\nabla f\cdot\nabla\Delta f. \tag{6.1}$$

N. Gigli, E. Pasqualetto, *Lectures on Nonsmooth Differential Geometry*, SISSA Springer Series 2, https://doi.org/10.1007/978-3-030-38613-9_6

Now consider any smooth Riemannian manifold (M, g). Recall that the *Riemann curvature tensor* is given by

$$\mathrm{R}(X, Y, Z, W) := \langle \nabla_X \nabla_Y Z - \nabla_Y \nabla_X Z - \nabla_{[X,Y]} Z, W \rangle,$$

while the *Ricci curvature tensor* is defined as

$$\mathrm{Ric}(X, Y) := \sum_{i=1}^{\dim M} \mathrm{R}(e_i, X, Y, e_i)$$

where $(e_i)_i$ is any (local) *frame*, i.e., a family of vector fields that form an orthonormal basis of the tangent space at all points.

Observe that in (6.1) three derivatives of f appear, thus an analogous formula for M should contain a correction term due to the presence of the curvature. Indeed, it turns out that for any $f \in C^\infty(M)$ we have

$$\Delta \frac{|\nabla f|^2}{2} = |\mathrm{H} f|^2_{\mathsf{HS}} + \nabla f \cdot \nabla \Delta f + \mathrm{Ric}(\nabla f, \nabla f). \tag{6.2}$$

Formula (6.2) is called *Bochner identity*. In order to generalise the notion of 'having Ricci curvature greater than or equal to K' to the framework of metric measure spaces, we need the following simple result:

Proposition 6.1.1 *Let (M, g) be a smooth Riemannian manifold and let $K \in \mathbb{R}$. Then the following are equivalent:*

i) $\mathrm{Ric}_M \geq K g$, *i.e., for any $p \in M$ and $v \in T_p M$ we have that* $\mathrm{Ric}_p(v, v) \geq K|v|^2$.
ii) *For any $f \in C^\infty(M)$ it holds that*

$$\Delta \frac{|\nabla f|^2}{2} \geq \nabla f \cdot \nabla \Delta f + K |\nabla f|^2, \tag{6.3}$$

which is called Bochner inequality.

Proof The implication i) $\Longrightarrow$ ii) is trivial by (6.2), then it just suffices to prove ii) $\Longrightarrow$ i). Suppose to have $p \in M$ and $v \in T_p M$ such that $\mathrm{Ric}_p(v, v) < K|v|^2$. Hence there is $f \in C^\infty(M)$ satisfying $\nabla f_p = v$ and $\mathrm{H} f_p = 0$. Then $\Delta \frac{|\nabla f|^2}{2}(p) < \nabla f_p \cdot \nabla \Delta f_p + K|\nabla f_p|^2$, which is in contradiction with (6.2). □

We are now in a position to give the definition of the $\mathsf{RCD}(K, \infty)$ condition:

Definition 6.1.2 ($\mathsf{RCD}(K, \infty)$ **Space**) Let $(\mathrm{X}, \mathsf{d}, \mathfrak{m})$ be a metric measure space and $K \in \mathbb{R}$. Then we say that $(\mathrm{X}, \mathsf{d}, \mathfrak{m})$ is an $\mathsf{RCD}(K, \infty)$ *space* provided:

i) There exist $C > 0$ and $\bar{x} \in \mathrm{X}$ such that $\mathfrak{m}\big(B_r(\bar{x})\big) \leq \exp(Cr^2)$ for all $r > 0$.
ii) If $f \in W^{1,2}(\mathrm{X})$ satisfies $|Df| \in L^\infty(\mathfrak{m})$, then there exists $\tilde{f} \in \mathrm{LIP}(\mathrm{X})$ such that $\tilde{f} = f$ holds $\mathfrak{m}$-a.e. and $\mathrm{Lip}(\tilde{f}) = \big\| |Df| \big\|_{L^\infty(\mathfrak{m})}$.

iii) (X, d, m) is infinitesimally Hilbertian.
iv) The *weak Bochner inequality* is satisfied, i.e.,

$$\int \Delta g \, \frac{|\nabla f|^2}{2} \, \mathrm{dm} \geq \int g \big[\nabla f \cdot \nabla \Delta f + K |\nabla f|^2 \big] \, \mathrm{dm} \tag{6.4}$$

for every choice of functions $f \in D(\Delta)$ and $g \in D(\Delta) \cap L^\infty(\mathfrak{m})^+$ with $\Delta f \in W^{1,2}(\mathrm{X})$ and $\Delta g \in L^\infty(\mathfrak{m})$.

Remark 6.1.3 Item ii) in Definition 6.1.2 is verified if and only if both these conditions hold:

a) If $f \in W^{1,2}(\mathrm{X})$ satisfies $|Df| \in L^\infty(\mathfrak{m})$, then there exists $\tilde{f} : \mathrm{X} \to \mathbb{R}$ locally Lipschitz such that $\tilde{f} = f$ holds m-a.e. in X and $\mathrm{lip}(\tilde{f}) \leq \big\| |Df| \big\|_{L^\infty(\mathfrak{m})}$.
b) If $\tilde{f} : \mathrm{X} \to \mathbb{R}$ is locally Lipschitz and $\mathrm{lip}(\tilde{f}) \leq L$, then $\tilde{f}$ is L-Lipschitz.

The role of ii) is to link the metric structure of the space with the Sobolev calculus. ■

6.1.2 *Heat Flow on RCD Spaces*

From now on, (X, d, m) will always be an $\mathsf{RCD}(K, \infty)$ space, for some $K \in \mathbb{R}$.

Theorem 6.1.4 (Bakry-Émery Estimate) *Consider* $f \in W^{1,2}(\mathrm{X})$ *and* $t \geq 0$. *Then*

$$|D\mathsf{h}_t f|^2 \leq e^{-2Kt} \, \mathsf{h}_t\big(|Df|^2\big) \quad \text{holds m-a.e. in X.} \tag{6.5}$$

Proof Fix $g \in D(\Delta) \cap L^\infty(\mathfrak{m})^+$ such that $\Delta g \in L^\infty(\mathfrak{m})$ and $t > 0$. Define $F : [0, t] \to \mathbb{R}$ as

$$F(s) := \int \mathsf{h}_s g \, |D\mathsf{h}_{t-s} f|^2 \, \mathrm{dm} \qquad \text{for every } s \in [0, t].$$

Since $t \mapsto \mathsf{h}_t f \in W^{1,2}(\mathrm{X})$ is of class C^1 by Proposition 5.2.12, we know that $t \mapsto |D\mathsf{h}_t f|^2 \in L^1(\mathfrak{m})$ is of class C^1 as well. Moreover, from the m-a.e. inequality

$$|\mathsf{h}_t g - \mathsf{h}_s g| = \left| \int_s^t \frac{\mathrm{d}}{\mathrm{d}r} \mathsf{h}_r g \, \mathrm{d}r \right| \leq \int_s^t |\Delta \mathsf{h}_r g| \, \mathrm{d}r = \int_s^t |\mathsf{h}_r \Delta g| \, \mathrm{d}r \leq |t - s| \, \|\Delta g\|_{L^\infty(\mathfrak{m})},$$

which is granted by Proposition 5.2.7 and the weak maximum principle, we immediately deduce that $\|\mathsf{h}_t g - \mathsf{h}_s g\|_{L^\infty(\mathfrak{m})} \leq |t - s| \, \|\Delta g\|_{L^\infty(\mathfrak{m})}$, in other words

$t \mapsto \mathsf{h}_t g \in L^\infty(\mathfrak{m})$ is Lipschitz. Therefore F is Lipschitz and it holds that

$$\begin{aligned}\frac{\mathrm{d}}{\mathrm{d}s} F(s) &= \int \Delta \mathsf{h}_s g\, |D\mathsf{h}_{t-s} f|^2 - 2\, \mathsf{h}_s g\, \nabla \mathsf{h}_{t-s} f \cdot \nabla \Delta \mathsf{h}_{t-s} f \,\mathrm{d}\mathfrak{m} \\ &\overset{(6.4)}{\geq} 2\,K \int \mathsf{h}_s g\, |D\mathsf{h}_{t-s} f|^2 \,\mathrm{d}\mathfrak{m} \\ &= 2\,K\, F(s) \quad \text{for a.e. } s \in [0,t].\end{aligned}$$

Hence Gronwall lemma grants that $F(t) \geq e^{2Kt} F(0)$, or equivalently

$$\int g\, |D\mathsf{h}_t f|^2 \,\mathrm{d}\mathfrak{m} \leq e^{-2Kt} \int g\, \mathsf{h}_t\big(|Df|^2\big) \,\mathrm{d}\mathfrak{m}.$$

Since the class of functions g under consideration is weakly*-dense in $\{g \in L^\infty(\mathfrak{m}) \,:\, g \geq 0\}$ as a consequence of Proposition 5.2.18, we finally conclude that (6.5) is satisfied. □

Lemma 6.1.5 *Let $f, g \in D(\Delta) \cap L^\infty(\mathfrak{m})$ be given. Then*

$$\int \Delta g \,\frac{f^2}{2} \,\mathrm{d}\mathfrak{m} = \int g\big(f\,\Delta f + |Df|^2\big) \,\mathrm{d}\mathfrak{m}. \tag{6.6}$$

Proof Since $fg \in W^{1,2}(\mathrm{X})$, we see that

$$\begin{aligned}\int fg\,\Delta f \,\mathrm{d}\mathfrak{m} &= -\int \nabla(fg) \cdot \nabla f \,\mathrm{d}\mathfrak{m} = -\int g\,|Df|^2 + f\,\nabla g \cdot \nabla f \,\mathrm{d}\mathfrak{m} \\ &= -\int g\,|Df|^2 + \nabla \frac{f^2}{2} \cdot \nabla g \,\mathrm{d}\mathfrak{m},\end{aligned}$$

which gives the statement. □

Proposition 6.1.6 (L^∞-Lip **Regularisation of the Heat Flow**) *Let $f \in L^\infty(\mathfrak{m})$ and $t > 0$ be given. Then $|D\mathsf{h}_t f| \in L^\infty(\mathfrak{m})$ and*

$$\big\||D\mathsf{h}_t f|\big\|_{L^\infty(\mathfrak{m})} \leq \frac{C(K)}{\sqrt{t}} \|f\|_{L^\infty(\mathfrak{m})} \quad \textit{for every } t \in (0,1). \tag{6.7}$$

In particular, the function $\mathsf{h}_t f$ admits a Lipschitz representative.

Proof It suffices to prove the statement for $f \in L^2(\mathfrak{m}) \cap L^\infty(\mathfrak{m})$. Fix any $g \in D(\Delta) \cap L^\infty(\mathfrak{m})^+$ such that $\Delta g \in L^\infty(\mathfrak{m})$. Take $t \in (0,1)$ and define $F : [0,t] \to \mathbb{R}$ as

$$F(s) := \int \mathsf{h}_s g\, |\mathsf{h}_{t-s} f|^2 \,\mathrm{d}\mathfrak{m} \quad \text{for every } s \in [0,t].$$

We already know that $F \in C([0,t]) \cap C^1((0,t))$ and that for a.e. $s \in [0,t]$ it holds

$$\frac{\mathrm{d}}{\mathrm{d}s} F(s) = \int \Delta \mathsf{h}_s g \, |\mathsf{h}_{t-s} f|^2 - 2\mathsf{h}_s g \, \mathsf{h}_{t-s} f \, \Delta \mathsf{h}_{t-s} f \, \mathrm{d}\mathfrak{m} \overset{(6.6)}{=} 2 \int \mathsf{h}_s g \, |D\mathsf{h}_{t-s} f|^2 \, \mathrm{d}\mathfrak{m}$$
$$= 2 \int g \, \mathsf{h}_s(|D\mathsf{h}_{t-s} f|^2) \, \mathrm{d}\mathfrak{m} \overset{(6.5)}{\geq} 2\,C(K) \int g \, |D\mathsf{h}_t f|^2 \, \mathrm{d}\mathfrak{m}.$$

By integrating the previous inequality on $[0,t]$, we obtain that

$$2\,C(K)\,t \int g \, |D\mathsf{h}_t f|^2 \, \mathrm{d}\mathfrak{m} \leq F(t) - F(0) \leq \int g \, \mathsf{h}_t(f^2) \, \mathrm{d}\mathfrak{m}.$$

By the weak*-density of such functions g, we see that the inequality $2\,C(K)\,t\,|D\mathsf{h}_t f|^2 \leq \mathsf{h}_t(f^2)$ holds $\mathfrak{m}$-a.e. in X. Therefore, the weak maximum principle grants that (6.7) is satisfied. Finally, the last statement immediately follows from item ii) of Definition 6.1.2. □

6.1.3 *Test Functions*

We now introduce the algebra $\mathrm{Test}^\infty(\mathrm{X})$ of *test functions* on $(\mathrm{X}, \mathsf{d}, \mathfrak{m})$. These represent the 'smoothest possible objects' on X and will be used (in place of C_c^∞) to define several differential operators via suitable integration-by-parts formulae.

Definition 6.1.7 (Test Function) Let us define

$$\mathrm{Test}^\infty(\mathrm{X}) := \Big\{ f \in \mathrm{LIP}(\mathrm{X}) \cap L^\infty(\mathfrak{m}) \cap D(\Delta) \;\Big|\; \Delta f \in W^{1,2}(\mathrm{X}) \cap L^\infty(\mathfrak{m}) \Big\},$$
$$\mathrm{Test}^\infty_+(\mathrm{X}) := \big\{ f \in \mathrm{Test}^\infty(\mathrm{X}) \;\big|\; f \geq 0 \text{ holds } \mathfrak{m}\text{-a.e. on } \mathrm{X} \big\}. \tag{6.8}$$

Proposition 6.1.8 *The space* $\mathrm{Test}^\infty_+(\mathrm{X})$ *is dense in* $W^{1,2}(\mathrm{X})^+$. *Moreover, the space* $\mathrm{Test}^\infty(\mathrm{X})$ *is dense in* $W^{1,2}(\mathrm{X})$.

Proof Let $f \in W^{1,2}(\mathrm{X})^+$ be fixed. Call $f_n := f \wedge n \in W^{1,2}(\mathrm{X})^+ \cap L^\infty(\mathfrak{m})$ for any $n \in \mathbb{N}$, so that $f_n \to f$ in $W^{1,2}(\mathrm{X})$. Then it suffices to prove that each f_n belongs to the $W^{1,2}(\mathrm{X})$-closure of $\mathrm{Test}^\infty_+(\mathrm{X})$. We now claim that

$$\mathsf{h}_\varphi f_n \in \mathrm{Test}^\infty_+(\mathrm{X}) \quad \text{for every } \varphi \in C_c^\infty(0,+\infty). \tag{6.9}$$

We have that $\mathsf{h}_\varphi f_n \geq 0$ holds $\mathfrak{m}$-a.e. by the weak maximum principle. By arguing as in Proposition 5.2.18, we also see that $\mathsf{h}_\varphi f_n \in D(\Delta) \cap L^\infty(\mathfrak{m})$. Choose $\varepsilon \in (0,1)$ so that the support of φ is contained in $[\varepsilon, \varepsilon^{-1}]$, then the fact that $\Delta \mathsf{h}_t f_n = \mathsf{h}_{t-\varepsilon/2} \Delta \mathsf{h}_{\varepsilon/2} f_n$ for all $t \geq \varepsilon$ can be used to prove that $\Delta \mathsf{h}_\varphi f_n \in W^{1,2}(\mathrm{X}) \cap L^\infty(\mathfrak{m})$. Finally, $\mathsf{h}_\varphi f_n \in \mathrm{LIP}(\mathrm{X})$ by Proposition 6.1.6. Hence, the claim (6.9) is proved. Now take any $(\varphi_k)_k \subseteq C_c^\infty(0,+\infty)$ such that $\varphi_k \rightharpoonup \delta_0$. Then $\mathsf{h}_{\varphi_k} f_n \to f_n$ strongly in $W^{1,2}(\mathrm{X})$, proving that each function f_n is in the closure of the space $\mathrm{Test}_+^\infty(\mathrm{X})$, as required.

The second statement follows from the first one by noticing that for every $f \in W^{1,2}(\mathrm{X})$ it holds that $f = f^+ - f^-$ and $f^\pm \in W^{1,2}(\mathrm{X})^+$. □

By making use of the assumed lower Ricci curvature bounds, we can prove the following regularity of minimal weak upper gradients of test functions:

Lemma 6.1.9 *Let $f \in \mathrm{Test}^\infty(\mathrm{X})$ be given. Then $|Df|^2 \in W^{1,2}(\mathrm{X})$.*

Proof Given any $g \in D(\Delta) \cap L^\infty(\mathfrak{m})^+$ and any sequence $(\varphi_k)_k \subseteq C_c^\infty(0,+\infty)$ with $\varphi_k \rightharpoonup \delta_0$, we deduce from Proposition 5.2.18 that $\mathsf{h}_{\varphi_k} g \rightharpoonup g$ weakly* in $L^\infty(\mathfrak{m})$ and $L^\infty(\mathfrak{m}) \ni \Delta \mathsf{h}_{\varphi_k} g \to \Delta g$ in $L^2(\mathfrak{m})$. Thus taking into account item iv) of Definition 6.1.2 and the fact that $|\nabla f|^2 \in L^2(\mathfrak{m})$, we see that

$$\frac{1}{2}\int \Delta g\, |\nabla f|^2 \,\mathrm{d}\mathfrak{m} \geq \int g\,\big(\nabla f \cdot \nabla \Delta f + K|\nabla f|^2\big)\,\mathrm{d}\mathfrak{m} \quad \text{for every } g \in D(\Delta) \cap L^\infty(\mathfrak{m})^+. \tag{6.10}$$

(Notice that in (6.10), differently from item iv) of Definition 6.1.2, the function Δg is not required to be essentially bounded.) Now we apply (6.10) with $g := \mathsf{h}_t\big(|\nabla f|^2\big)$, which satisfies the inequality $g \leq \mathrm{Lip}^2(f)$ in the $\mathfrak{m}$-a.e. sense by the weak maximum principle, obtaining

$$\begin{aligned}
E\big(|\nabla f|^2\big) &\leq \varliminf_{t\searrow 0} E\big(\mathsf{h}_{t/2}(|\nabla f|^2)\big) = \varliminf_{t\searrow 0} \frac{1}{2}\int \big|\nabla \mathsf{h}_{t/2}(|\nabla f|^2)\big|^2\,\mathrm{d}\mathfrak{m} \\
&= -\varlimsup_{t\searrow 0} \frac{1}{2}\int \mathsf{h}_{t/2}\big(|\nabla f|^2\big)\,\Delta \mathsf{h}_{t/2}\big(|\nabla f|^2\big)\,\mathrm{d}\mathfrak{m} \\
&= -\varlimsup_{t\searrow 0} \frac{1}{2}\int \Delta \mathsf{h}_t\big(|\nabla f|^2\big)\,|\nabla f|^2\,\mathrm{d}\mathfrak{m} \\
&\leq -\varlimsup_{t\searrow 0} \int \mathsf{h}_t\big(|\nabla f|^2\big)\,\big(\nabla f \cdot \nabla \Delta f + K|\nabla f|^2\big)\,\mathrm{d}\mathfrak{m} \\
&\leq \mathrm{Lip}(f)^2 \int \big|\nabla f \cdot \nabla \Delta f + K|\nabla f|^2\big|\,\mathrm{d}\mathfrak{m} < +\infty,
\end{aligned}$$

whence $|Df|^2 \in W^{1,2}(\mathrm{X})$, as required. □

Remark 6.1.10 Given any $f \in \mathrm{Test}^\infty(\mathrm{X})$, it holds that

$$E\big(|Df|^2\big) \leq \mathrm{Lip}(f)^2\, \|f\|_{W^{1,2}(\mathrm{X})}\big(\|\Delta f\|_{W^{1,2}(\mathrm{X})} + |K|\,\|f\|_{W^{1,2}(\mathrm{X})}\big), \tag{6.11}$$

as a consequence of the estimates in the proof of Lemma 6.1.9. ■

Theorem 6.1.11 *The space* $\mathrm{Test}^\infty(\mathrm{X})$ *is an algebra.*

Proof It is clear that $\mathrm{Test}^\infty(\mathrm{X})$ is a vector space. Now fix $f, g \in \mathrm{Test}^\infty(\mathrm{X})$. We aim to prove that $fg \in \mathrm{Test}^\infty(\mathrm{X})$ as well. It is immediate to check that $fg \in \mathrm{LIP}(\mathrm{X}) \cap L^\infty(\mathfrak{m})$. Moreover, we already know from item iii) of Proposition 5.2.3 that $fg \in D(\Delta)$ and

$$\Delta(fg) = f\,\Delta g + g\,\Delta f + 2\,\nabla f \cdot \nabla g,$$

in particular $\Delta(fg) \in L^\infty(\mathfrak{m})$. Finally, given that $f\,\Delta g, g\,\Delta f \in W^{1,2}(\mathrm{X})$ by the Leibniz rule (i.e. item C) of Theorem 2.1.28), while $\nabla f \cdot \nabla g \in W^{1,2}(\mathrm{X})$ by Lemma 6.1.9 and a polarisation argument, we conclude that $\Delta(fg) \in W^{1,2}(\mathrm{X})$. Hence $fg \in \mathrm{Test}^\infty(\mathrm{X})$, as required. □

6.2 Hessian

6.2.1 Definition and Basic Properties

We briefly recall the notion of Hessian on a smooth Riemannian manifold (M, g).

Given any two smooth vector fields X, Y on M, we consider the *covariant derivative* $\nabla_Y X$ *of* X *in the direction of* Y, which is characterised by the following result:

Theorem 6.2.1 *There exists a unique bilinear map* $(X, Y) \mapsto \nabla_Y X$ *with these properties:*

1) *It is an* affine connection*:*

   1a) *It is* tensorial *with respect to* Y*, i.e.,* $\nabla_{fY} X = f\,\nabla_Y X$ *holds for all* $f \in C^\infty(M)$ *and* X, Y *smooth vector fields on* M.

   1b) *It holds that* $\nabla_Y(fX) = Y(f)X + f\,\nabla_Y X$ *for all* $f \in C^\infty(M)$ *and* X, Y *smooth vector fields on* M.

2) *It is the* Levi-Civita connection*:*

   2a) *It is* torsion-free*, i.e.,* $\nabla_X Y - \nabla_Y X = [X, Y]$ *holds for all* X, Y *smooth vector fields on* M.

   2b) *It is* compatible with the metric*, i.e.,* $X\big(\langle Y, Z\rangle\big) = \langle \nabla_X Y, Z\rangle + \langle Y, \nabla_X Z\rangle$ *holds for all* X, Y, Z *smooth vector fields on* M.

Proof Properties (1), (2) imply that *Koszul's formula*

$$\langle \nabla_X Y, Z\rangle = X\big(\langle Y, Z\rangle\big) + Y\big(\langle X, Z\rangle\big) - Z\big(\langle X, Y\rangle\big) + \big\langle [X, Y], Z\big\rangle - \big\langle [X, Z], Y\big\rangle - \big\langle [Y, Z], X\big\rangle$$

holds for any smooth vector fields X, Y, Z. This formula characterises $\nabla_X Y$ in terms of scalar product and Lie brackets only, thus showing uniqueness of the bilinear map satisfying (1), (2). As for existence, we use again Koszul's formula to define $\nabla_X Y$ as the only vector field for which the formula is valid for any Z: it is easy to see that the definition is well-posed and simple computations show that the resulting object satisfies (1), (2), thus concluding the proof. □

Given a smooth vector field X on M, we define the *covariant derivative* ∇X of X as

$$\nabla X(Y, Z) := \langle \nabla_Y X, Z\rangle \quad \text{for all } Y, Z \text{ smooth vector fields on } M. \tag{6.12}$$

Then we define the *Hessian* $\mathrm{H}f$ of a function $f \in C^\infty(M)$ as

$$\mathrm{H}f := \nabla(\nabla f). \tag{6.13}$$

It can be readily proved that the Hessian is a symmetric tensor, i.e.,

$$\mathrm{H}f(X, Y) = \mathrm{H}f(Y, X) \quad \text{for all } f \in C^\infty(M) \text{ and } X, Y \text{ smooth vector fields on } M. \tag{6.14}$$

In order to prove it, just observe that item 2b) of Theorem 6.2.1 yields

$$\begin{aligned} \mathrm{H}f(X, Y) &= \langle \nabla_X \nabla f, Y\rangle = X\big(\langle \nabla f, Y\rangle\big) - \langle \nabla f, \nabla_X Y\rangle = X\big(Y(f)\big) - (\nabla_X Y)(f), \\ \mathrm{H}f(Y, X) &= \langle \nabla_Y \nabla f, X\rangle = Y\big(\langle \nabla f, X\rangle\big) - \langle \nabla f, \nabla_Y X\rangle = Y\big(X(f)\big) - (\nabla_Y X)(f). \end{aligned}$$

By subtracting the second line from the first one, we thus obtain that

$$\mathrm{H}f(X, Y) - \mathrm{H}f(Y, X) = (XY - YX)(f) - \underbrace{(\nabla_X Y - \nabla_Y X)}_{=[X,Y] \text{ by 2a)}}(f) = 0,$$

proving the claim (6.14).

Lemma 6.2.2 *Let $f \in C^\infty(M)$ be given. Then*

$$\nabla \frac{|\nabla f|^2}{2} = \mathrm{H}f(\nabla f, \cdot). \tag{6.15}$$

Proof Just observe that for any smooth vector field X on M it holds

$$\Big\langle \nabla \frac{|\nabla f|^2}{2}, X\Big\rangle = \frac{1}{2} X\big(|\nabla f|^2\big) \overset{\text{2b)}}{=} \langle \nabla_X \nabla f, \nabla f\rangle = \nabla(\nabla f)(X, \nabla f) \overset{(6.14)}{=} \mathrm{H}f(\nabla f, X),$$

whence the statement follows. □

Remark 6.2.3 By polarisation, starting from (6.15) and with simple computations we get that the identity

$$2\,\mathrm{H}f(\nabla g_1, \nabla g_2) = \nabla(\nabla f \cdot \nabla g_1) \cdot \nabla g_2 + \nabla(\nabla f \cdot \nabla g_2) \cdot \nabla g_1 - \nabla f \cdot \nabla(\nabla g_1 \cdot \nabla g_2) \tag{6.16}$$

is satisfied for every $f, g_1, g_2 \in C^\infty(M)$. ■

Definition 6.2.4 Let $(\mathrm{X}, \mathsf{d}, \mathfrak{m})$ be an $\mathsf{RCD}(K, \infty)$ space. Then we define

$$L^2\big((T^*)^{\otimes 2}\mathrm{X}\big) := L^2(T^*\mathrm{X}) \otimes L^2(T^*\mathrm{X}). \tag{6.17}$$

Given any $A \in L^2\big((T^*)^{\otimes 2}\mathrm{X}\big)$, we define

$$A(X, Y) := A(X \otimes Y) \in L^0(\mathfrak{m}) \quad \text{for every } X, Y \in L^2(T\mathrm{X}). \tag{6.18}$$

Clearly $L^2\big((T^*)^{\otimes 2}\mathrm{X}\big)$ can be identified with the dual of $L^2(T^{\otimes 2}\mathrm{X}) := L^2(T\mathrm{X}) \otimes L^2(T\mathrm{X})$, the duality mapping being given by

$$(\omega \otimes \eta)(X \otimes Y) := \omega(X)\,\eta(Y) \quad \mathfrak{m}\text{-a.e.}$$

for all $\omega, \eta \in L^2(T^*\mathrm{X})$ and $X, Y \in L^2(T\mathrm{X})$, then extended by linearity and continuity. We also point out that

$$\big|A(X, Y)\big| \le |A|_{\mathsf{HS}} |X| |Y| \quad \text{holds } \mathfrak{m}\text{-a.e. on } \mathrm{X} \tag{6.19}$$

for every $A \in L^2\big((T^*)^{\otimes 2}\mathrm{X}\big)$ and $X, Y \in L^2(T\mathrm{X})$.

Lemma 6.2.5 *Let* $(\mathrm{X}, \mathsf{d}, \mathfrak{m})$ *be an* $\mathsf{RCD}(K, \infty)$ *space. Then*

$$\left\{ \sum_{i=1}^n h_i \nabla g_i \ :\ h_i, g_i \in \mathrm{Test}^\infty(\mathrm{X}) \right\} \quad \text{is dense in } L^2(T\mathrm{X}). \tag{6.20}$$

In particular, it holds that

$$\left\{ \sum_{i=1}^n h_i \nabla g_{1,i} \otimes \nabla g_{2,i} \ :\ h_i, g_{1,i}, g_{2,i} \in \mathrm{Test}^\infty(\mathrm{X}) \right\} \quad \text{is dense in } L^2(T^{\otimes 2}\mathrm{X}). \tag{6.21}$$

Proof To get (6.20), recall that $\mathrm{Test}^\infty(\mathrm{X})$ is dense in $W^{1,2}(\mathrm{X})$ and weakly* dense in $L^\infty(\mathfrak{m})$. To deduce (6.21) from (6.20), it suffices to apply Lemma 3.2.21 and Theorem 6.1.11. □

Having formula (6.16) in mind, we thus give the following definition:

Definition 6.2.6 (The Space $W^{2,2}(\mathrm{X})$) Let $(\mathrm{X}, \mathsf{d}, \mathfrak{m})$ be an $\mathsf{RCD}(K,\infty)$ space, with $K \in \mathbb{R}$. Let $f \in W^{1,2}(\mathrm{X})$. Then we say that $f \in W^{2,2}(\mathrm{X})$ provided there exists $A \in L^2\big((T^*)^{\otimes 2}\mathrm{X}\big)$ such that for every choice of $h, g_1, g_2 \in \mathrm{Test}^\infty(\mathrm{X})$ it holds that

$$2\int h\, A(\nabla g_1, \nabla g_2)\,\mathrm{d}\mathfrak{m} = -\int \nabla f \cdot \nabla g_1 \,\mathrm{div}(h\nabla g_2) + \nabla f \cdot \nabla g_2 \,\mathrm{div}(h\nabla g_1) + h\,\nabla f \cdot \nabla(\nabla g_1 \cdot \nabla g_2)\,\mathrm{d}\mathfrak{m}.$$

Such tensor A, which is uniquely determined by (6.21), will be unambiguously denoted by $\mathrm{H}f$ and called *Hessian* of f. Moreover, the resulting vector space $W^{2,2}(\mathrm{X})$ is naturally endowed with the norm $\|\cdot\|_{W^{2,2}(\mathrm{X})}$, defined as

$$\|f\|_{W^{2,2}(\mathrm{X})} := \sqrt{\|f\|^2_{L^2(\mathfrak{m})} + \|\mathrm{d}f\|^2_{L^2(T^*\mathrm{X})} + \|\mathrm{H}f\|^2_{L^2((T^*)^{\otimes 2}\mathrm{X})}} \quad \text{for every } f \in W^{2,2}(\mathrm{X}).$$

Theorem 6.2.7 *The space $\mathrm{W}^{2,2}(\mathrm{X})$ is a separable Hilbert space and the Hessian is a closed operator, i.e.,*

$$\big\{(f, \mathrm{H}f)\,:\, f \in \mathrm{W}^{2,2}(\mathrm{X})\big\} \quad \textit{is closed in } \mathrm{W}^{1,2}(\mathrm{X}) \times L^2\big((T^*)^{\otimes 2}\mathrm{X}\big). \tag{6.22}$$

Proof Proving (6.22) amounts to showing that $f \in W^{2,2}(\mathrm{X})$ and $\mathrm{H}f = A$ whenever a given sequence $(f_n)_n \subseteq W^{2,2}(\mathrm{X})$ satisfies $f_n \to f$ in $W^{1,2}(\mathrm{X})$ and $\mathrm{H}f_n \to A$ in $L^2\big((T^*)^{\otimes 2}\mathrm{X}\big)$. This can be achieved by writing the integral formula characterising $\mathrm{H}f_n$ and letting $n \to \infty$. Completeness of $W^{2,2}(\mathrm{X})$ is then a direct consequence of (6.22). Finally, we deduce the separability of $W^{2,2}(\mathrm{X})$ from the fact that the operator $f \mapsto (f, \mathrm{d}f, \mathrm{H}f)$ is an isometry from the space $W^{2,2}(\mathrm{X})$ to the separable space $L^2(\mathfrak{m}) \times L^2(T^*\mathrm{X}) \times L^2\big((T^*)^{\otimes 2}\mathrm{X}\big)$, provided the latter is endowed with the product norm. □

6.2.2 *Measure-Valued Laplacian*

Definition 6.2.8 (Measure-Valued Laplacian) Let $(\mathrm{X}, \mathsf{d}, \mathfrak{m})$ be an infinitesimally Hilbertian metric measure space. Let $f \in W^{1,2}(\mathrm{X})$. Then we say that f has *measure-valued Laplacian*, briefly $f \in D(\mathbf{\Delta})$, provided there exists a finite (signed) Radon measure μ on X such that

$$\int g\,\mathrm{d}\mu = -\int \nabla g \cdot \nabla f\,\mathrm{d}\mathfrak{m} \quad \text{for every } g \in \mathrm{LIP}_{bs}(\mathrm{X}). \tag{6.23}$$

The measure μ, which is uniquely determined by the density of $\mathrm{LIP}_{bs}(\mathrm{X})$ in $C_b(\mathrm{X})$, will be unambiguously denoted by $\mathbf{\Delta} f$.

It holds that $D(\mathbf{\Delta})$ is a vector space and that $\mathbf{\Delta} : D(\mathbf{\Delta}) \to \{$*finite Radon measures on* $\mathrm{X}\}$ is a linear map. Both properties immediately follow from (6.23).

Remark 6.2.9 Suppose that (X, d) is bounded. Then

$$\mathbf{\Delta} f(\mathrm{X}) = 0 \quad \text{for every } f \in D(\mathbf{\Delta}). \tag{6.24}$$

Indeed, $g \equiv 1$ trivially belongs to $\mathrm{LIP}_{bs}(\mathrm{X})$, whence (6.23) yields $\mathbf{\Delta} f(\mathrm{X}) = \int \mathrm{d}\mathbf{\Delta} f = 0$. ■

Example 6.2.10 Let $\mathrm{X} := [0, 1]$ and $\mathfrak{m} := \mathcal{L}^1|_{[0,1]}$. Then the identity function $f(x) := x$ belongs to $D(\mathbf{\Delta})$ and $\mathbf{\Delta} f = \delta_0 - \delta_1$. ■

Remark 6.2.11 In this framework, the Laplacian is not necessarily the trace of the Hessian. ■

Lemma 6.2.12 *Let* $(\mathrm{X}, \mathsf{d}, \mathfrak{m})$ *be an* $\mathsf{RCD}(K, \infty)$ *space. Then* $\mathrm{LIP}_{bs}(\mathrm{X})$ *is dense in* $W^{1,2}(\mathrm{X})$.

Proof We already know that $\mathrm{Test}^\infty(\mathrm{X})$ is dense in $W^{1,2}(\mathrm{X})$ (cf. Proposition 6.1.8). Then it suffices to prove that $\mathrm{LIP}_{bs}(\mathrm{X})$ is $W^{1,2}(\mathrm{X})$-dense in $\mathrm{Test}^\infty(\mathrm{X})$. To this aim, fix $f \in \mathrm{Test}^\infty(\mathrm{X})$ and define $\chi_n := \big(1 - \mathsf{d}(\cdot, B_n(\bar{x}))\big)^+$ for all $n \in \mathbb{N}$, where $\bar{x} \in \mathrm{X}$ is any fixed point. Now let us call $f_n := \chi_n f \in \mathrm{LIP}_{bs}(\mathrm{X})$ for every $n \in \mathbb{N}$. Then the dominated convergence theorem gives

$$\left.\begin{aligned} |f_n - f| &= |1 - \chi_n|\,|f| \longrightarrow 0, \\ |\mathrm{d}f_n - \mathrm{d}f| &\le |1 - \chi_n|\,|\mathrm{d}f| + |\mathrm{d}\chi_n|\,|f| \longrightarrow 0, \end{aligned}\right\} \quad \text{in } L^2(\mathfrak{m}),$$

thus proving that $f_n \to f$ in $W^{1,2}(\mathrm{X})$, as required. □

Proposition 6.2.13 (Compatibility of Δ and $\mathbf{\Delta}$) *The following properties hold:*

i) *Let* $f \in D(\mathbf{\Delta})$ *satisfy* $\mathbf{\Delta} f = \rho\,\mathfrak{m}$ *for some* $\rho \in L^2(\mathfrak{m})$. *Then* $f \in D(\Delta)$ *and* $\Delta f = \rho$.
ii) *Let* $f \in D(\Delta)$ *satisfy* $\Delta f \in L^1(\mathfrak{m})$. *Then* $f \in D(\mathbf{\Delta})$ *and* $\mathbf{\Delta} f = \Delta f\,\mathfrak{m}$.

Proof

i) We know that $\int g\,\rho\,\mathrm{d}\mathfrak{m} = -\int \nabla g \cdot \nabla f\,\mathrm{d}\mathfrak{m}$ holds for every $g \in \mathrm{LIP}_{bs}(\mathrm{X})$, whence also for every $g \in W^{1,2}(\mathrm{X})$ by Lemma 6.2.12. This proves that $f \in D(\Delta)$ and $\Delta f = \rho$.
ii) Since $\int g\,\mathrm{d}(\Delta f\,\mathfrak{m}) = \int g\,\Delta f\,\mathrm{d}\mathfrak{m} = -\int \nabla g \cdot \nabla f\,\mathrm{d}\mathfrak{m}$ for every $g \in \mathrm{LIP}_{bs}(\mathrm{X}) \subseteq W^{1,2}(\mathrm{X})$, we see that $f \in D(\mathbf{\Delta})$ and $\mathbf{\Delta} f = \Delta f\,\mathfrak{m}$.

□

In the sequel we shall need the following result, whose proof we omit:

Lemma 6.2.14 *Let* $(\mathrm{X}, \mathsf{d}, \mathfrak{m})$ *be a given* $\mathsf{RCD}(K, \infty)$ *space. Let* $\Omega \subseteq \mathrm{X}$ *be an open set and let* $K \subseteq \Omega$ *be a compact set such that* $\mathrm{dist}(K, \partial\Omega) > 0$. *Then there exists* $h \in \mathrm{Test}^\infty(\mathrm{X})$ *with* $0 \leq h \leq 1$ *such that* $h = 1$ *on* K *and* $\mathrm{spt}(h) \subseteq \Omega$.

Lemma 6.2.15 (Good Cut-Off Functions) *Let* $(\mathrm{X}, \mathsf{d}, \mathfrak{m})$ *be a proper* $\mathsf{RCD}(K, \infty)$ *space, i.e. all bounded closed subsets of* X *are compact. Then there exists a sequence* $(\chi_n)_n \subseteq \mathrm{Test}^\infty(\mathrm{X})$ *such that*

i) $\chi_n(x) \nearrow 1$ *for every* $x \in \mathrm{spt}(\mathfrak{m})$,
ii) $\Delta\chi_n$ *converges to* 0 *in the weak* topology of* $L^\infty(\mathfrak{m})$.

Proof Choose any $(g_n)_n \subseteq \mathrm{LIP}_{bs}(\mathrm{X})^+$ such that $g_n(x) \nearrow 1$ for every $x \in \mathrm{X}$. We claim that

$$\mathsf{h}_t g_n(x) \nearrow 1 \text{ as } n \to \infty \quad \text{for every } x \in \mathrm{spt}(\mathfrak{m}) \text{ and } t > 0. \tag{6.25}$$

Since $g_n - g_{n+1} \leq 0$ holds $\mathfrak{m}$-a.e., we deduce from item i) of Proposition 5.2.14 that $\mathsf{h}_t g_n \leq \mathsf{h}_t g_{n+1}$ holds $\mathfrak{m}$-a.e., thus also everywhere on $\mathrm{spt}(\mathfrak{m})$ because each $\mathsf{h}_t g_n$ is continuous (by the Sobolev-to-Lipschitz property). Given any $t > 0$ and $n \in \mathbb{N}$, it holds that the function $\mathsf{h}_t g_n$ is Lipschitz with $\mathrm{Lip}(\mathsf{h}_t g_n) \leq C(K)/\sqrt{t}$ by Proposition 6.1.6 and item ii) of Definition 6.1.2, whence the limit function $\ell_t := \sup_n \mathsf{h}_t g_n \leq 1$ is Lipschitz as well with $\mathrm{Lip}(\ell_t) \leq C(K)/\sqrt{t}$. By dominated convergence theorem it is immediate to see that g_n converges to 1 in the weak* topology of $L^\infty(\mathfrak{m})$, so for any $f \in L^1(\mathfrak{m})$ we have

$$\lim_{n\to\infty} \int f\, \mathsf{h}_t g_n \, \mathrm{d}\mathfrak{m} = \lim_{n\to\infty} \int \mathsf{h}_t f \, g_n \, \mathrm{d}\mathfrak{m} = \int \mathsf{h}_t f \, \mathrm{d}\mathfrak{m} = \int f \, \mathrm{d}\mathfrak{m},$$

which shows that for any $t > 0$ the functions $\mathsf{h}_t g_n$ converge to 1 with respect to the weak* topology of $L^\infty(\mathfrak{m})$. We can now prove (6.25) arguing by contradiction: if $\{\ell_t < 1\} \neq \emptyset$ for some $t > 0$, then there exists a Borel set $E \subseteq \mathrm{spt}(\mathfrak{m})$ with $0 < \mathfrak{m}(E) < +\infty$ such that $\ell_t(x) < 1$ for every $x \in E$. Then $\int_E \mathsf{h}_t g_n \, \mathrm{d}\mathfrak{m} \to \int_E \ell_t \, \mathrm{d}\mathfrak{m} < \mathfrak{m}(E)$ by monotone convergence theorem, which contradicts the weak* convergence of $\mathsf{h}_t g_n$ to 1. Therefore (6.25) is achieved.

Fix any function $\varphi \in C_c^\infty(0,1)^+$ with $\int_0^1 \varphi(t)\, \mathrm{d}t = 1$ and put $\chi_n := \int_0^1 \varphi(t)\, \mathsf{h}_t g_n \, \mathrm{d}t \in L^2(\mathfrak{m})$. By recalling Proposition 5.2.18 we see that $(\chi_n)_n \subseteq \mathrm{Test}^\infty(\mathrm{X})$ and that the sequence $(\Delta\chi_n)_n$ is bounded in $L^\infty(\mathfrak{m})$. Given any $x \in \mathrm{spt}(\mathfrak{m})$, we know from (6.25) that $\varphi(t)\, \mathsf{h}_t g_n(x) \nearrow \varphi(t)$ for all $t \in (0,1)$, thus accordingly

$$\chi_n(x) = \int_0^1 \varphi(t)\, \mathsf{h}_t g_n(x)\, \mathrm{d}t \nearrow \int_0^1 \varphi(t)\, \mathrm{d}t = 1,$$

which proves i). Moreover, from the bounded sequence $(\Delta\chi_n)_n \subseteq L^\infty(\mathfrak{m})$ we can extract a (not relabeled) subsequence converging to some limit function $G \in L^\infty(\mathfrak{m})$ in the weak* topology of $L^\infty(\mathfrak{m})$. In order to conclude it suffices to show that $G = 0$.

Fix any $\psi \in \mathrm{Test}^\infty(\mathrm{X})$ with compact support. Lemma 6.2.14 grants the existence of a function $\eta \in \mathrm{Test}^\infty(\mathrm{X})$ with compact support that equals 1 on a neighbourhood of $\mathrm{spt}(\psi)$. Since $\psi = 0$ on $\mathrm{X}\setminus\mathrm{spt}(\psi)$ we have that $\Delta\psi = 0$ holds $\mathfrak{m}$-a.e. on $\mathrm{X}\setminus\mathrm{spt}(\psi)$, therefore

$$\int \Delta\psi \,\mathrm{d}\mathfrak{m} = \int \eta\,\Delta\psi \,\mathrm{d}\mathfrak{m} = -\int \nabla\eta\cdot\nabla\psi \,\mathrm{d}\mathfrak{m} = 0,$$

where the last equality follows from the fact that $\chi_{\mathrm{spt}(\psi)}\nabla\eta = 0$ and $\chi_{\mathrm{X}\setminus\mathrm{spt}(\psi)}\nabla\psi = 0$ by locality of ∇. By dominated convergence theorem and i) one has $\int \Delta\psi\,\chi_n \,\mathrm{d}\mathfrak{m} \to \int \Delta\psi \,\mathrm{d}\mathfrak{m}$, thus

$$\lim_{n\to\infty}\int \psi\,\Delta\chi_n \,\mathrm{d}\mathfrak{m} = \lim_{n\to\infty}\int \Delta\psi\,\chi_n \,\mathrm{d}\mathfrak{m} = \int \Delta\psi \,\mathrm{d}\mathfrak{m} = 0.$$

Since test functions having compact support are dense in $L^1(\mathfrak{m})$ (by Lemma 6.2.14), this is enough to conclude that $G = 0$. Hence also item ii) is proved. □

Proposition 6.2.16 *Let* $(\mathrm{X}, \mathsf{d}, \mathfrak{m})$ *be a proper* $\mathsf{RCD}(K,\infty)$ *space. Let* $f \in W^{1,2}(\mathrm{X}) \cap L^1(\mathfrak{m})$ *and let* μ *be a finite Radon measure on* X *such that*

$$-\int \nabla g\cdot\nabla f \,\mathrm{d}\mathfrak{m} \geq \int g\,\mathrm{d}\mu \quad \textit{for every } g \in \mathrm{LIP}_{bs}(\mathrm{X})^+. \tag{6.26}$$

Then $f \in D(\mathbf{\Delta})$ *and* $\mathbf{\Delta} f \geq \mu$.

Proof Fix a sequence $(\chi_n)_n$ as in Lemma 6.2.15. Define $V_n := \{g \in \mathrm{LIP}(\mathrm{X}) : \mathrm{spt}(g) \subseteq \Omega_n\}$ for all $n \in \mathbb{N}$, where we set $\Omega_n := \{\chi_n > 1/2\}$. The elements of $\mathrm{LIP}_{bs}(\mathrm{X})$ have compact support (as the space is supposed to be proper), the sets Ω_n are open (by continuity of χ_n) and $\bigcup_n \Omega_n = \mathrm{X}$ (as $\chi_n \nearrow 1$ by Lemma 6.2.15). Therefore $\mathrm{LIP}_{bs}(\mathrm{X}) = \bigcup_n V_n$. We define the linear map $L : \mathrm{LIP}_{bs}(\mathrm{X}) \to \mathbb{R}$ as

$$L(g) := -\int \nabla g\cdot\nabla f \,\mathrm{d}\mathfrak{m} - \int g\,\mathrm{d}\mu \quad \text{for every } g \in \mathrm{LIP}_{bs}(\mathrm{X}).$$

Note that $L(g) \geq 0$ whenever $g \geq 0$. Given $n \in \mathbb{N}$ and $g \in V_n$, we have $2\,\|g\|_{L^\infty(\mathfrak{m})}\,\chi_n \pm g \geq 0$, so that $\pm L(g) \leq 2\,\|g\|_{L^\infty(\mathfrak{m})}\,L(\chi_n)$, or equivalently $|L(g)| \leq 2\,\|g\|_{L^\infty(\mathfrak{m})}\,L(\chi_n)$. This grants that L can be uniquely extended to a linear continuous map $L : C_c(\mathrm{X}) \to \mathbb{R}$ by Lemma 6.2.12. Since L is positive, by applying the Riesz representation theorem we deduce that there exists a Radon measure $\nu \geq 0$ on X such that $L(g) = \int g\,\mathrm{d}\nu$ for all $g \in C_c(\mathrm{X})$, thus in particular

$$-\int \nabla f\cdot\nabla g \,\mathrm{d}\mathfrak{m} = \int g\,\mathrm{d}(\mu+\nu) \quad \text{for every } g \in \mathrm{LIP}_{bs}(\mathrm{X}). \tag{6.27}$$

Now fix $n \in \mathbb{N}$ and pick a sequence $(\eta_k)_k \subseteq \mathrm{LIP}_{bs}(\mathrm{X})^+$ of cut-off functions with $\mathrm{Lip}(\eta_k) \leq 1$ such that $\eta_k \nearrow 1$. It holds that $(\eta_k \chi_n)_k \subseteq \mathrm{LIP}_{bs}(\mathrm{X})$. Given that $\eta_k \chi_n \to \chi_n$ holds pointwise $\mathfrak{m}$-a.e. and $|D(\eta_k \chi_n)| \leq |D\chi_n| + \chi_n \in L^2(\mathfrak{m})$, we can extract a (not relabeled) subsequence of $(\eta_k \chi_n)_k$ for which $\nabla(\eta_k \chi_n) \rightharpoonup \nabla \chi_n$ in the weak topology of $L^2(T\mathrm{X})$ (as ∇ is a closed operator). Moreover, one has that $\int \eta_k \chi_n \, \mathrm{d}\mu \to \int \chi_n \, \mathrm{d}\mu$ by dominated convergence theorem, while $\int \eta_k \chi_n \, \mathrm{d}\nu \to \int \chi_n \, \mathrm{d}\nu$ by monotone convergence theorem. Hence by choosing $g = \eta_k \chi_n$ in (6.27) and letting $k \to \infty$, we obtain that

$$-\int \nabla f \cdot \nabla \chi_n \, \mathrm{d}\mathfrak{m} = \int \chi_n \, \mathrm{d}(\mu + \nu) \quad \text{for every } n \in \mathbb{N}. \tag{6.28}$$

By applying (6.28) and recalling that the functions $\Delta \chi_n$ weakly* converge in $L^\infty(\mathfrak{m})$ to the null function, we see that

$$\int \chi_n \, \mathrm{d}(\mu + \nu) = -\int \nabla \chi_n \cdot \nabla f \, \mathrm{d}\mathfrak{m} = \int f \, \Delta \chi_n \, \mathrm{d}\mathfrak{m} \longrightarrow 0.$$

We thus deduce that

$$\nu(\mathrm{X}) = \lim_{n \to \infty} \int \chi_n \, \mathrm{d}\nu = -\lim_{n \to \infty} \int \chi_n \, \mathrm{d}\mu = -\mu(\mathrm{X}) < +\infty,$$

whence accordingly ν is a finite measure. In particular, one has that $\mu + \nu$ is a finite measure as well, so that (6.27) yields $f \in D(\boldsymbol{\Delta})$ and $\boldsymbol{\Delta} f = \mu + \nu \geq \mu$. □

Corollary 6.2.17 *Let* $(\mathrm{X}, \mathsf{d}, \mathfrak{m})$ *be a proper* $\mathsf{RCD}(K, \infty)$ *space. Fix* $f \in \mathrm{Test}^\infty(\mathrm{X})$. *Then it holds that* $|\nabla f|^2 \in D(\boldsymbol{\Delta})$ *and*

$$\boldsymbol{\Delta} \frac{|\nabla f|^2}{2} \geq \left(\nabla f \cdot \nabla \Delta f + K |\nabla f|^2\right)\mathfrak{m}. \tag{6.29}$$

Proof Denote by μ the right hand side of (6.29). We know from (6.4) that

$$-\int \nabla g \cdot \nabla \left(\frac{|\nabla f|^2}{2}\right) \mathrm{d}\mathfrak{m} = \int \Delta g \, \frac{|\nabla f|^2}{2} \, \mathrm{d}\mathfrak{m} \geq \int g \, \mathrm{d}\mu \quad \text{for every } g \in \mathrm{Test}^\infty_+(\mathrm{X}).$$

By regularisation via the mollified heat flow (cf. Proposition 5.2.18), we see that the previous inequality is verified for every $g \in \mathrm{LIP}_{bs}(\mathrm{X})^+$, so that Proposition 6.2.16 gives the statement. □

6.2.3 Presence of Many $W^{2,2}$-Functions

Given any $f_1, f_2 \in \mathrm{Test}^\infty(\mathrm{X})$, let us define

$$\Gamma_2(f_1, f_2) := \frac{1}{2}\Big[\boldsymbol{\Delta}(\nabla f_1 \cdot \nabla f_2) - \big(\nabla f_1 \cdot \nabla\Delta f_2 + \nabla f_2 \cdot \nabla\Delta f_1\big)\mathfrak{m}\Big]. \qquad (6.30)$$

Notice that $\Gamma_2(f_1, f_2)$ is a finite Radon measure on X and that Γ_2 is bilinear. Then the inequality (6.29) can be restated in the following compact form:

$$\Gamma_2(f, f) \geq K|\nabla f|^2\mathfrak{m} \quad \text{for every } f \in \mathrm{Test}^\infty(\mathrm{X}). \qquad (6.31)$$

Moreover, given any $f, g, h \in \mathrm{Test}^\infty(\mathrm{X})$ we define

$$[\mathrm{H}f](g, h) := \frac{1}{2}\big(\nabla(\nabla f\cdot\nabla g)\cdot\nabla h+\nabla(\nabla f\cdot\nabla h)\cdot\nabla g-\nabla f\cdot\nabla(\nabla g\cdot\nabla h)\big). \qquad (6.32)$$

Clearly $(f, g, h) \mapsto [\mathrm{H}f](g, h)$ is a trilinear map.

Given two non-negative Radon measures μ, ν on X, we define the Radon measure $\sqrt{\mu\nu}$ as

$$\sqrt{\mu\nu} := \sqrt{\frac{\mathrm{d}\mu}{\mathrm{d}\sigma}\frac{\mathrm{d}\nu}{\mathrm{d}\sigma}}\,\sigma \quad \text{for any Radon measure } \sigma \geq 0 \text{ with } \mu, \nu \ll \sigma. \qquad (6.33)$$

Its well-posedness stems from the fact that the function $(a, b) \mapsto \sqrt{ab}$ is 1-homogeneous.

Lemma 6.2.18 *Let μ_1, μ_2, μ_3 be (finite) Radon measures on* X. *Assume $\lambda^2\mu_1 + 2\lambda\mu_2 + \mu_3 \geq 0$ for every $\lambda \in \mathbb{R}$. Then $\mu_1, \mu_3 \geq 0$ and $\mu_2 \leq \sqrt{\mu_1\mu_3}$.*

Proof By choosing $\lambda = 0$ we see that $\mu_3 \geq 0$. Given any Borel set $E \subseteq \mathrm{X}$ and $\lambda > 0$, we have that $\mu_1(E) + 2\mu_2(E)/\lambda + \mu_3(E)/\lambda^2 \geq 0$, so that $\mu_1(E) \geq -\lim_{\lambda\to+\infty} 2\mu_2(E)/\lambda + \mu_3(E)/\lambda^2 = 0$, which shows that $\mu_1 \geq 0$. Now take any Radon measure $\nu \geq 0$ such that $\mu_1, \mu_2, \mu_3 \ll \nu$. Write $\mu_i = f_i\,\nu$ for $i = 1, 2, 3$. Then $\lambda^2 f_1 + 2\lambda f_2 + f_3 \geq 0$ holds ν-a.e., whence accordingly we have that the inequality $f_2 \leq \sqrt{f_1 f_3}$ holds ν-a.e. as well, concluding the proof. □

Lemma 6.2.19 *Let $n \in \mathbb{N}$ and let $\Phi : \mathbb{R}^n \to \mathbb{R}$ be a polynomial with no constant term. Let us fix $f_1, \ldots, f_n \in \mathrm{Test}^\infty(\mathrm{X})$, briefly $\boldsymbol{f} = (f_1, \ldots, f_n)$. Denote by Φ_i the partial derivative of Φ with respect to its ith-entry. Then $\Phi(\boldsymbol{f}) \in \mathrm{Test}^\infty(\mathrm{X})$ and*

$$\Gamma_2\big(\Phi(\boldsymbol{f}), \Phi(\boldsymbol{f})\big) = A + (B + C)\,\mathfrak{m}, \quad \big|\nabla\Phi(\boldsymbol{f})\big|^2 = D, \qquad (6.34)$$

where we set

$$
\begin{aligned}
A &:= \sum_{i,j=1}^{n} \Phi_i(\boldsymbol{f})\,\Phi_j(\boldsymbol{f})\,\Gamma_2(f_i, f_j),\\
B &:= 2 \sum_{i,j,k=1}^{n} \Phi_i(\boldsymbol{f})\,\Phi_{jk}(\boldsymbol{f})\,[\mathrm{H}f_i](f_j, f_k),\\
C &:= \sum_{i,j,k,h=1}^{n} \Phi_{ik}(\boldsymbol{f})\,\Phi_{jh}(\boldsymbol{f})\,\langle \nabla f_i, \nabla f_j\rangle\,\langle \nabla f_k, \nabla f_h\rangle,\\
D &:= \sum_{i,j=1}^{n} \Phi_i(\boldsymbol{f})\,\Phi_j(\boldsymbol{f})\,\langle \nabla f_i, \nabla f_j\rangle.
\end{aligned}
\tag{6.35}
$$

Proof The fact that $\Phi(\boldsymbol{f}) \in \mathrm{Test}^\infty(\mathrm{X})$ follows from Theorem 6.1.11. To prove that (6.34) is satisfied it suffices to manipulate the calculus rules described so far; for instance, it can be readily checked that $\mathrm{d}\Phi(\boldsymbol{f}) = \sum_{i=1}^n \Phi_i(\boldsymbol{f})\,\mathrm{d}f_i$ as a consequence of the Leibniz rule. □

Before stating and proving Theorem 6.2.21 below in its full generality, we illustrate the ideas by treating a simpler case (the following approach is due to Bakry):

Proposition 6.2.20 *Let M be a smooth Riemannian manifold with $\Delta\frac{|\nabla f|^2}{2} \geq \nabla f \cdot \nabla\Delta f$ for every $f \in C^\infty(M)$. Then $\Delta\frac{|\nabla f|^2}{2} \geq \nabla f \cdot \nabla\Delta f + |\mathrm{H}f|^2_{\mathsf{op}}$.*

Proof Let $\Phi(x_1, x_2) := \lambda x_1 + (x_2 - c)^2 - c^2$ for some $\lambda, c \in \mathbb{R}$. Then for arbitrary $h \in C^\infty(M)$ Lemma 6.2.19 yields

$$
\begin{aligned}
0 &\leq \Gamma_2\big(\lambda f + (h-c)^2, \lambda f + (h-c)^2\big)\\
&= \lambda^2\,\Gamma_2(f, f) + 4\lambda(h-c)\Gamma_2(f, h) + 4(h-c)^2\,\Gamma_2(h, h)\\
&\quad + 4\lambda\,\mathrm{H}f(\nabla h, \nabla h) + 8(h-c)\,\mathrm{H}h(\nabla h, \nabla h) + 4|\nabla h|^4.
\end{aligned}
$$

Since c is arbitrary, we can for every point $x \in M$ choose $c = h(x)$, thus getting that the inequality $\lambda^2\,\Gamma_2(f, f) + 4\lambda\,\mathrm{H}f(\nabla h, \nabla h) + 4|\nabla h|^4 \geq 0$ holds for all $\lambda \in \mathbb{R}$, whence accordingly one has $\big|\mathrm{H}f(\nabla h, \nabla h)\big| \leq \sqrt{\Gamma_2(f, f)}\,|\nabla h|^2$. Since $\mathrm{H}f$ is symmetric, for all $x \in M$ we have

$$
|\mathrm{H}f|_{\mathsf{op}}(x) = \sup\Big\{\big|\mathrm{H}f(\nabla h, \nabla h)\big| \;:\; h \in C^\infty(M),\ |\nabla h|(x) = 1\Big\} \leq \sqrt{\Gamma_2(f, f)(x)},
$$

getting the statement. □

We now state and prove the following fundamental result:

Theorem 6.2.21 (Key Lemma) *Let $f_i, g_i, h_j \in \mathrm{Test}^\infty(\mathrm{X})$ for $i = 1, \dots, n$ and $j = 1, \dots, m$. We define the Radon measure μ on X as*

$$\mu := \sum_{i,i'} g_i\, g_{i'} \big(\Gamma_2(f_i, f_{i'}) - K \langle \nabla f_i, \nabla f_{i'} \rangle\, \mathfrak{m}\big) + \sum_{i,i'} 2\, g_i\, [\mathrm{H} f_i](f_{i'}, g_{i'})\, \mathfrak{m}$$
$$+ \sum_{i,i'} \frac{\langle \nabla f_i, \nabla f_{i'} \rangle \langle \nabla g_i, \nabla g_{i'} \rangle + \langle \nabla f_i, \nabla g_{i'} \rangle \langle \nabla f_{i'}, \nabla g_i \rangle}{2}\, \mathfrak{m}. \tag{6.36}$$

Let us write $\mu = \rho\, \mathfrak{m} + \mu^s$, with $\mu^s \perp \mathfrak{m}$. Then $\mu^s \geq 0$ and

$$\left| \sum_{i,j} \langle \nabla f_i, \nabla h_j \rangle \langle \nabla g_i, \nabla h_j \rangle + g_i\, [\mathrm{H} f_i](h_j, h_j) \right|^2 \leq \rho \sum_{j,j'} \big| \langle \nabla h_j, \nabla h_{j'} \rangle \big|^2 \quad \mathfrak{m}\text{-a.e.}. \tag{6.37}$$

Proof Given any $\lambda, a_i, b_i, c_j \in \mathbb{R}$, let us define

$$\Phi(x_1, \dots, x_n, y_1, \dots, y_n, z_1, \dots, z_m) := \sum_{i=1}^n (\lambda y_i x_i + a_i x_i - b_i y_i) + \sum_{j=1}^m \big((z_j - c_j)^2 - c_j^2\big).$$

Simple computations show that the only non-vanishing derivatives are

$$\partial_{x_i} \Phi = \lambda y_i + a_i, \quad \partial_{y_i} \Phi = \lambda x_i - b_i, \quad \partial_{x_i y_i} \Phi = \lambda, \quad \partial_{z_j} \Phi = 2(z_j - c_j), \quad \partial_{z_j z_j} \Phi = 2.$$

Let $f := (f_1, \dots, f_n, g_1, \dots, g_n, h_1, \dots, h_m) \in \big[\mathrm{Test}^\infty(\mathrm{X})\big]^{2n+m}$, so that $\Phi(f) \in \mathrm{Test}^\infty(\mathrm{X})$ by Lemma 6.2.19. Note that $\Gamma_2\big(\Phi(f), \Phi(f)\big) \geq K \big|\nabla \Phi(f)\big|^2 \mathfrak{m}$ by (6.31). Moreover, in this case the objects A, B, C, D defined in Lemma 6.2.19 read as

$$A(\lambda, a, b, c) = \sum_{i,i'} (\lambda g_i + a_i)(\lambda g_{i'} + a_{i'}) \Gamma_2(f_i, f_{i'}) + o.t.,$$

$$B(\lambda, a, b, c) = 4 \sum_{i,i'} (\lambda g_i + a_i) \lambda [\mathrm{H} f_i](f_{i'}, g_{i'}) + 4 \sum_{i,j} (\lambda g_i + a_i) [\mathrm{H} f_i](h_j, h_j) + o.t.,$$

$$C(\lambda, a, b, c) = 2 \sum_{i,i'} \lambda^2 \big(\langle \nabla f_i, \nabla f_{i'} \rangle \langle \nabla g_i, \nabla g_{i'} \rangle + \langle \nabla f_i, \nabla g_{i'} \rangle \langle \nabla g_i, \nabla f_{i'} \rangle\big)$$
$$+ 8\lambda \sum_{i,j} \langle \nabla f_i, \nabla h_j \rangle \langle \nabla g_i, \nabla h_j \rangle + 4 \sum_{j,j'} \big| \langle \nabla h_j, \nabla h_{j'} \rangle \big|^2 + o.t.,$$

$$D(\lambda, a, b, c) = \sum_{i,i'} (\lambda g_i + a_i)(\lambda g_{i'} + a_{i'}) \langle \nabla f_i, \nabla f_{i'} \rangle + o.t.,$$

where each o.t.='other terms' contains either a factor $\lambda f_i - b_i$ or a factor $h_j - c_j$. Therefore Lemma 6.2.19 grants that for any $\lambda \in \mathbb{R}$, $a, b \in \mathbb{R}^n$ and $c \in \mathbb{R}^m$ we have

$$A(\lambda, a, b, c) + \big(B(\lambda, a, b, c) + C(\lambda, a, b, c)\big)\,\mathfrak{m} \geq K D(\lambda, a, b, c)\,\mathfrak{m}. \tag{6.38}$$

Now choose a Radon measure $\sigma \geq 0$ such that $\mathfrak{m}, \Gamma_2(f_i, f_{i'}) \ll \sigma$ for all i, i'. Write $\mathfrak{m} = \eta\,\sigma$. Then property (6.38) gives the σ-a.e. inequality $\frac{\mathrm{d}A}{\mathrm{d}\sigma} + (B + C)\eta \geq K D\,\eta$. Now let us choose a sequence $m \mapsto (E^m_\ell)_\ell$ of Borel partitions of X and uniformly bounded $a^{m\ell}_i, b^{m\ell}_i, c^{m\ell}_j \in \mathbb{R}$ with

$$\sum_{\ell \in \mathbb{N}} a^{m\ell}_i \chi_{E^m_\ell} \xrightarrow{m} \lambda g_i, \quad \sum_{\ell \in \mathbb{N}} b^{m\ell}_i \chi_{E^m_\ell} \xrightarrow{m} \lambda f_i, \quad \sum_{\ell \in \mathbb{N}} c^{m\ell}_j \chi_{E^m_\ell} \xrightarrow{m} h_j$$

with respect to the strong topology of $L^\infty(\sigma)$, for every i, j. Therefore we deduce that

$$\begin{aligned} &\sum_{\ell \in \mathbb{N}} \chi_{E^m_\ell} \left[\frac{\mathrm{d}A(\lambda, a^{m\ell}, b^{m\ell}, c^{m\ell})}{\mathrm{d}\sigma} + \big(B(\lambda, a^{m\ell}, b^{m\ell}, c^{m\ell}) + C(\lambda, a^{m\ell}, b^{m\ell}, c^{m\ell})\big)\eta \right] \\ &\geq K \sum_{\ell \in \mathbb{N}} \chi_{E^m_\ell} D(\lambda, a^{m\ell}, b^{m\ell}, c^{m\ell})\eta. \end{aligned} \tag{6.39}$$

Since both sides of (6.39) are converging in $L^1(\sigma)$, we conclude that $\lambda^2\mu + 2\lambda F + G \geq 0$ for all $\lambda \in \mathbb{R}$, where μ is defined as in (6.36), while

$$F := \sum_{i,j} \langle \nabla f_i, \nabla h_j \rangle \langle \nabla g_i, \nabla h_j \rangle\,\mathfrak{m} + g_i\,[\mathrm{H}f_i](h_j, h_j)\,\mathfrak{m}, \quad G := \sum_{j,j'} \big|\langle \nabla h_j, \nabla h_{j'} \rangle\big|^2 \mathfrak{m}.$$

Hence Lemma 6.2.18 grants that $\mu \geq 0$, so in particular $\mu^s \geq 0$, and that $F \leq \sqrt{(\rho\mathfrak{m})G}$, which is nothing but (6.37). This proves the statement. □

Theorem 6.2.22 *It holds that* $\mathrm{Test}^\infty(\mathrm{X}) \subseteq W^{2,2}(\mathrm{X})$. *Moreover, if we take* $f \in \mathrm{Test}^\infty(\mathrm{X})$ *and we write* $\Gamma_2(f, f) = \gamma_2\,\mathfrak{m} + \Gamma^s_2$ *with* $\Gamma^s_2 \perp \mathfrak{m}$, *then* $\Gamma^s_2 \geq 0$ *and for all* $g_1, g_2 \in \mathrm{Test}^\infty(\mathrm{X})$ *we have that*

$$\left.\begin{aligned} |\mathrm{H}f|^2_{\mathsf{HS}} &\leq \gamma_2 - K\,|\nabla f|^2, \\ \mathrm{H}f(\nabla g_1, \nabla g_2) &= [\mathrm{H}f](g_1, g_2) \end{aligned}\right\} \quad \textit{hold } \mathfrak{m}\textit{-a.e. in } \mathrm{X}. \tag{6.40}$$

Proof Apply Theorem 6.2.21 with $n = 1$. We thus get the $\mathfrak{m}$-a.e. inequality

$$\begin{aligned} &\left| \sum_{j=1}^m \langle \nabla f, \nabla h_j \rangle \langle \nabla g, \nabla h_j \rangle + g\,[\mathrm{H}f](h_j, h_j) \right|^2 \\ &\leq \Big(g^2(\gamma_2 - K\,|\nabla f|^2) + 2\,g\,[\mathrm{H}f](f, g) \Big) \sum_{j,j'=1}^m \big|\langle \nabla h_j, \nabla h_{j'} \rangle\big|^2 \end{aligned} \tag{6.41}$$

for any choice of $f, g, h_1, \ldots, h_m \in \mathrm{Test}^\infty(\mathrm{X})$. Define μ as in (6.36) for this choice of test functions; since μ is the sum of $g^2\,\Gamma_2(f,f)$ and a measure that is absolutely continuous with respect to $\mathfrak{m}$, we see that $\mu^s = g^2\,\Gamma_2^s$, thus accordingly the fact that $\mu^s \geq 0$ grants that $\Gamma_2^s \geq 0$ as well. Moreover, notice that both sides of (6.41) are $W^{1,2}(\mathrm{X})$-continuous with respect to the entry g with values in $L^1(\mathfrak{m})$, so the inequality (6.41) is actually verified for any $g \in W^{1,2}(\mathrm{X})$. Then by choosing suitable g's, namely identically equal to 1 on an arbitrarily big ball, we deduce that

$$\begin{aligned}\left|\sum_{j=1}^m g_j\,[\mathrm{H}f](h_j,h_j)\right|^2 &\leq \left(\gamma_2 - K\,|\nabla f|^2\right)\sum_{j,j'=1}^m g_j\,g_{j'}\,\langle\nabla h_j,\nabla h_{j'}\rangle^2\\ &= \left(\gamma_2 - K\,|\nabla f|^2\right)\left\langle\sum_{j=1}^m g_j\,\nabla h_j\otimes\nabla h_j,\ \sum_{j'=1}^m g_{j'}\,\nabla h_{j'}\otimes\nabla h_{j'}\right\rangle\\ &= \left(\gamma_2 - K\,|\nabla f|^2\right)\left|\sum_{j=1}^m g_j\,\nabla h_j\otimes\nabla h_j\right|^2\end{aligned} \tag{6.42}$$

for all $f, g_1, \ldots, g_m, h_1, \ldots, h_m \in \mathrm{Test}^\infty(\mathrm{X})$. Now note that for $f, g, h, h' \in \mathrm{Test}^\infty(\mathrm{X})$ one has

$$\begin{aligned}2\,[\mathrm{H}f](h,h') &= [\mathrm{H}f](h+h',h+h') - [\mathrm{H}f](h,h) - [\mathrm{H}f](h',h'),\\ g\,(\nabla h\otimes\nabla h' + \nabla h'\otimes\nabla h) &= g\,\big(\nabla(h+h')\otimes\nabla(h+h') - \nabla h\otimes\nabla h - \nabla h'\otimes\nabla h'\big).\end{aligned}$$

By combining these two identities with (6.42) and the $\mathfrak{m}$-a.e. inequality $\left|\frac{A+A^t}{2}\right|_{\mathsf{HS}}^2 \leq |A|_{\mathsf{HS}}^2$, which is trivially verified for any $A \in L^2(T^{\otimes 2}\mathrm{X})$, we obtain that

$$\left|\sum_{j=1}^m g_j\,[\mathrm{H}f](h_j,h'_j)\right| \leq \sqrt{\gamma_2 - K\,|\nabla f|^2}\,\left|\sum_{j=1}^m g_j\,\nabla h_j\otimes\nabla h'_j\right| \tag{6.43}$$

holds $\mathfrak{m}$-a.e. for any $f, g_j, h_j, h'_j \in \mathrm{Test}^\infty(\mathrm{X})$. Define $\mathcal{V} \subseteq L^2(T^{\otimes 2}\mathrm{X})$ as the linear span of the tensors of the form $g\,\nabla h\otimes\nabla h'$, with $g, h, h' \in \mathrm{Test}^\infty(\mathrm{X})$. Then the operator $L : \mathcal{V} \to L^1(\mathfrak{m})$, which is given by

$$L\left(\sum_{j=1}^m g_j\,\nabla h_j\otimes\nabla h'_j\right) := \sum_{j=1}^m g_j\,[\mathrm{H}f](h_j,h'_j) \quad \text{for every } \sum_{j=1}^m g_j\,\nabla h_j\otimes\nabla h'_j \in \mathcal{V},$$

is well-defined, linear and continuous by (6.43). Since $\mathcal{V}$ is dense in $L^2(T^{\otimes 2}\mathrm{X})$ by Lemma 6.2.5, there exists a unique linear and continuous extension of L to the whole $L^2(T^{\otimes 2}\mathrm{X})$. Such extension is $L^\infty(\mathfrak{m})$-linear by construction, whence it can be viewed as an element B of the space $L^2\big((T^*)^{\otimes 2}\mathrm{X}\big)$. Notice that (6.43) gives

$\left|L(A)\right| \leq \sqrt{\gamma_2 - K\,|\nabla f|^2}\,|A|_{\mathsf{HS}}$ for all $A \in \mathcal{V}$, so that $|L|_{\mathsf{HS}} \leq \sqrt{\gamma_2 - K\,|\nabla f|^2}$ and accordingly $|B|_{\mathsf{HS}} \leq \sqrt{\gamma_2 - K\,|\nabla f|^2}$ as well. Finally, for any $g, h \in \mathrm{Test}^\infty(\mathrm{X})$ we have

$$\begin{aligned} 2\int g\,B(\nabla h \otimes \nabla h)\,\mathrm{d}\mathfrak{m} &= 2\int L\big(g\,\nabla h \otimes \nabla h\big)\,\mathrm{d}\mathfrak{m} \\ &= \int g\,\big(2\,\nabla(\nabla f \cdot \nabla h) \cdot \nabla h - \nabla f \cdot \nabla|\nabla h|^2\big)\,\mathrm{d}\mathfrak{m} \\ &= -\int \nabla f \cdot \nabla h\,\mathrm{div}(\nabla g \cdot \nabla h) + \nabla f \cdot \nabla|\nabla h|^2\,\mathrm{d}\mathfrak{m}. \end{aligned}$$

Therefore $f \in W^{2,2}(\mathrm{X})$ and (6.40) can be easily checked to hold true; the first line of (6.40) is a consequence of (6.43), while the second one follows from the very definition of the involved objects. □

Corollary 6.2.23 *It holds that $D(\Delta) \subseteq W^{2,2}(\mathrm{X})$. Moreover, we have that*

$$\int |\mathrm{H}f|^2_{\mathsf{HS}}\,\mathrm{d}\mathfrak{m} \leq \int |\Delta f|^2 - K\,|\nabla f|^2\,\mathrm{d}\mathfrak{m} \quad \textit{for every } f \in D(\Delta). \tag{6.44}$$

Proof Formula (6.44) holds for all $f \in \mathrm{Test}^\infty(\mathrm{X})$ as a consequence of Theorem 6.2.21. The general case $f \in D(\Delta)$ follows by approximating f with a sequence $(f_n)_n \subseteq \mathrm{Test}^\infty(\mathrm{X})$. □

Let us define the space $H^{2,2}(\mathrm{X})$ as the $W^{2,2}(\mathrm{X})$-closure of $\mathrm{Test}^\infty(\mathrm{X})$. An important open problem is the following: is it true that $H^{2,2}(\mathrm{X}) = W^{2,2}(\mathrm{X})$?

6.2.4 Calculus Rules

Let us consider the functional

$$L^2(\mathfrak{m}) \ni f \longmapsto \begin{cases} \int |\mathrm{H}f|^2_{\mathsf{HS}}\,\mathrm{d}\mathfrak{m} & \text{if } f \in W^{2,2}(\mathrm{X}), \\ +\infty & \text{otherwise.} \end{cases} \tag{6.45}$$

An open problem is the following: is such functional lower semicontinuous?

It is known that such functional is convex and lower semicontinuous when its domain is replaced by $W^{1,2}(\mathrm{X})$.

Proposition 6.2.24 (Leibniz Rule for H) *Let $f_1, f_2 \in W^{2,2}(\mathrm{X}) \cap \mathrm{LIP}(\mathrm{X}) \cap L^\infty(\mathfrak{m})$ be given. Then $f_1 f_2 \in W^{2,2}(\mathrm{X})$ and*

$$\mathrm{H}(f_1 f_2) = f_1\,\mathrm{H}f_2 + f_2\,\mathrm{H}f_1 + \mathrm{d}f_1 \otimes \mathrm{d}f_2 + \mathrm{d}f_2 \otimes \mathrm{d}f_1. \tag{6.46}$$

Proof By polarisation, it holds that an element $A \in L^2\big((T^*)^{\otimes 2}\mathrm{X}\big)$ coincides with $\mathrm{H}(f_1 f_2)$ if and only if $A^{\mathrm{t}} = A$ and

$$-\int h\, A(\nabla g, \nabla g)\,\mathrm{d}\mathfrak{m} = \int \nabla(f_1 f_2)\cdot\nabla g\,\mathrm{div}(h\nabla g) + h\,\nabla(f_1 f_2)\cdot\nabla\frac{|\nabla g|^2}{2}\,\mathrm{d}\mathfrak{m} \tag{6.47}$$

holds for all $g, h \in \mathrm{Test}^\infty(\mathrm{X})$. By using the Leibniz rule for gradients, we see that the right hand side of (6.47) can be rewritten as

$$\int f_1\,\nabla f_2\cdot\nabla g\,\mathrm{div}(h\nabla g) + f_2\,\nabla f_1\cdot\nabla g\,\mathrm{div}(h\nabla g) + hf_1\,\nabla f_2\cdot\nabla\frac{|\nabla g|^2}{2} + hf_2\,\nabla f_1\cdot\nabla\frac{|\nabla g|^2}{2}\,\mathrm{d}\mathfrak{m}. \tag{6.48}$$

Moreover, since $f_1, f_2 \in W^{2,2}(\mathrm{X}) \cap \mathrm{LIP}(\mathrm{X}) \cap L^\infty(\mathfrak{m})$, we also have that

$$\int hf_2\,\mathrm{H}f_1(\nabla g, \nabla g)\,\mathrm{d}\mathfrak{m} = -\int \nabla f_1\cdot\nabla g\,\mathrm{div}(hf_2\nabla g) + hf_2\,\nabla f_1\cdot\nabla\frac{|\nabla g|^2}{2}\,\mathrm{d}\mathfrak{m},$$
$$\int hf_1\,\mathrm{H}f_2(\nabla g, \nabla g)\,\mathrm{d}\mathfrak{m} = -\int \nabla f_2\cdot\nabla g\,\mathrm{div}(hf_1\nabla g) + hf_1\,\nabla f_2\cdot\nabla\frac{|\nabla g|^2}{2}\,\mathrm{d}\mathfrak{m}. \tag{6.49}$$

Therefore (6.48) and (6.49) yield (6.47) for $A := f_1\,\mathrm{H}f_2 + f_2\,\mathrm{H}f_1 + \mathrm{d}f_1 \otimes \mathrm{d}f_2 + \mathrm{d}f_2 \otimes \mathrm{d}f_1$. Since such A defines a symmetric tensor, the statement is achieved. □

Proposition 6.2.25 (Chain Rule for H) *Let $f \in W^{2,2}(\mathrm{X}) \cap \mathrm{LIP}(\mathrm{X})$. Suppose $\varphi \in C^{1,1}(\mathbb{R})$ has bounded derivative and satisfies $\varphi(0) = 0$ if $\mathfrak{m}(\mathrm{X}) = \infty$. Then $\varphi \circ f \in W^{2,2}(\mathrm{X})$ and*

$$\mathrm{H}(\varphi\circ f) = \varphi''\circ f\,\mathrm{d}f\otimes\mathrm{d}f + \varphi'\circ f\,\mathrm{H}f. \tag{6.50}$$

Proof The statement can be achieved by using the chain rule for gradients, similarly to how the Leibniz rule for gradients gives (6.46). □

Lemma 6.2.26 *Let $(\mathrm{X}, \mathsf{d}, \mathfrak{m})$ be infinitesimally Hilbertian. Let $f \in L^2(\mathfrak{m})$. Then $f \in W^{1,2}(\mathrm{X})$ if and only if there exists $\omega \in L^2(T^*\mathrm{X})$ such that*

$$\int f\,\mathrm{div}(X)\,\mathrm{d}\mathfrak{m} = -\int \omega(X)\,\mathrm{d}\mathfrak{m} \quad \textit{for every } X \in D(\mathrm{div}). \tag{6.51}$$

In this case, it holds that $\omega = \mathrm{d}f$. Moreover, if $(\mathrm{X}, \mathsf{d}, \mathfrak{m})$ is an $\mathsf{RCD}(K, \infty)$ space for some constant $K \in \mathbb{R}$, then it suffices to check this property for $X = \nabla g$ with $g \in \mathrm{Test}^\infty(\mathrm{X})$.

Proof Sufficiency follows from the definition of divergence. To prove necessity, let $X := \nabla \mathsf{h}_t f$ for $t > 0$. Notice that $\mathrm{div}(X) = \Delta \mathsf{h}_t f$. Moreover, since the Cheeger energy decreases along the heat flow, it holds that

$$\begin{aligned}\int |\nabla \mathsf{h}_{t/2} f|^2 \, \mathrm{d}\mathfrak{m} &= -\int f \, \Delta \mathsf{h}_t f \, \mathrm{d}\mathfrak{m} \\ &= \int \omega(\nabla \mathsf{h}_t f) \, \mathrm{d}\mathfrak{m} \leq \|\omega\|_{L^2(T^*\mathrm{X})} \bigg(\int |\nabla \mathsf{h}_{t/2} f|^2 \, \mathrm{d}\mathfrak{m} \bigg)^{1/2},\end{aligned}$$

whence accordingly $\int |\nabla \mathsf{h}_{t/2} f|^2 \, \mathrm{d}\mathfrak{m} \leq \int |\omega|^2 \, \mathrm{d}\mathfrak{m}$. Since the Cheeger energy is lower semicontinuous, we conclude that $f \in W^{1,2}(\mathrm{X})$ and $\omega = \mathrm{d}f$. Finally, the last statement follows from a density argument (noticing that in the argument just given we only used X gradient). □

Proposition 6.2.27 *Let* $(\mathrm{X}, \mathsf{d}, \mathfrak{m})$ *be an* $\mathsf{RCD}(K, \infty)$ *space. Let* $f_1, f_2 \in H^{2,2}(\mathrm{X}) \cap \mathrm{LIP}(\mathrm{X})$ *be given. Then* $\langle \nabla f_1, \nabla f_2 \rangle \in W^{1,2}(\mathrm{X})$ *and*

$$\mathrm{d}\langle \nabla f_1, \nabla f_2 \rangle = \mathrm{H} f_1(\nabla f_2, \cdot) + \mathrm{H} f_2(\nabla f_1, \cdot). \tag{6.52}$$

Proof By polarisation and by density of test functions in $H^{2,2}(\mathrm{X})$, it is sufficient to show that one has $|\nabla f|^2 \in W^{1,2}(\mathrm{X})$ and $\mathrm{d}|\nabla f|^2 = 2\,\mathrm{H} f(\nabla f, \cdot)$ for every $f \in \mathrm{Test}^\infty(\mathrm{X})$. Given that we have $2 \int h \, \mathrm{H} f(\nabla f, \nabla g) \, \mathrm{d}\mathfrak{m} = -\int |\nabla f|^2 \, \mathrm{div}(h \nabla g) \, \mathrm{d}\mathfrak{m}$ for all $g, h \in \mathrm{Test}^\infty(\mathrm{X})$, we know that

$$\int |\nabla f|^2 \, \mathrm{div}(\nabla g) \, \mathrm{d}\mathfrak{m} = -2 \int \mathrm{H} f(\nabla f, \nabla g) \, \mathrm{d}\mathfrak{m} \quad \text{for every } g \in \mathrm{Test}^\infty(\mathrm{X}),$$

whence Lemma 6.2.26 yields $|\nabla f|^2 \in W^{1,2}(\mathrm{X})$ and $\mathrm{d}|\nabla f|^2 = 2\,\mathrm{H} f(\nabla f, \cdot)$, as required. □

Corollary 6.2.28 (Locality of H) *Let* $f, g \in H^{2,2}(\mathrm{X}) \cap \mathrm{LIP}(\mathrm{X})$ *be given. Then*

$$\mathrm{H} f = \mathrm{H} g \quad \textit{holds } \mathfrak{m}\textit{-a.e. on } \{f = g\}. \tag{6.53}$$

Proof By linearity of H, it suffices to prove that $\mathrm{H} f = 0$ holds $\mathfrak{m}$-a.e. on the set $\{f = 0\}$. Given any $g \in \mathrm{Test}^\infty(\mathrm{X})$, we know from Proposition 6.2.27 that $\langle \nabla f, \nabla g \rangle \in W^{1,2}(\mathrm{X})$ and

$$\mathrm{H} f(\nabla g, \cdot) = \mathrm{d}\langle \nabla f, \nabla g \rangle - \mathrm{H} g(\nabla f, \cdot). \tag{6.54}$$

Since $\nabla f = 0$ holds $\mathfrak{m}$-a.e. on $\{f = 0\}$, we see that the right hand side of (6.54) vanishes $\mathfrak{m}$-a.e. on $\{f = 0\}$. Hence $\mathrm{H} f(\nabla g, \cdot) = 0$ $\mathfrak{m}$-a.e. on $\{f = 0\}$ for all $g \in \mathrm{Test}^\infty(\mathrm{X})$, which implies that $\mathrm{H} f = 0$ $\mathfrak{m}$-a.e. on $\{f = 0\}$, proving the statement. □

Given a Borel subset E of X, we define its *essential interior* as

$$\operatorname{ess\,int}(E) := \bigcup \{\Omega : \Omega \subseteq \mathrm{X} \text{ open}, \mathfrak{m}(\Omega \setminus E) = 0\}. \tag{6.55}$$

By using Lemma 6.2.14, we can prove that functions in $W^{2,2}(\mathrm{X})$ (but not necessarily in $H^{2,2}(\mathrm{X})$) satisfy a weaker form of locality:

Proposition 6.2.29 *Let $f \in W^{2,2}(\mathrm{X})$. Then $\mathrm{H}f = 0$ holds $\mathfrak{m}$-a.e. on ess int$\big(\{f = 0\}\big)$.*

Proof Let us denote by Ω the essential interior of $\{f = 0\}$. Given any $g_1, g_2, h \in \mathrm{Test}^\infty(\mathrm{X})$ with $\mathrm{spt}(h) \subseteq \Omega$, we have that $\int h\, \mathrm{H}f(\nabla g_1, \nabla g_2)\, \mathrm{d}\mathfrak{m}$ equals

$$-\int \nabla f \cdot \nabla g_1 \operatorname{div}(h \nabla g_2) + \nabla f \cdot \nabla g_2 \operatorname{div}(h \nabla g_1) + h\, \nabla f \cdot \nabla(\nabla g_1 \cdot \nabla g_2)\, \mathrm{d}\mathfrak{m}, \tag{6.56}$$

which vanishes as a consequence of the fact that $f = 0$ $\mathfrak{m}$-a.e. on Ω and $h = 0$ on $\mathrm{X} \setminus \Omega$. We thus deduce that $\int h\, \mathrm{H}f(\nabla g_1, \nabla g_2)\, \mathrm{d}\mathfrak{m} = 0$, which grants that $\mathrm{H}f = 0$ holds $\mathfrak{m}$-a.e. on Ω, as required. □

6.3 Covariant Derivative

On a Riemannian manifold M, we have for any vector field X and any $f, g \in C^\infty(M)$ that

$$\langle \nabla_{\nabla f} X, \nabla g \rangle = \big\langle \nabla \langle X, \nabla g \rangle, \nabla f \big\rangle - \mathrm{H}g(X, \nabla f). \tag{6.57}$$

Such formula motivates the following definition of covariant derivative on RCD spaces.

Definition 6.3.1 (Covariant Derivative) Let $(\mathrm{X}, \mathsf{d}, \mathfrak{m})$ be an $\mathsf{RCD}(K, \infty)$ space. Then a vector field $X \in L^2(T\mathrm{X})$ belongs to $W^{1,2}_{\mathrm{C}}(T\mathrm{X})$ provided there exists $T \in L^2(T^{\otimes 2}\mathrm{X})$ such that

$$\int h\, T : (\nabla f \otimes \nabla g)\, \mathrm{d}\mathfrak{m} = -\int \langle X, \nabla g \rangle \operatorname{div}(h \nabla f) + h\, \mathrm{H}g(X, \nabla f)\, \mathrm{d}\mathfrak{m} \tag{6.58}$$

holds for every $f, g, h \in \mathrm{Test}^\infty(\mathrm{X})$. The element T, which is uniquely determined by (6.58), is called *covariant derivative* of X and denoted by ∇X. The Sobolev norm of X is defined as

$$\|X\|_{W^{1,2}_{\mathrm{C}}(T\mathrm{X})} := \Big(\|X\|^2_{L^2(T\mathrm{X})} + \|\nabla X\|^2_{L^2(T^{\otimes 2}\mathrm{X})} \Big)^{1/2}. \tag{6.59}$$

It turns out that the operator $\nabla : W^{1,2}_{\mathrm{C}}(T\mathrm{X}) \to L^2(T^{\otimes 2}\mathrm{X})$ is linear.

In the sequel, we shall denote by $\sharp : L^2((T^*)^{\otimes 2}X) \to L^2(T^{\otimes 2}X)$ the Riesz isomorphism.

Theorem 6.3.2 *The following hold:*

i) $W_C^{1,2}(TX)$ *is a separable Hilbert space.*
ii) *The unbounded operator* $\nabla : L^2(TX) \to L^2(T^{\otimes 2}X)$ *is closed.*
iii) *If* $f \in H^{2,2}(X) \cap \mathrm{LIP}(X)$, *then* $\nabla f \in W_C^{1,2}(TX)$ *and* $\nabla(\nabla f) = (\mathrm{H}f)^\sharp$.

Proof The proof goes as follows:

i) Separability follows from the following facts: $X \mapsto (X, \nabla X)$ is an isometry from $W_C^{1,2}(TX)$ to $L^2(TX) \times L^2(T^{\otimes 2}X)$ and the latter space is separable. Moreover, it directly stems from the construction that the norm $\|\cdot\|_{W_C^{1,2}(TX)}$ satisfies the parallelogram identity. Finally, the completeness of $W_C^{1,2}(TX)$ is an immediate consequence of ii).
ii) Let $(X_n)_n \subseteq W_C^{1,2}(TX)$ satisfy $X_n \to X$ in $L^2(TX)$ and $\nabla X_n \to T$ in $L^2(T^{\otimes 2}X)$. Therefore by writing equation (6.58) for X_n and letting $n \to \infty$, we conclude that $X \in W_C^{1,2}(TX)$ and that $\nabla X = T$. This proves that ∇ is a closed unbounded operator.
iii) This can be readily checked by direct computations, by using Proposition 6.2.27. □

Proposition 6.3.3 (Leibniz Rule) *Let* $X \in W_C^{1,2}(TX) \cap L^\infty(TX)$ *and* $f \in W^{1,2}(X) \cap L^\infty(\mathfrak{m})$. *Then* $fX \in W_C^{1,2}(TX)$ *and* $\nabla(fX) = \nabla f \otimes X + f\nabla X$.

Proof Direct computation. □

We define the class of *test vector fields* as

$$\mathrm{TestV}(X) := \left\{ \sum_{i=1}^n g_i \nabla f_i \ : \ f_i, g_i \in \mathrm{Test}^\infty(X) \right\}. \tag{6.60}$$

Then we can formulate an important consequence of Proposition 6.3.3 in the following way:

Corollary 6.3.4 *It holds that* $\mathrm{TestV}(X) \subseteq W_C^{1,2}(TX)$. *Given any* $X = \sum_{i=1}^n g_i \nabla f_i$, *we have*

$$\nabla X = \sum_{i=1}^n \nabla g_i \otimes \nabla f_i + g_i\, (\mathrm{H}f_i)^\sharp. \tag{6.61}$$

Definition 6.3.5 We define the space $H_C^{1,2}(TX)$ as the $W_C^{1,2}(TX)$-closure of $\mathrm{TestV}(X)$.

Given any $X \in W_C^{1,2}(T\mathrm{X})$ and $Z \in L^0(T\mathrm{X})$, we define the vector field $\nabla_Z X \in L^0(T\mathrm{X})$ as the unique element such that

$$\langle \nabla_Z X, Y \rangle = \nabla X(Z, Y) \quad \text{for every } Y \in L^0(T\mathrm{X}). \tag{6.62}$$

Observe that $\nabla_Z X \in L^2(T\mathrm{X})$ whenever $Z \in L^\infty(T\mathrm{X})$.

Proposition 6.3.6 (Compatibility with the Metric) *Let $X, Y \in H_C^{1,2}(T\mathrm{X}) \cap L^\infty(T\mathrm{X})$ be given. Then $\langle X, Y \rangle \in W^{1,2}(\mathrm{X})$ and*

$$\mathrm{d}\langle X, Y \rangle(Z) = \langle \nabla_Z X, Y \rangle + \langle X, \nabla_Z Y \rangle \quad \textit{for every } Z \in L^0(T\mathrm{X}). \tag{6.63}$$

Proof First of all, the statement can be obtained for $X = g\nabla f$ and $Y = \tilde{g}\nabla\tilde{f}$ by direct computation. By linearity we get it for $X, Y \in \mathrm{TestV}(\mathrm{X})$. Then the general case follows by approximation. □

Given any $X, Y \in H_C^{1,2}(T\mathrm{X}) \cap L^\infty(T\mathrm{X})$ and $f \in W^{1,2}(\mathrm{X})$, we define

$$\begin{aligned} X(f) &:= \nabla f \cdot X = \mathrm{d}f(X), \\ [X, Y] &:= \nabla_X Y - \nabla_Y X. \end{aligned} \tag{6.64}$$

We call $[X, Y]$ the *commutator*, or *Lie brackets*, between X and Y.

Proposition 6.3.7 (Torsion-Free Identity) *Let $X, Y \in H_C^{1,2}(T\mathrm{X}) \cap L^\infty(T\mathrm{X})$. Then*

$$X\big(Y(f)\big) - Y\big(X(f)\big) = [X, Y](f) \quad \textit{for every } f \in H^{2,2}(\mathrm{X}) \cap \mathrm{LIP}(\mathrm{X}). \tag{6.65}$$

Proof Observe that

$$\begin{aligned} \nabla(\nabla f \cdot Y) \cdot X &= \nabla_X(\nabla f) \cdot Y + \nabla f \cdot \nabla_X Y = \mathrm{H}f(X, Y) + \nabla f \cdot \nabla_X Y, \\ \nabla(\nabla f \cdot X) \cdot Y &= \nabla_Y(\nabla f) \cdot X + \nabla f \cdot \nabla_Y X = \mathrm{H}f(Y, X) + \nabla f \cdot \nabla_Y X. \end{aligned} \tag{6.66}$$

Since $\mathrm{H}f$ is symmetric, by subtracting the second equation of (6.66) from the first one we obtain precisely (6.65). □

Remark 6.3.8 Since $\{\mathrm{d}f : f \in H^{2,2}(\mathrm{X}) \cap \mathrm{LIP}(\mathrm{X})\}$ generates the module $L^2(T^*\mathrm{X})$, we deduce that $[X, Y]$ is the unique element satisfying (6.65). ■

6.4 Exterior Derivative

6.4.1 *Sobolev Differential Forms*

We now want to introduce the notion of exterior differential on RCD spaces.

Given a Riemannian manifold M and a smooth k-form ω, it is well-known that $\mathrm{d}\omega$ is given by the following formula: given $X_0, \ldots, X_k$ smooth vector fields on M, one has

$$\mathrm{d}\omega(X_0, \ldots, X_k) = \sum_{i=0}^{k}(-1)^i X_i\big(\omega(\ldots, \hat{X}_i, \ldots)\big) + \sum_{i<j}(-1)^{i+j}\omega\big([X_i, X_j], \ldots, \hat{X}_i, \ldots, \hat{X}_j, \ldots\big). \tag{6.67}$$

Such formula actually defines a $(k+1)$-form, because it is alternating, functorial and linear in each entry.

Definition 6.4.1 Let $(\mathrm{X}, \mathsf{d}, \mathfrak{m})$ be an $\mathsf{RCD}(K, \infty)$ space. Then we denote the kth-exterior power of the cotangent module $L^0(T^*\mathrm{X})$ by

$$L^0(\Lambda^k T^*\mathrm{X}) := \Lambda^k L^0(T^*\mathrm{X}), \tag{6.68}$$

while we denote by $L^2(\Lambda^k T^*\mathrm{X})$ the subspacè of $L^0(\Lambda^k T^*\mathrm{X})$ consisting of those elements having pointwise norm in $L^2(\mathfrak{m})$.

Then formula (6.67) suggests the following definition:

Definition 6.4.2 (Exterior Derivative) Let $(\mathrm{X}, \mathsf{d}, \mathfrak{m})$ be an $\mathsf{RCD}(K, \infty)$ space and $k \in \mathbb{N}$. Then we say that a k-form $\omega \in L^2(\Lambda^k T^*\mathrm{X})$ belongs to $W^{1,2}_{\mathrm{d}}(\Lambda^k T^*\mathrm{X})$ provided there exists a $(k+1)$-form $\eta \in L^2(\Lambda^{k+1} T^*\mathrm{X})$ such that for any $X_0, \ldots, X_k \in \mathrm{TestV}(\mathrm{X})$ it holds

$$\begin{aligned}\int \eta(X_0, \ldots, X_k)\,\mathrm{d}\mathfrak{m} &= \sum_{i=0}^{k}\int (-1)^{i+1}\omega(\ldots, \hat{X}_i, \ldots)\,\mathrm{div}(X_i)\,\mathrm{d}\mathfrak{m} \\ &\quad + \sum_{i<j}\int (-1)^{i+j}\omega\big([X_i, X_j], \ldots, \hat{X}_i, \ldots, \hat{X}_j, \ldots\big)\,\mathrm{d}\mathfrak{m}.\end{aligned} \tag{6.69}$$

The element η, which is uniquely determined, is called *exterior differential* of ω and denoted by $\mathrm{d}\omega$. Its norm is defined as

$$\|\omega\|_{W^{1,2}_{\mathrm{d}}(\Lambda^k T^*\mathrm{X})} := \Big(\|\omega\|^2_{L^2(\Lambda^k T^*\mathrm{X})} + \|\mathrm{d}\omega\|^2_{L^2(\Lambda^{k+1} T^*\mathrm{X})}\Big)^{1/2}. \tag{6.70}$$

Much like in Theorem 6.3.2, one can prove that $W_{\mathrm{d}}^{1,2}(\Lambda^k T^*\mathrm{X})$ is a separable Hilbert space and that the unbounded operator $\mathrm{d} : L^2(\Lambda^k T^*\mathrm{X}) \to L^2(\Lambda^{k+1} T^*\mathrm{X})$ is closed.

Proposition 6.4.3 *Let $f_0, \ldots, f_k \in \mathrm{Test}^\infty(\mathrm{X})$ be given. Then both elements $f_0 \,\mathrm{d} f_1 \wedge \ldots \wedge \mathrm{d} f_k$ and $\mathrm{d} f_1 \wedge \ldots \wedge \mathrm{d} f_k$ belong to $W_{\mathrm{d}}^{1,2}(\Lambda^k T^*\mathrm{X})$ and it holds*

$$\begin{aligned} \mathrm{d}(f_0 \,\mathrm{d} f_1 \wedge \ldots \wedge \mathrm{d} f_k) &= \mathrm{d} f_0 \wedge \ldots \wedge \mathrm{d} f_k, \\ \mathrm{d}(\mathrm{d} f_1 \wedge \ldots \wedge \mathrm{d} f_k) &= 0. \end{aligned} \tag{6.71}$$

Proof Direct computation. □

Definition 6.4.4 Given any $k \in \mathbb{N}$, we define the space of *test k-forms* on $(\mathrm{X}, \mathsf{d}, \mathfrak{m})$ as

$$\mathrm{TestForm}_k(\mathrm{X}) := \text{linear span of the } f_0 \,\mathrm{d} f_1 \wedge \ldots \wedge \mathrm{d} f_k, \text{ with } f_0, \ldots, f_k \in \mathrm{Test}^\infty(\mathrm{X}). \tag{6.72}$$

It turns out that $\mathrm{TestForm}_k(\mathrm{X})$ is dense in $L^2(\Lambda^k T^*\mathrm{X})$ for all $k \in \mathbb{N}$. We define $H_{\mathrm{d}}^{1,2}(\Lambda^k T^*\mathrm{X})$ as the $W_{\mathrm{d}}^{1,2}(\Lambda^k T^*\mathrm{X})$-closure of $\mathrm{TestForm}_k(\mathrm{X})$.

Proposition 6.4.5 *Let $\omega \in H_{\mathrm{d}}^{1,2}(\Lambda^k T^*\mathrm{X})$. Then $\mathrm{d}\omega \in H_{\mathrm{d}}^{1,2}(\Lambda^{k+1} T^*\mathrm{X})$ and $\mathrm{d}(\mathrm{d}\omega) = 0$.*

Proof The statement holds for any test k-form by Proposition 6.4.3. The general case follows from the closure of the exterior differential. □

6.4.2 de Rham Cohomology and Hodge Theorem

Definition 6.4.6 (Closed/Exact Forms) Let $\omega \in H_{\mathrm{d}}^{1,2}(\Lambda^k T^*\mathrm{X})$. Then we say that ω is *closed* provided $\mathrm{d}\omega = 0$, while it is said to be *exact* if there exists $\alpha \in H_{\mathrm{d}}^{1,2}(\Lambda^{k-1} T^*\mathrm{X})$ such that $\omega = \mathrm{d}\alpha$.

We point out that any exact form is also closed by Proposition 6.4.5.

By the closure of d, the space of all closed k-forms is strongly closed in $L^2(\Lambda^k T^*\mathrm{X})$. Accordingly, the closed k-forms, endowed with the $L^2(\Lambda^k T^*\mathrm{X})$-norm, constitute a Hilbert space. In general, the same fails if we replace 'closed k-forms' with 'exact k-forms', but we point out that the $L^2(\Lambda^k T^*\mathrm{X})$-closure of the space of exact k-forms is a Hilbert space.

Definition 6.4.7 (de Rham Cohomology) Let $(\mathrm{X}, \mathsf{d}, \mathfrak{m})$ be any $\mathsf{RCD}(K, \infty)$ space. Then the *de Rham cohomology* is the quotient Hilbert space defined as follows:

$$\mathsf{H}^k_{\mathrm{dR}}(\mathrm{X}) := \frac{\text{closed } k\text{-forms}}{L^2(\Lambda^k T^*\mathrm{X})\text{-closure of exact } k\text{-forms}}. \tag{6.73}$$

Exercise 6.4.8 Let H_1, H_2 be Hilbert spaces. Let $\varphi : H_1 \to H_2$ be a linear and continuous operator. Then there exists a unique linear and continuous operator $\Lambda^k\varphi : \Lambda^k H_1 \to \Lambda^k H_2$ such that $\Lambda^k\varphi(v_1 \wedge \ldots \wedge v_k) = \varphi(v_1) \wedge \ldots \wedge \varphi(v_k)$ is satisfied for every $v_1, \ldots, v_k \in H_1$. Prove that $\|\Lambda^k\varphi\|_{\mathrm{op}} \leq \|\varphi\|^k_{\mathrm{op}}$. ■

Lemma 6.4.9 *Let* $(\mathrm{X}, \mathsf{d}_\mathrm{X}, \mathfrak{m}_\mathrm{X})$ *and* $(\mathrm{Y}, \mathsf{d}_\mathrm{Y}, \mathfrak{m}_\mathrm{Y})$ *be infinitesimally Hilbertian metric measure spaces. Let* $\varphi : \mathrm{X} \to \mathrm{Y}$ *be a map of bounded deformation. Then there exists a unique linear and continuous operator* $\varphi^* : L^2(\Lambda^k T^*\mathrm{Y}) \to L^2(\Lambda^k T^*\mathrm{X})$ *such that*

$$\varphi^*(\omega_1 \wedge \ldots \wedge \omega_k) = (\varphi^*\omega_1) \wedge \ldots \wedge (\varphi^*\omega_k) \quad \text{for every } \omega_1, \ldots, \omega_k \in L^2(\Lambda^k T^*\mathrm{Y}). \tag{6.74}$$

Moreover, $|\varphi^* A| \leq \mathrm{Lip}(\varphi)^k\, |A| \circ \varphi$ *holds* $\mathfrak{m}_\mathrm{X}$*-a.e. for every* $A \in L^2(\Lambda^k T^*\mathrm{Y})$.

Proof It follows from Exercise 6.4.8 by making use of an 'Hilbertian basis' (as in the definition of $|\cdot|_{\mathsf{HS}}$). □

Proposition 6.4.10 *Let* $(\mathrm{X}, \mathsf{d}_\mathrm{X}, \mathfrak{m}_\mathrm{X})$ *and* $(\mathrm{Y}, \mathsf{d}_\mathrm{Y}, \mathfrak{m}_\mathrm{Y})$ *be* $\mathsf{RCD}(K, \infty)$ *spaces. Let* $\varphi : \mathrm{X} \to \mathrm{Y}$ *be a map of bounded deformation and* $\omega \in H^{1,2}_\mathrm{d}(\Lambda^k T^*\mathrm{Y})$*. Then* $\varphi^*\omega \in H^{1,2}_\mathrm{d}(\Lambda^k T^*\mathrm{X})$ *and it holds that* $\varphi^*(\mathrm{d}\omega) = \mathrm{d}(\varphi^*\omega)$.

Proof For any test k-form $\omega = f_0\, \mathrm{d}f_1 \wedge \ldots \wedge \mathrm{d}f_k$, we have that

$$\varphi^*\omega = f_0 \circ \varphi\, (\varphi^*\mathrm{d}f_1) \wedge \ldots \wedge (\varphi^*\mathrm{d}f_k) = f_0 \circ \varphi\, \mathrm{d}(f_1 \circ \varphi) \wedge \ldots \wedge \mathrm{d}(f_k \circ \varphi),$$

whence Proposition 6.4.3 grants that $\varphi^*(\mathrm{d}\omega) = \mathrm{d}(\varphi^*\omega)$. The general case follows from the closure of the exterior differential by an approximation argument. □

Corollary 6.4.11 *Let* $k \in \mathbb{N}$ *be given. Then the map* φ^* *as in Proposition 6.4.10 canonically induces a linear and continuous operator from* $\mathsf{H}^k_{\mathrm{dR}}(\mathrm{Y})$ *to* $\mathsf{H}^k_{\mathrm{dR}}(\mathrm{X})$.

Proof Direct consequence of Proposition 6.4.10 and the closure of d. □

We briefly recall the Hodge theory for smooth Riemannian manifolds. With abuse of notation, we will sometimes identify tangent and cotangent objects, via the musical isomorphisms.

Let (M, g) be a smooth Riemannian manifold. Then for any $k \in \mathbb{N}$ we can define the de Rham cohomology $\mathsf{H}^k_{\mathrm{dR}}(M)$ as the quotient of closed k-forms over exact k-forms. Observe that this construction makes use only of the smooth structure of the manifold M, in other words the metric g plays no role. For brevity, we denote by L^2_k the space of all L^2 k-forms on the manifold M, which is a Hilbert space if endowed with the scalar product induced by g. Then we define $\delta : L^2_{k+1} \to L^2_k$ as the adjoint of the unbounded operator $\mathrm{d} : L^2_k \to L^2_{k+1}$, i.e. satisfying $\int \langle \delta\omega, \eta \rangle_k \,\mathrm{dVol} = \int \langle \omega, \mathrm{d}\eta \rangle_{k+1} \,\mathrm{dVol}$. Observe that $\mathrm{d}^2 = 0$, whence $\delta^2 = 0$ as well.

Given any 1-form ω, it holds that $\delta\omega = -\mathrm{div}(X)$, where the vector field X corresponds to ω via the musical isomorphism.

Definition 6.4.12 We define the *Hodge Laplacian* as the unbounded operator $\Delta_{\mathrm{H}} : L^2_k \to L^2_k$, which is given by

$$\Delta_{\mathrm{H}}\omega := (\delta\mathrm{d} + \mathrm{d}\delta)\omega = (\mathrm{d} + \delta)^2\omega. \tag{6.75}$$

A k-form ω is said to be *coexact* provided there exists $\eta \in L^2_{k+1}$ such that $\omega = \delta\eta$, while it is said to be *harmonic* if $\Delta_{\mathrm{H}}\omega = 0$.

Remark 6.4.13 Given any smooth 0-form f, i.e. any smooth function $f \in C^\infty(M)$, it holds that $\Delta f = -\Delta_{\mathrm{H}} f$. Moreover, one has that

$$\int \langle \eta, \Delta_{\mathrm{H}}\omega \rangle_k \,\mathrm{dVol} = \int \langle \mathrm{d}\eta, \mathrm{d}\omega \rangle_{k+1} \,\mathrm{dVol} + \int \langle \delta\eta, \delta\omega \rangle_{k-1} \,\mathrm{dVol} \tag{6.76}$$

is verified for $\eta, \omega \in L^2_k$. ■

The following result is due to W. V. D. Hodge:

Theorem 6.4.14 *The following properties hold:*

i) $L^2_k = \{$*exact k-forms*$\} \oplus \{$*coexact k-forms*$\} \oplus \{$*harmonic k-forms*$\}$.
ii) *For any* $[\omega] \in \mathsf{H}^k_{\mathrm{dR}}(M)$ *there exists a unique* $\eta \in [\omega]$ *such that* $\Delta_{\mathrm{H}}\eta = 0$.
iii) *One has* $\Delta_{\mathrm{H}}\eta = 0$ *if and only if* $\mathrm{d}\eta = 0$ *and* $\delta\eta = 0$.

Proof The proof goes as follows:

iii) If $\mathrm{d}\eta = 0$ and $\delta\eta = 0$, then trivially $\Delta_{\mathrm{H}}\eta = 0$. Conversely, suppose that $\Delta_{\mathrm{H}}\eta = 0$. Then (6.76) yields $0 = \int \langle \eta, \Delta_{\mathrm{H}}\eta \rangle_k \,\mathrm{dVol} = \int |\mathrm{d}\eta|^2 + |\delta\eta|^2 \,\mathrm{dVol}$, whence $\mathrm{d}\eta = 0$ and $\delta\eta = 0$.

i) Let $\omega = \mathrm{d}\omega'$, $\alpha = \delta\alpha'$ and $\Delta_{\mathrm{H}}\eta = 0$. We have $\int \langle \mathrm{d}\omega', \delta\alpha' \rangle_k \,\mathrm{dVol} = \int \langle \mathrm{d}^2\omega', \alpha' \rangle_{k-1} \,\mathrm{dVol} = 0$. Moreover, it holds that

$$\int \langle \mathrm{d}\omega', \eta \rangle_k \,\mathrm{dVol} = \int \langle \omega', \delta\eta \rangle_{k-1} \,\mathrm{dVol} = 0,$$

$$\int \langle \delta\alpha', \eta \rangle_k \,\mathrm{dVol} = \int \langle \alpha', \mathrm{d}\eta \rangle_{k+1} \,\mathrm{dVol} = 0$$

by item iii). Hence exact, coexact and harmonic k-forms are in direct sum. Now let $\omega \in L^2_k$ be fixed. Choose $\omega' \in L^2_{k-1}$ that minimises the quantity $\|\omega - \mathrm{d}\alpha\|_{L^2_k}$ among all $\alpha \in L^2_{k-1}$. (We omit the proof of the existence of such minimiser.) Then the Euler-Lagrange equation yields $\int \langle \omega - \mathrm{d}\omega', \mathrm{d}\alpha \rangle_k \,\mathrm{dVol} = 0$ for all $\alpha \in L^2_{k-1}$, whence we have that $\delta(\omega - \mathrm{d}\omega') = 0$. Now let $\beta' \in L^2_{k+1}$ be the minimiser of $\|\omega - \delta\alpha'\|_{L^2_k}$ among all $\alpha' \in L^2_{k+1}$. Then the Euler-Lagrange equation yields $\int \langle \omega - \delta\beta', \delta\alpha' \rangle_k \,\mathrm{dVol} = 0$ for all $\alpha' \in L^2_{k+1}$, whence we have $\mathrm{d}(\omega - \delta\beta') = 0$. Therefore we can write ω as

$$\omega = \underbrace{\mathrm{d}\omega'}_{\text{exact}} + \underbrace{\delta\beta'}_{\text{coexact}} + \underbrace{(\omega - \mathrm{d}\omega' - \delta\beta')}_{\text{harmonic}},$$

thus proving that i) holds.

ii) Let ω be a closed k-form. Since the space of closed k-forms is orthogonal to that of coexact k-forms, there exists a unique $\eta \in L^2_k$ harmonic such that $\omega - \eta$ is an exact k-form. Then it holds that $[\eta] = [\omega] \in \mathsf{H}^k_{\mathrm{dR}}(M)$, thus proving ii).

□

In the language of Hodge theory, we can state a sharper form of the Bochner inequality:

$$\Delta \frac{|\omega|^2}{2} \geq -\langle \omega, \Delta_{\mathrm{H}}\omega \rangle + K|\omega|^2 \quad \text{for every smooth 1-form } \omega. \tag{6.77}$$

Actually, the Bochner identity can be written as follows:

$$\Delta \frac{|\omega|^2}{2} = |\nabla\omega|^2_{\mathsf{HS}} - \langle \omega, \Delta_{\mathrm{H}}\omega \rangle + \mathrm{Ric}(\omega, \omega) \quad \text{for every smooth 1-form } \omega. \tag{6.78}$$

Moreover, we define the *connection Laplacian* $\Delta_{\mathrm{C}} X$ of a smooth vector field X as

$$\int \langle \Delta_{\mathrm{C}} X, Y \rangle \,\mathrm{dVol} = -\int \nabla X : \nabla Y \,\mathrm{dVol} \quad \text{for every smooth vector field } Y. \tag{6.79}$$

One can prove that $\Delta\big(|X|^2/2\big) = |\nabla X|^2_{\mathsf{HS}} + \langle X, \Delta_{\mathrm{C}} X \rangle$ holds for any smooth vector field X. We also have that

$$\Delta_{\mathrm{C}} X + \Delta_{\mathrm{H}} X = \mathrm{Ric}(X, \cdot) \quad \text{for every smooth vector field } X, \tag{6.80}$$

which is known as the *Weitzenböck identity*.

Theorem 6.4.15 (Bochner) *Suppose that* $\mathrm{Ric}_M \geq 0$. *Then*

$$\dim \mathsf{H}^1_{\mathrm{dR}}(M) \leq \dim M, \tag{6.81}$$

with equality if and only if M is a flat torus.

Proof We know from Theorem 6.4.14 that the dimension of $\mathsf{H}^1_{\mathrm{dR}}(M)$ coincides with that of the space of all harmonic 1-forms. Then fix a harmonic 1-form ω. We thus have that

$$0 = \int \Delta \frac{|\omega|^2}{2}\, \mathrm{dVol} \overset{(6.78)}{\geq} \int |\nabla\omega|^2_{\mathsf{HS}}\, \mathrm{dVol} - \int \langle \omega, \Delta_{\mathsf{H}}\omega \rangle\, \mathrm{dVol} = \int |\nabla\omega|^2_{\mathsf{HS}}\, \mathrm{dVol}.$$

Therefore $\int |\nabla\omega|^2_{\mathsf{HS}}\, \mathrm{dVol} = 0$, so by using the parallel transport we conclude that the dimension of the space of harmonic 1-forms is smaller than or equal to $\dim M$, proving (6.81). We omit the proof of the last part of the statement. □

We now introduce the Hodge theory for RCD spaces. Hereafter, the space $(\mathrm{X}, \mathsf{d}, \mathfrak{m})$ will be a fixed $\mathsf{RCD}(K, \infty)$ space.

Definition 6.4.16 (Codifferential) We denote by $D(\delta)$ the family of all k-forms $\omega \in L^2(\Lambda^k T^*\mathrm{X})$ such that there exists $\eta \in L^2(\Lambda^{k-1}T^*\mathrm{X})$ for which

$$\int \langle \omega, \mathrm{d}\alpha \rangle\, \mathrm{d}\mathfrak{m} = \int \langle \eta, \alpha \rangle\, \mathrm{d}\mathfrak{m} \quad \text{holds for every } \alpha \in \mathrm{TestForm}_{k-1}(\mathrm{X}). \tag{6.82}$$

The element η, which is uniquely determined, is denoted by $\delta\omega$ and called *codifferential* of ω.

It is easy to see that δ is a closed unbounded operator.

Proposition 6.4.17 *It holds that* $\mathrm{TestForm}_k(\mathrm{X}) \subseteq D(\delta)$ *for all* $k \in \mathbb{N}$. *More explicitly,*

$$\begin{aligned}\delta(\mathrm{d}f_1 \wedge \ldots \wedge \mathrm{d}f_k) &= \sum_{i=1}^{k} (-1)^i\, \Delta f_i\, \mathrm{d}f_1 \wedge \ldots \wedge \widehat{\mathrm{d}f_i} \wedge \ldots \wedge \mathrm{d}f_k \\ &\quad + \sum_{i<j} (-1)^{i+j}\, [\nabla f_i, \nabla f_j] \wedge \ldots \wedge \widehat{\mathrm{d}f_i} \wedge \ldots \wedge \widehat{\mathrm{d}f_j} \wedge \ldots \wedge \mathrm{d}f_k\end{aligned} \tag{6.83}$$

is verified for every $f_1, \ldots, f_k \in \mathrm{Test}^\infty(\mathrm{X})$.

Proof Direct computation. □

Definition 6.4.18 Let us define $W^{1,2}_{\mathrm{H}}(\Lambda^k T^*\mathrm{X}) := W^{1,2}_{\mathrm{d}}(\Lambda^k T^*\mathrm{X}) \cap D(\delta)$ for every $k \in \mathbb{N}$. The norm of an element $\omega \in W^{1,2}_{\mathrm{H}}(\Lambda^k T^*\mathrm{X})$ is given by

$$\|\omega\|_{W^{1,2}_{\mathrm{H}}(\Lambda^k T^*\mathrm{X})} := \left(\|\omega\|^2_{L^2(\Lambda^k T^*\mathrm{X})} + \|\mathrm{d}\omega\|^2_{L^2(\Lambda^{k+1} T^*\mathrm{X})} + \|\delta\omega\|^2_{L^2(\Lambda^{k-1} T^*\mathrm{X})} \right)^{1/2}. \tag{6.84}$$

Finally, let us define $H^{1,2}_{\mathrm{H}}(\Lambda^k T^*\mathrm{X})$ as the $W^{1,2}_{\mathrm{H}}(\Lambda^k T^*\mathrm{X})$-closure of $\mathrm{TestForm}_k(\mathrm{X})$.

We have that $W^{1,2}_{\mathrm{H}}(\Lambda^k T^*\mathrm{X})$ and $H^{1,2}_{\mathrm{H}}(\Lambda^k T^*\mathrm{X})$ are separable Hilbert spaces.

Definition 6.4.19 (Hodge Laplacian) Let $\omega \in H^{1,2}_{\mathrm{H}}(\Lambda^k T^*\mathrm{X})$ be given. Then we declare that $\omega \in D(\Delta_{\mathrm{H}})$ provided there exists $\eta \in L^2(\Lambda^k T^*\mathrm{X})$ such that

$$\int \langle \eta, \alpha \rangle \,\mathrm{dm} = \int \langle \mathrm{d}\omega, \mathrm{d}\alpha \rangle + \langle \delta\omega, \delta\alpha \rangle \,\mathrm{dm} \quad \text{for every } \alpha \in \mathrm{TestForm}_k(\mathrm{X}). \tag{6.85}$$

The element η, which is uniquely determined, is denoted by $\Delta_{\mathrm{H}}\omega$ and called *Hodge Laplacian*.

Definition 6.4.20 (Harmonic k-Forms) Let $k \in \mathbb{N}$. Then we define $\mathrm{Harm}_k(\mathrm{X})$ as the set of all $\omega \in H^{1,2}_{\mathrm{H}}(\Lambda^k T^*\mathrm{X})$ such that $\Delta_{\mathrm{H}}\omega = 0$. The elements of $\mathrm{Harm}_k(\mathrm{X})$ are called *harmonic*.

Remark 6.4.21 It holds that Δ_{H} is a closed unbounded operator. Indeed, suppose $\omega_n \to \omega$ and $\Delta_{\mathrm{H}}\omega_n \to \eta$ in $L^2(\Lambda^k T^*\mathrm{X})$. Observe that

$$\sup_{n\in\mathbb{N}} \int |\mathrm{d}\omega_n|^2 + |\delta\omega_n|^2 \,\mathrm{dm} = \sup_{n\in\mathbb{N}} \int \langle \omega_n, \Delta_{\mathrm{H}}\omega_n \rangle \,\mathrm{dm} < +\infty,$$

whence it easily follows that $\omega \in D(\Delta_{\mathrm{H}})$ and $\eta = \Delta_{\mathrm{H}}\omega$, since d and δ are closed. ■

Corollary 6.4.22 *The space* $\big(\mathrm{Harm}_k(\mathrm{X}), \|\cdot\|_{L^2(\Lambda^k T^*\mathrm{X})}\big)$ *is Hilbert.*

Proof Direct consequence of the closure of Δ_{H}. □

Theorem 6.4.23 (Hodge Theorem for RCD Spaces) *Let $k \in \mathbb{N}$ be given. Then the map*

$$\mathrm{Harm}_k(\mathrm{X}) \ni \omega \longmapsto [\omega] \in \mathsf{H}^k_{\mathrm{dR}}(\mathrm{X}) \tag{6.86}$$

is an isomorphism of Hilbert spaces.

Proof First of all, observe that any element of $\mathrm{Harm}_k(\mathrm{X})$ is a closed k-form. In analogy with item iii) of Theorem 6.4.14, we also have that for any $\omega \in H^{1,2}_{\mathrm{H}}(\Lambda^k T^*\mathrm{X})$ it holds

$$\omega \in \mathrm{Harm}_k(\mathrm{X}) \iff \mathrm{d}\omega = 0 \text{ and } \delta\omega = 0. \tag{6.87}$$

Moreover, we recall the following general functional analytic fact:

$$H \text{ Hilbert space, } V \subseteq H \text{ linear subspace} \implies \begin{cases} V^\perp \ni \omega \mapsto \omega + \overline{V} \in H/\overline{V} \\ \text{is an isomorphism.} \end{cases} \tag{6.88}$$

Now let us apply (6.88) with $H := \{\text{closed } k\text{-forms}\}$ and $V := \{\text{exact } k\text{-forms}\}$. Since it holds that $V^\perp = \mathrm{Harm}_k(\mathrm{X})$ by (6.87), we get the statement. □

Remark 6.4.24 Let us define the energy functional $\mathcal{E}_{\mathrm{H}} : L^2(\Lambda^k T^*\mathrm{X}) \to [0, +\infty]$ as follows:

$$\mathcal{E}_{\mathrm{H}}(\omega) := \begin{cases} \frac{1}{2}\int |\mathrm{d}\omega|^2 + |\delta\omega|^2 \,\mathrm{dm} & \text{if } \omega \in H^{1,2}_{\mathrm{H}}(\Lambda^k T^*\mathrm{X}), \\ +\infty & \text{otherwise.} \end{cases} \tag{6.89}$$

Then $\mathcal{E}_{\mathrm{H}}$ is convex and lower semicontinuous. Moreover, we have that $\omega \in D(\Delta_{\mathrm{H}})$ if and only if $\partial^-\mathcal{E}_{\mathrm{H}}(\omega) \neq \emptyset$. In this case, $\Delta_{\mathrm{H}}\omega$ is the only element of $\partial^-\mathcal{E}_{\mathrm{H}}(\omega)$. ■

Definition 6.4.25 (Heat Flow of Forms) Let $\omega \in L^2(\Lambda^k T^*\mathrm{X})$. Then we denote by $t \mapsto \mathsf{h}_{\mathrm{H},t}\omega$ the unique gradient flow of $\mathcal{E}_{\mathrm{H}}$ starting from ω.

Exercise 6.4.26 Prove that

$$\mathsf{h}_{\mathrm{H},t}(\mathrm{d}\omega) = \mathrm{d}\mathsf{h}_{\mathrm{H},t}\omega \quad \text{for every } \omega \in W^{1,2}_{\mathrm{d}}(\Lambda^k T^*\mathrm{X}) \text{ and } t \geq 0. \tag{6.90}$$

Moreover, an analogous property is satisfied by the codifferential δ. ■

Given any closed k-form ω, its (unique) harmonic representative is $\lim_{t\to\infty} \mathsf{h}_{\mathrm{H},t}\,\omega$.

Definition 6.4.27 (Connection Laplacian) Let $X \in H^{1,2}_{\mathrm{C}}(T\mathrm{X})$ be given. Then we declare that $X \in D(\Delta_{\mathrm{C}})$ provided there exists $Z \in L^2(T\mathrm{X})$ such that

$$\int \langle Z, X\rangle \,\mathrm{dm} = -\int \langle \nabla X, \nabla Y\rangle \,\mathrm{dm} \quad \text{for every } Y \in \mathrm{TestV}(\mathrm{X}). \tag{6.91}$$

The element Z is denoted by $\Delta_{\mathrm{C}}X$ and called *connection Laplacian* of ω.

Remark 6.4.28 We define the *connection energy* $\mathcal{E}_{\mathrm{C}} : L^2(T\mathrm{X}) \to [0, +\infty]$ as

$$\mathcal{E}_{\mathrm{C}}(X) := \begin{cases} \frac{1}{2}\int |\nabla X|^2_{\mathsf{HS}} \,\mathrm{dm} & \text{if } X \in H^{1,2}_{\mathrm{C}}(T\mathrm{X}), \\ +\infty & \text{otherwise.} \end{cases} \tag{6.92}$$

Then $\mathcal{E}_{\mathrm{C}}$ is a convex and lower semicontinuous functional. Moreover, we have that $X \in D(\Delta_{\mathrm{C}})$ if and only if $\partial^-\mathcal{E}_{\mathrm{C}}(X) \neq \emptyset$. In this case, $-\Delta_{\mathrm{C}}X$ is the unique element of $\partial^-\mathcal{E}_{\mathrm{C}}(X)$. ■

Proposition 6.4.29 *Let $X \in D(\Delta_{\mathrm{C}}) \cap L^\infty(T\mathrm{X})$ be given. Then $|X|^2/2 \in D(\mathbf{\Delta})$ and*

$$\mathbf{\Delta}\frac{|X|^2}{2} = \big(|\nabla X|^2_{\mathsf{HS}} + \langle X, \Delta_{\mathrm{C}}X\rangle\big)\mathrm{m}. \tag{6.93}$$

Proof We know that $|X|^2 \in W^{1,2}(\mathrm{X})$ and $\nabla|X|^2 = 2\nabla X(\cdot, X)$. Hence the equalities

$$\begin{aligned}\int f\big(|\nabla X|^2_{\mathsf{HS}} + \langle X, \Delta_{\mathsf{C}} X\rangle\big)\,\mathrm{dm} &= \int f\,|\nabla X|^2_{\mathsf{HS}} - \nabla(fX) : \nabla X\,\mathrm{dm}\\ &= \int f\,|\nabla X|^2_{\mathsf{HS}} - (f\,\nabla X + \nabla f \otimes \nabla X) : \nabla X\,\mathrm{dm}\\ &= -\int \nabla X(\nabla f, X)\,\mathrm{dm}\\ &= -\int \nabla f \cdot \nabla\frac{|X|^2}{2}\,\mathrm{dm}\end{aligned}$$

hold for every $f \in \mathrm{LIP}_{bs}(\mathrm{X})$, thus obtaining (6.93). □

Definition 6.4.30 (Heat Flow of Vector Fields) Let $X \in L^2(T\mathrm{X})$ be given. Then we denote by $t \mapsto \mathsf{h}_{\mathsf{C},t}X$ the unique gradient flow of $\mathcal{E}_{\mathsf{C}}$ starting from X.

Proposition 6.4.31 *Let $X \in L^2(T\mathrm{X})$. Then it holds that*

$$|\mathsf{h}_{\mathsf{C},t}X|^2 \leq \mathsf{h}_t\big(|X|^2\big) \quad \mathfrak{m}\text{-a.e.} \quad \text{for every } t \geq 0. \tag{6.94}$$

Proof Fix $t > 0$ and set $F_s := \mathsf{h}_s\big(|\mathsf{h}_{\mathsf{C},t-s}X|^2\big)$ for all $s \in [0,t]$. Then for a.e. $s \in [0,t]$ one has

$$F_s' = \mathsf{h}_s\big(\Delta|\mathsf{h}_{\mathsf{C},t-s}X|^2 - 2\,\langle \mathsf{h}_{\mathsf{C},t-s}X, \Delta_{\mathsf{C}}\mathsf{h}_{\mathsf{C},t-s}X\rangle\big) = \mathsf{h}_s\big(|\nabla \mathsf{h}_{\mathsf{C},t-s}X|^2\big) \geq 0,$$

whence (6.94) immediately follows. □

With the terminology introduced so far, we can restate Theorem 6.2.21 as follows:

$$\frac{|X|^2}{2} \in D(\mathbf{\Delta}) \quad \text{and} \quad \mathbf{\Delta}\frac{|X|^2}{2} \geq \big(|\nabla X|^2_{\mathsf{HS}} - \langle X, \Delta_{\mathsf{H}}X\rangle + K|\nabla X|^2\big)\mathfrak{m} \tag{6.95}$$

are verified for every $X \in \mathrm{TestV}(\mathrm{X})$.

Lemma 6.4.32 *It holds that $H^{1,2}_{\mathsf{H}}(T\mathrm{X}) \subseteq H^{1,2}_{\mathsf{C}}(T\mathrm{X})$. More precisely, we have that*

$$\mathcal{E}_{\mathsf{C}}(X) \leq \mathcal{E}_{\mathsf{H}}(X) - \frac{K}{2}\int |X|^2\,\mathrm{dm} \quad \text{for every } X \in H^{1,2}_{\mathsf{H}}(T\mathrm{X}). \tag{6.96}$$

Proof The statement can be proved by integrating the Bochner inequality (6.95). □

6.5 Ricci Curvature Operator

In light of the Bochner identity (6.2), it is natural to give the following definition:

$$\mathbf{Ric}(X, Y) := \mathbf{\Delta}\frac{\langle X, Y\rangle}{2} - \left(\langle\nabla X, \nabla Y\rangle - \frac{\langle X, \Delta_{\mathrm{H}} Y\rangle}{2} - \frac{\langle Y, \Delta_{\mathrm{H}} X\rangle}{2}\right)\mathfrak{m} \tag{6.97}$$

for every $X, Y \in \mathrm{TestV}(\mathrm{X})$. We can thus introduce the *Ricci curvature* operator:

Theorem 6.5.1 (Ricci Curvature) *There exists a unique bilinear and continuous extension of* **Ric** *to an operator (still denoted by* **Ric***) from* $H^{1,2}_{\mathrm{H}}(T\mathrm{X}) \times H^{1,2}_{\mathrm{H}}(T\mathrm{X})$ *to the space of finite Radon measures on* X. *Moreover, it holds that*

$$\begin{aligned}
&\mathbf{Ric}(X, X) \geq K|X|^2\mathfrak{m},\\
&\left\|\mathbf{Ric}(X, Y)\right\|_{\mathsf{TV}} \leq 2\left(\mathcal{E}_{\mathrm{H}}(X) + K^-\|X\|^2_{L^2(T\mathrm{X})}\right)^{1/2}\left(\mathcal{E}_{\mathrm{H}}(Y) + K^-\|Y\|^2_{L^2(T\mathrm{X})}\right)^{1/2}\\
&\mathbf{Ric}(X, Y)(\mathrm{X}) = \int \langle \mathrm{d}X, \mathrm{d}Y\rangle + \langle\delta X, \delta Y\rangle - \nabla X : \nabla Y \,\mathrm{d}\mathfrak{m},
\end{aligned} \tag{6.98}$$

for every $X, Y \in H^{1,2}_{\mathrm{H}}(T\mathrm{X})$.

Proof The first line and the third line in (6.98) are verified for every $X \in \mathrm{TestV}(\mathrm{X})$ by (6.95) and (6.97). In order to prove the second line (for test vector fields), we first consider the case in which $X = Y$ and $K = 0$: since $\mathbf{Ric}(X, X) \geq 0$, we have that

$$\left\|\mathbf{Ric}(X, X)\right\|_{\mathsf{TV}} = \mathbf{Ric}(X, X)(\mathrm{X}) = 2\,\mathcal{E}_{\mathrm{H}}(X) - 2\,\mathcal{E}_{\mathrm{C}}(X) \leq 2\,\mathcal{E}_{\mathrm{H}}(X),$$

which is precisely the second line in (6.98). Its polarised version—for $X, Y \in \mathrm{TestV}(\mathrm{X})$—can be achieved by noticing that for all $\lambda \in \mathbb{R}$ one has

$$\lambda^2\,\mathbf{Ric}(X, X) + 2\,\lambda\,\mathbf{Ric}(X, Y) + \mathbf{Ric}(Y, Y) = \mathbf{Ric}(\lambda\,X + Y, \lambda\,X + Y) \geq 0,$$

whence $\left|\mathbf{Ric}(X, Y)\right| \leq \left(\mathbf{Ric}(X, X)\,\mathbf{Ric}(Y, Y)\right)^{1/2}$ by Lemma 6.2.18 and accordingly

$$\left\|\mathbf{Ric}(X, Y)\right\|_{\mathsf{TV}} \leq \left(\left\|\mathbf{Ric}(X, X)\right\|_{\mathsf{TV}}\left\|\mathbf{Ric}(Y, Y)\right\|_{\mathsf{TV}}\right)^{1/2},$$

which proves the second in (6.98) for $K = 0$. The general case $K \in \mathbb{R}$ can be shown by repeating the same argument with $\widetilde{\mathbf{Ric}}$ instead of **Ric**, where we set

$$\widetilde{\mathbf{Ric}}(X, Y) := \mathbf{Ric}(X, Y) - K\langle X, Y\rangle\mathfrak{m} \quad \text{for every } X, Y \in \mathrm{TestV}(\mathrm{X}).$$

Finally, once (6.98) is proven for test vector fields, the full statement easily follows. □

The next result shows that the Ricci curvature is 'tensorial':

Proposition 6.5.2 *Let* $X, Y \in H^{1,2}_{\mathrm{H}}(TX)$ *and* $f \in \mathrm{Test}^{\infty}(\mathrm{X})$. *Then* $fX \in H^{1,2}_{\mathrm{H}}(TX)$ *and*

$$\mathbf{Ric}(fX, Y) = f\,\mathbf{Ric}(X, Y). \tag{6.99}$$

Proof Immediate consequence of the defining property (6.97) of **Ric** and a direct computation based on the calculus rules developed so far. □

Proposition 6.5.3 (Refined Bakry-Émery Estimate) *Let* $\omega \in L^2(T^*\mathrm{X})$. *Then it holds*

$$|\mathsf{h}_{\mathrm{H},t}\,\omega|^2 \leq e^{-2Kt}\,\mathsf{h}_t\big(|\omega|^2\big) \quad \mathfrak{m}\text{-}a.e. \quad \textit{for every } t \geq 0. \tag{6.100}$$

Proof Fix $t > 0$ and set $F_s := \mathsf{h}_s\big(|\mathsf{h}_{\mathrm{H},t-s}\,\omega|^2\big)$ for all $s \in [0, t]$. Then for a.e. $s \in [0, t]$ one has

$$F_s' = \mathsf{h}_s\big(\Delta|\mathsf{h}_{\mathrm{H},t-s}\,\omega|^2 + 2\,\langle \mathsf{h}_{\mathrm{H},t-s}\,\omega, \Delta_{\mathrm{H}}\mathsf{h}_{\mathrm{H},t-s}\,\omega\rangle\big) \geq 2\,\mathsf{h}_s\big(K\,|\mathsf{h}_{\mathrm{H},t-s}\,\omega|^2\big),$$

i.e. $F_s' \geq 2K\,F_s$ for a.e. $s \in [0, t]$. Then (6.100) follows by Gronwall lemma. □

Bibliographical Remarks

The original curvature-dimension condition for metric measure spaces, called CD condition, has been independently proposed by Sturm and Lott-Villani in [29, 30] and [24], respectively. Such formulation, which is based upon an optimal transport language, is related to the convexity properties of certain entropy functionals along Wasserstein geodesics. Its Riemannian counterpart, namely the RCD condition, has been introduced in [5] in the infinite dimensional case and in [18] in the finite dimensional one. The approach we adopted in these notes, that fits into the framework of the Bakry-Émery theory [9, 10], has been proposed by Ambrosio-Gigli-Savaré in [6]. As seen in Definition 6.1.2, it consists of a weak formulation of the Bochner inequality; the proof of the equivalence of the resulting notion with the above-mentioned RCD condition can be found in [8, 15]. We refer to the surveys [1, 31, 32] for a detailed account of the curvature-dimension conditions.

Section 6.1 is subdivided as follows: the definition of $\mathsf{RCD}(K, \infty)$ space in Sect. 6.1.1 is taken from [6], but is formulated in terms of the language proposed in [19]; the results in Sect. 6.1.2, concerning the properties of the heat flow on RCD spaces, can be found in the paper [5]; the material of Sect. 6.1.3 about test functions on RCD spaces is basically extracted from [27], where in particular Theorem 6.1.11 has been proved.

The remaining part of the chapter—from Sects. 6.2 to 6.5—is almost entirely taken from [17] (and [19]). The only exceptions are given by Lemma 6.2.14 (that is proved in [7, Lemma 6.7]), by Lemma 6.2.15 (that constitutes a new result) and by the equality statement in Theorem 6.4.15 (proven in [20]).

Appendix A
Functional Analytic Tools

Let us state and prove two well-known fundamental results of functional analysis:

Lemma A.1 *Let $\mathbb{E}_1, \mathbb{E}_2$ be Banach spaces. Let $i : \mathbb{E}_1 \to \mathbb{E}_2$ be a linear and continuous injection. Suppose that $\mathbb{E}_1$ is reflexive and that $\mathbb{E}_2$ is separable. Then $\mathbb{E}_1$ is separable as well.*

Proof Recall that any continuous bijection f from a compact topological space X to a Hausdorff topological space Y is a homeomorphism (each closed subset $C \subseteq$ X is compact because X is compact, hence $f(C)$, being compact in the Hausdorff space Y, is closed). Call

X the closed unit ball in $\mathbb{E}_1$ endowed with the (restriction of the) weak topology of $\mathbb{E}_1$,

Y the image $i(\mathrm{X})$ endowed with the (restriction of the) weak topology of $\mathbb{E}_2$,

f the map $i_{|\mathrm{X}}$ from X to Y.

Since X is compact (by reflexivity of $\mathbb{E}_1$), Y is Hausdorff and f is continuous (as i is linear and continuous), we thus deduce that f is a homeomorphism. In particular, the separability of Y grants that X is separable as well, i.e. the closed unit ball B of $\mathbb{E}_1$ is weakly separable. Now fix a countable weakly dense subset D of such ball. Denote by Q the set of all finite convex combinations with coefficients in $\mathbb{Q}$ of elements of D. It is clear that the set Q, which is countable by construction, is strongly dense in the convex hull C of D. Since C is convex, we have that the weak closure and the strong closure of C coincide. Moreover, such closure contains B. Hence Q is strongly dense in the set B, which accordingly turns out to be strongly separable. Finally, we conclude that $\mathbb{E}_1 = \bigcup_{n\in\mathbb{N}} nB$ is strongly separable as well, thus achieving the statement. □

Theorem A.2 (Mazur's Lemma) *Let $\mathbb{B}$ be a Banach space. Let $(v_n)_n \subseteq \mathbb{B}$ be a sequence that weakly converges to some limit $v \in \mathbb{B}$. Then there exist*

N. Gigli, E. Pasqualetto, *Lectures on Nonsmooth Differential Geometry*,
SISSA Springer Series 2, https://doi.org/10.1007/978-3-030-38613-9

$(N_n)_n \subseteq \mathbb{N}$ *and* $(\alpha_{n,i})_{i=n}^{N_n} \subseteq [0,1]$ *such that* $\sum_{i=n}^{N_n} \alpha_{n,i} = 1$ *for all* $n \in \mathbb{N}$ *and* $\tilde{v}_n := \sum_{i=n}^{N_n} \alpha_{n,i}\, v_i \to v$ *in the strong topology of* $\mathbb{B}$.

Proof Given any $n \in \mathbb{N}$, let us denote by K_n the strong closure of the set of all (finite) convex combinations of the $(v_i)_{i \geq n}$. Each set K_n, being strongly closed and convex, is weakly closed by Hahn-Banach theorem. Given that $v \in \bigcap_{n \in \mathbb{N}} K_n$, for every $n \in \mathbb{N}$ we can choose $N_n \geq n$ and some $\alpha_{n,n}, \ldots, \alpha_{n,N_n} \in [0,1]$ such that $\sum_{i=n}^{N_n} \alpha_{n,i} = 1$ and $\|\tilde{v}_n - v\|_{\mathbb{B}} < 1/n$, where we put $\tilde{v}_n := \sum_{i=n}^{N_n} \alpha_{n,i}\, v_i$. This proves the claim. □

Appendix B
Solutions to the Exercises

Exercise 1.1.5 Suppose that X is compact. Prove that if a sequence $(f_n)_n \subseteq C(\mathrm{X})$ satisfies $f_n(x) \searrow 0$ for every $x \in \mathrm{X}$, then $f_n \to 0$ uniformly on X.

Solution First of all, we claim that

$$(f_n)_n \subseteq C(\mathrm{X}) \quad \text{is equicontinuous.} \tag{B.1}$$

We argue by contradiction: if not, there exist $\bar{x} \in \mathrm{X}$ and $\varepsilon > 0$ such that for any $\delta > 0$ there are $n \in \mathbb{N}$ and $y \in B_\delta(\bar{x})$ satisfying $\left|f_n(y) - f_n(\bar{x})\right| \geq \varepsilon$. Choose $\bar{n} \in \mathbb{N}$ for which $f_{\bar{n}}(\bar{x}) < \varepsilon/2$, then take any $\bar{\delta} > 0$ such that $\left|f_{\bar{n}}(y) - f_{\bar{n}}(\bar{x})\right| < \varepsilon/2$ for every $y \in B_{\bar{\delta}}(\bar{x})$. This clearly grants that $f_n(y) < \varepsilon$ for every $n \geq \bar{n}$ and $y \in B_{\bar{\delta}}(\bar{x})$, thus in particular

$$\left|f_n(y) - f_n(\bar{x})\right| < \varepsilon \quad \text{for every } n \geq \bar{n} \text{ and } y \in B_{\bar{\delta}}(\bar{x}). \tag{B.2}$$

Now choose any sequence $(\delta_k)_k \subseteq (0, \bar{\delta})$ such that $\delta_k \searrow 0$. For any $k \in \mathbb{N}$ there exist $n_k \in \mathbb{N}$ and $y_k \in B_{\delta_k}(\bar{x})$ that satisfy $\left|f_{n_k}(y_k) - f_{n_k}(\bar{x})\right| \geq \varepsilon$. Observe that (B.2) forces $n_k < \bar{n}$ for every $k \in \mathbb{N}$. Up to passing to a not relabeled subsequence, one has that there exists $n' < \bar{n}$ such that $n_k = n'$ for all $k \in \mathbb{N}$. Since $\lim_k \mathsf{d}(y_k, \bar{x}) = 0$ and the map $f_{n'}$ is continuous, we have that $\lim_k \left|f_{n'}(y_k) - f_{n'}(\bar{x})\right| = 0$, which is a contradiction. Therefore (B.1) is proved.

Take any subsequence $(f_{n_k})_k$ of $(f_n)_n$. Given that $\sup_k \|f_{n_k}\|_{C_b(\mathrm{X})} < +\infty$ by hypothesis and $(f_{n_k})_k$ is equicontinuous by (B.1), we conclude that a subsequence of $(f_{n_k})_k$ uniformly converges to some map $f \in C_b(\mathrm{X})$ by Arzelà-Ascoli theorem. Since $f_{n_k} \searrow 0$ pointwise, we have that $f = 0$. Therefore the whole sequence $(f_n)_n$ is uniformly converging to 0, thus proving the statement. □

Exercise 1.1.7 Let (X, d) be a complete and separable metric space. Prove that if $C_b(\mathrm{X})$ is separable, then the space X is compact.

N. Gigli, E. Pasqualetto, *Lectures on Nonsmooth Differential Geometry*,
SISSA Springer Series 2, https://doi.org/10.1007/978-3-030-38613-9

Solution Suppose that (X, d) is not compact, or equivalently that it is not totally bounded. Then there exists $r > 0$ such that X cannot be covered by finitely many balls of radius $2r$. Choose any family $\{x_i\}_{i\in I}$ of distinct points in X such that $\{B_r(x_i)\}_{i\in I}$ is a maximal r-net in X—thus in particular the set I is at most countable. Since the family $\{B_{2r}(x_i)\}_{i\in I}$ is a cover of X, we know that I must be infinite. For any index $i \in I$, let us pick any continuous function $g_i : \mathrm{X} \to [0, 1]$ such that $g_i(x_i) = 1$ and $\mathrm{spt}(g_i) \subseteq B_r(x_i)$. Given any subset $S \subseteq I$, we define the function $f_S \in C_b(\mathrm{X})$ as $f_S := \sum_{i\in S} g_i$. Hence $\{f_S\}_{S\subseteq I}$ is an uncountable family of elements of $C_b(\mathrm{X})$ such that $\|f_S - f_T\|_{C_b(\mathrm{X})} = 1$ whenever $S, T \subseteq I$ satisfy $S \neq T$. This shows that the space $C_b(\mathrm{X})$ is not separable, as desired. □

Exercise 1.1.23 Prove that $L^p(\mathfrak{m})$ is dense in $L^0(\mathfrak{m})$ for every $p \in [1, \infty]$.

Solution Let $f \in L^0(\mathfrak{m})$ be fixed. Pick any $\bar{x} \in \mathrm{X}$ and define

$$f_n := \chi_{B_n(\bar{x})}\,(f \wedge n) \vee (-n) \in L^1(\mathfrak{m}) \cap L^\infty(\mathfrak{m}) \quad \text{for every } n \in \mathbb{N}.$$

Fix any Borel probability measure $\mathfrak{m}'$ on X with $\mathfrak{m} \ll \mathfrak{m}' \ll \mathfrak{m}$. Given that the $\mathfrak{m}'$-measure of $\{f \neq f_n\} = \big(\mathrm{X} \setminus B_n(\bar{x})\big) \cup \big\{|f| > n\big\}$ goes to 0 as $n \to \infty$, we see that $f_n \to f$ in $L^0(\mathfrak{m})$. Since $L^1(\mathfrak{m}) \cap L^\infty(\mathfrak{m}) = \bigcap_{p\in[1,\infty]} L^p(\mathfrak{m})$, the statement is achieved. □

Exercise 1.1.26 Suppose that the measure $\mathfrak{m}$ has no atoms. Let $L : L^0(\mathfrak{m}) \to \mathbb{R}$ be linear and continuous. Then $L = 0$.

Solution We argue by contradiction: suppose that there exists $f \in L^0(\mathfrak{m})$ such that $L(f) = 1$. Since $\mathfrak{m}$ is atomless and outer regular, any point of X is center of some ball having arbitrarily small $\mathfrak{m}$-measure. In particular, by using the Lindelöf property of (X, d) we can provide, for any $n \in \mathbb{N}$, a Borel partition $(A_n^k)_{k\in\mathbb{N}}$ of X such that $\mathfrak{m}(A_n^k) \leq 1/n$ for every $k \in \mathbb{N}$. Since the limit $f = \lim_{N\to\infty} \sum_{k=1}^N \chi_{A_n^k} f$ holds in $L^0(\mathfrak{m})$ and L is linear continuous, we see that

$$\sum_{k\in\mathbb{N}} L(\chi_{A_n^k} f) = \lim_{N\to\infty} \sum_{k=1}^N L(\chi_{A_n^k} f) = L\bigg(\lim_{N\to\infty} \sum_{k=1}^N \chi_{A_n^k} f\bigg) = L(f) = 1,$$

whence there exists $k_n \in \mathbb{N}$ such that $L(\chi_{A_n^{k_n}} f) > 0$. Now let us define

$$f_n := f + \frac{\chi_{A_n^{k_n}} f}{L(\chi_{A_n^{k_n}} f)} \in L^0(\mathfrak{m}) \quad \text{for every } n \in \mathbb{N}.$$

Since $\mathfrak{m}\big(\{f \neq f_n\}\big) \leq \mathfrak{m}(A_n^{k_n}) \to 0$ as $n \to \infty$, we deduce that $f_n \to f$ in $L^0(\mathfrak{m})$. On the other hand, one has

$$L(f_n) = L(f) + L\bigg(\frac{\chi_{A_n^{k_n}} f}{L(\chi_{A_n^{k_n}} f)}\bigg) = 2 \quad \text{for every } n \in \mathbb{N},$$

so that $L(f_n)$ does not converge to $L(f) = 1$. This contradicts the continuity of L. □

Exercise 1.1.27 Let $(\mathrm{X}, \mathsf{d}, \mathfrak{m})$ be any metric measure space. Then the topology of $L^0(\mathfrak{m})$ comes from a norm if and only if $\mathfrak{m}$ has finite support.

Solution If the support of $\mathfrak{m}$ has cardinality $n \in \mathbb{N}$, then $L^0(\mathfrak{m})$ can be identified with the Euclidean space $\mathbb{R}^n$ (as a topological vector space), whence its topology comes from a norm.

Conversely, suppose that $\mathfrak{m}$ does not have finite support. We distinguish two cases:

(i) $\mathfrak{m}$ is purely atomic,
(ii) $\mathfrak{m}$ is not purely atomic.

In case (i), we can write $\mathfrak{m} = \sum_{n\in\mathbb{N}} \lambda_n\, \delta_{x_n}$ for some constants $(\lambda_n)_n \subseteq (0, +\infty)$ and some distinct points $(x_n)_n \subseteq \mathrm{X}$. Then $L^0(\mathfrak{m})$ can be identified (as a vector space) with the space ℓ^0 of all real-valued sequences, via the map $I : \sum_{n\in\mathbb{N}} a_n\, \chi_{\{x_n\}} \mapsto (a_n)_n$. Call $(e_n)_n$ the canonical basis of ℓ^0, i.e. $e_n := (\delta_{nk})_k$ for all $n \in \mathbb{N}$. Let $\|\cdot\|$ be any norm on ℓ^0. It can be readily checked that $I^{-1}\big(e_n/\|e_n\|\big) \to 0$ with respect to the $L^0(\mathfrak{m})$-topology. Since all vectors $e_n/\|e_n\|$ have $\|\cdot\|$-norm equal to 1, we conclude that the $L^0(\mathfrak{m})$-topology does not come from a norm.

In case (ii), we can find two Radon measures $\mu, \nu \geq 0$ on X with $\mu \perp \nu$ such that $\mu \neq 0$ has no atoms and $\mathfrak{m} = \mu + \nu$. Notice that $L^0(\mu)$ is a vector subspace of $L^0(\mathfrak{m})$ and that its topology coincides with the restriction of the $L^0(\mathfrak{m})$-topology. We argue by contradiction: suppose that some norm $\|\cdot\|$ on $L^0(\mathfrak{m})$ induces its usual topology, thus in particular the restriction of $\|\cdot\|$ to $L^0(\mu)$ induces the $L^0(\mu)$-topology. By Hahn-Banach theorem we know that there exists a non-null linear continuous operator $L : L^0(\mu) \to \mathbb{R}$, which contradicts Exercise 1.1.26. Hence the $L^0(\mathfrak{m})$-topology is not induced by any norm, as required. □

Exercise 1.2.2 Any open subset of a Polish space is a Polish space.

Solution Let (X, d) be a complete separable metric space and $\emptyset \neq \Omega \subseteq \mathrm{X}$ an open set. The product space $\mathbb{R} \times \mathrm{X}$ is a complete separable metric space if endowed with the distance

$$(\mathsf{d}_{\mathrm{Eucl}} \times \mathsf{d})^2\big((\lambda_1, x_1), (\lambda_2, x_2)\big) := |\lambda_1 - \lambda_2|^2 + \mathsf{d}^2(x_1, x_2)$$

and the map $f : \mathbb{R} \times \mathrm{X} \to \mathbb{R}$, defined as $(\lambda, x) \mapsto \lambda\, \mathsf{d}(x, \mathrm{X} \setminus \Omega)$, is continuous. This grants that the set $C := \big\{(\lambda, x) \in \mathbb{R} \times \mathrm{X} : f(\lambda, x) = 1\big\}$ is closed in $\mathbb{R} \times \mathrm{X}$. Moreover, it is easy to prove that the projection $\mathbb{R} \times \mathrm{X} \ni (\lambda, x) \mapsto x \in \mathrm{X}$ is a homeomorphism between C and Ω—here the openness of Ω enters into play. Therefore Ω (with the topology induced by d) is proven to be a Polish space, as required. □

Exercise 1.2.8 Prove that

$$\mathrm{KE}(\gamma)=\sup_{0=t_0<\ldots<t_n=1}\sum_{i=0}^{n-1}\frac{\mathsf{d}(\gamma_{t_{i+1}},\gamma_{t_i})^2}{t_{i+1}-t_i}\quad\text{holds for every }\gamma\in C([0,1],\mathrm{X}).\tag{B.3}$$

Solution Fix a partition $0=t_0<t_1<\ldots<t_n=1$. By Hölder inequality, we get that

$$\sum_{i=0}^{n-1}\frac{\mathsf{d}(\gamma_{t_{i+1}},\gamma_{t_i})^2}{t_{i+1}-t_i}\leq\sum_{i=0}^{n-1}\frac{1}{t_{i+1}-t_i}\left(\int_{t_i}^{t_{i+1}}|\dot\gamma_s|\,\mathrm{d}s\right)^2\leq\sum_{i=0}^{n-1}\int_{t_i}^{t_{i+1}}|\dot\gamma_s|^2\,\mathrm{d}s=\mathrm{KE}(\gamma),$$

showing that the sup in (B.3) is smaller than or equal to $\mathrm{KE}(\gamma)$.

Conversely, fix any curve $\gamma\in C([0,1],\mathrm{X})$. By Proposition 1.2.12, we can isometrically embed (X,d) into a complete, separable and geodesic metric space $(\tilde{\mathrm{X}},\tilde{\mathsf{d}})$. Denote by $\widetilde{\mathrm{KE}}$ the kinetic energy associated to $(\tilde{\mathrm{X}},\tilde{\mathsf{d}})$. Then γ can be viewed as an element of $C([0,1],\tilde{\mathrm{X}})$ and it holds that $\widetilde{\mathrm{KE}}(\gamma)=\mathrm{KE}(\gamma)$. Now fix $n\in\mathbb{N}$. Since the curve γ is uniformly continuous, there exist $k(n)\in\mathbb{N}$ and a partition $0=t_0^n<\ldots<t_{k(n)}^n=1$ such that

$$\mathsf{d}(\gamma_{t_i^n},\gamma_s)\leq\frac{1}{2n}\quad\text{for every } i=1,\ldots,k(n)\text{ and } s\in[t_{i-1}^n,t_i^n].\tag{B.4}$$

Given that $(\tilde{\mathrm{X}},\tilde{\mathsf{d}})$ is a geodesic space, there exists $\gamma^n\in C([0,1],\tilde{\mathrm{X}})$ such that $\mathrm{Restr}_{t_{i-1}^n}^{t_i^n}(\gamma^n)$ is a $\tilde{\mathsf{d}}$-geodesic joining $\gamma_{t_{i-1}^n}$ to $\gamma_{t_i^n}$ for any $i=1,\ldots,k(n)$. Hence (B.4) gives $\underline{\mathsf{d}}(\gamma,\gamma^n)\leq 1/n$ for every $n\in\mathbb{N}$. The functional $\widetilde{\mathrm{KE}}$ is $\underline{\mathsf{d}}$-lower semicontinuous by Proposition 1.2.7, whence

$$\mathrm{KE}(\gamma)=\widetilde{\mathrm{KE}}(\gamma)\leq\varliminf_{n\to\infty}\widetilde{\mathrm{KE}}(\gamma^n)=\varliminf_{n\to\infty}\sum_{i=1}^{k(n)}\frac{\mathsf{d}(\gamma_{t_i^n},\gamma_{t_{i-1}^n})^2}{t_i^n-t_{i-1}^n},$$

which proves that (B.3) is verified, as required. □

Exercise 1.3.3 Show that the integral in (1.40) is well-posed, i.e. it does not depend on the particular way of writing f, and that it is linear.

Solution Say $f=\sum_i\chi_{E_i}v_i=\sum_j\chi_{F_j}w_j$. Then it holds that

$$\sum_i\mu(E_i\cap E)\,v_i-\sum_j\mu(F_j\cap E)\,w_j=\sum_{i,j}\mu(E_i\cap F_j\cap E)\,(v_i-w_j)=0,$$

which proves that $\int_E f\,\mathrm{d}\mu$ is well-defined. Hence linearity follows by construction. □

Exercise 1.3.13 Prove Examples 1.3.11 and 1.3.12.

Solution About Example 1.3.11, we prove only i): let us fix a sequence $(f_n)_n \subseteq C^1([0,1])$ such that $f_n \to f$ and $f_n' \to g$ in $C([0,1])$, for suitable $f, g \in C([0,1])$. Therefore

$$f_n(t) - f_n(s) = \int_s^t f_n'(r)\,\mathrm{d}r \quad \text{for every } n \in \mathbb{N} \text{ and } t, s \in [0,1] \text{ with } s \le t. \tag{B.5}$$

Then by letting $n \to \infty$ in (B.5), we deduce that $f(t) - f(s) = \int_s^t g(r)\,\mathrm{d}r$ for every $t, s \in [0,1]$ with $s \le t$. Since g is continuous, we conclude that f is differentiable, with derivative g. This proves that $\big(D(T_1), T_1\big)$ is a closed operator, getting i).

To prove Example 1.3.12, fix any sequence $(g_k)_k \subseteq W^{1,2}(\mathbb{R})$ that $L^2(\mathbb{R})$-converges to some limit function $g \in L^2(\mathbb{R}) \setminus W^{1,2}(\mathbb{R})$. Now define $f_k := (0, \ldots, 0, g_k)$ for every $k \in \mathbb{N}$. Then the sequence $(f_k)_k \subseteq W^{1,2}(\mathbb{R}^n)$ converges to $(0, \ldots, 0, g)$ in $L^2(\mathbb{R}^n)$ and $T_4(f_k) \to 0$ in $L^2(\mathbb{R}^n)$, but the function $(0, \ldots, 0, g)$ does not belong to $W^{1,2}(\mathbb{R}^n)$, showing that $\big(D(T_4), T_4\big)$ is not a closed operator. □

Exercise 2.1.5 Prove that the map Restr_t^s is continuous.

Solution Fix $\gamma, \sigma \in C([0,1], \mathrm{X})$ and $t, s \in [0,1]$. Then

$$\underline{\mathsf{d}}\big(\mathsf{Restr}_s^t(\gamma), \mathsf{Restr}_s^t(\sigma)\big) = \max_{r \in [0,1]} \mathsf{d}(\gamma_{(1-r)t+rs}, \sigma_{(1-r)t+rs}) \le \max_{r \in [0,1]} \mathsf{d}(\gamma_r, \sigma_r) = \underline{\mathsf{d}}(\gamma, \sigma),$$

which shows that Restr_s^t is 1-Lipschitz. □

Exercise 2.1.14 Given a metric space (X, d) and $\alpha \in (0,1)$, we set the distance d_α on X as

$$\mathsf{d}_\alpha(x, y) := \mathsf{d}(x, y)^\alpha \quad \text{for every } x, y \in \mathrm{X}.$$

Prove that the metric space $(\mathrm{X}, \mathsf{d}_\alpha)$, which is called the *snowflaking* of (X, d), has the following property: if a curve γ is d_α-absolutely continuous, then it is constant.

Now consider any Borel measure $\mathfrak{m}$ on (X, d). Since d and d_α induce the same topology on X, we have that $\mathfrak{m}$ is also a Borel measure on $(\mathrm{X}, \mathsf{d}_\alpha)$. Prove that any Borel map on X belongs to $\mathrm{S}^2(\mathrm{X}, \mathsf{d}_\alpha, \mathfrak{m})$ and has null minimal weak upper gradient.

Solution Let $\gamma : [0,1] \to \mathrm{X}$ be d_α-absolutely continuous, say that $\mathsf{d}_\alpha(\gamma_t, \gamma_s) \le \int_s^t f(r)\,\mathrm{d}r$ for every $0 \le s < t \le 1$, for a suitable $f \in L^1(0,1)$. Define $C := \max\big\{\mathsf{d}(\gamma_t, \gamma_s) : t, s \in [0,1]\big\}$. Therefore one has

$$\mathsf{d}(\gamma_t, \gamma_s) = \mathsf{d}(\gamma_t, \gamma_s)^{1-\alpha}\,\mathsf{d}_\alpha(\gamma_t, \gamma_s) \le C^{1-\alpha}\,\mathsf{d}_\alpha(\gamma_t, \gamma_s) \le C^{1-\alpha} \int_s^t f(r)\,\mathrm{d}r,$$

which shows that γ is d-absolutely continuous. Moreover, given that $\lim_{h\to 0}$ $\mathsf{d}_\alpha(\gamma_{t+h},\gamma_t)/|h|$ exists finite for a.e. $t\in[0,1]$, we deduce that

$$\lim_{h\to 0}\frac{\mathsf{d}(\gamma_{t+h},\gamma_t)}{|h|}=\lim_{h\to 0}\left(\frac{\mathsf{d}_\alpha(\gamma_{t+h},\gamma_t)}{|h|}\right)^{1/\alpha}|h|^{(1-\alpha)/\alpha}=0\quad\text{for a.e. } t\in[0,1],$$

which grants that the curve γ is constant, as required.

To prove the last statement, simply notice that any test plan on $(\mathrm{X},\mathsf{d}_\alpha,\mathfrak{m})$ must be concentrated on the set of all constant curves in X. □

Exercise 2.2.1 Prove that $\mathrm{lip}_a(f)$ is an upper semicontinuous function.

Solution Fix $x\in\mathrm{X}$ and a sequence $(x_n)_n\subseteq\mathrm{X}$ such that $x_n\to x$. Given any $r>0$, we can find $\bar n\in\mathbb{N}$ such that $x_n\in B_r(x)$ for all $n\geq\bar n$ and accordingly there exists $(r_n)_{n\geq\bar n}\subseteq(0,1)$ such that $B_{r_n}(x_n)\subseteq B_r(x)$ for all $n\geq\bar n$. Therefore

$$\mathrm{lip}_a(f)(x_n)\leq\mathrm{Lip}\big(f_{|B_{r_n}(x_n)}\big)\leq\mathrm{Lip}\big(f_{|B_r(x)}\big)\quad\text{for all } n\geq\bar n. \tag{B.6}$$

By passing to the limit as $n\to\infty$ in (B.6), we get that $\overline{\lim}_n\,\mathrm{lip}_a(f)(x_n)\leq\mathrm{Lip}\big(f_{|B_r(x)}\big)$. By letting $r\searrow 0$, we finally conclude that $\overline{\lim}_n\,\mathrm{lip}_a(f)(x_n)\leq\mathrm{lip}_a(f)(x)$, which shows that the function $\mathrm{lip}_a(f)$ is upper semicontinuous, as required. □

Exercise 2.2.4 Prove that $\mathsf{E}_{*,a}$ is $L^2(\mathfrak{m})$-lower semicontinuous and is the maximal $L^2(\mathfrak{m})$-lower semicontinuous functional E such that $\mathsf{E}(f)\leq\frac{1}{2}\int\mathrm{lip}_a^2(f)\,\mathrm{d}\mathfrak{m}$ holds for every $f\in\mathrm{LIP}(\mathrm{X})$. Actually, the same properties are verified by E_* if we replace $\mathrm{lip}_a(f)$ with $\mathrm{lip}(f)$.

Solution First of all, observe that $\mathsf{E}_{*,a}(f)\leq\frac{1}{2}\int\mathrm{lip}_a^2(f)\,\mathrm{d}\mathfrak{m}$ for all $f\in L^2(\mathfrak{m})$: if f is not Lipschitz then $\frac{1}{2}\int\mathrm{lip}_a^2(f)\,\mathrm{d}\mathfrak{m}$ is set to be equal to $+\infty$ by convention, while if f is Lipschitz then the choice of the sequence constantly equal to f shows the above inequality.

Now we prove that the functional $\mathsf{E}_{*,a}$ is $L^2(\mathfrak{m})$-lower semicontinuous. Fix $f\in L^2(\mathfrak{m})$ and a sequence $(f_n)_n\subseteq\mathrm{LIP}(\mathrm{X})\cap L^2(\mathfrak{m})$ that $L^2(\mathfrak{m})$-converges to f. We aim to show the validity of the inequality $\mathsf{E}_{*,a}(f)\leq\underline{\lim}_n\,\mathsf{E}_{*,a}(f_n)$. Possibly passing to a subsequence, we can suppose that the liminf is actually a limit. Moreover, if $\lim_n\mathsf{E}_{*,a}(f_n)=+\infty$ then the claim is trivially satisfied, so we can also assume that $\lim_n\mathsf{E}_{*,a}(f_n)$ is finite and accordingly that $\mathsf{E}_{*,a}(f_n)<+\infty$ for all $n\in\mathbb{N}$. Given any $n\in\mathbb{N}$, we can find a sequence $(f_n^k)_k\subseteq\mathrm{LIP}(\mathrm{X})\cap L^2(\mathfrak{m})$ such that

$$\underline{\lim}_{k\to\infty}\frac{1}{2}\int\mathrm{lip}_a^2(f_n^k)\,\mathrm{d}\mathfrak{m}=\lim_{k\to\infty}\frac{1}{2}\int\mathrm{lip}_a^2(f_n^k)\,\mathrm{d}\mathfrak{m}\leq\mathsf{E}_{*,a}(f_n)+\frac{1}{n}.$$

A diagonalisation argument yields an increasing sequence $(k_n)_n\subseteq\mathbb{N}$ such that $g_n:=f_n^{k_n}\to f$ in $L^2(\mathfrak{m})$ and $\frac{1}{2}\int\mathrm{lip}_a^2(g_n)\,\mathrm{d}\mathfrak{m}\leq\mathsf{E}_{*,a}(f_n)+2/n$ for all $n\in\mathbb{N}$.

Therefore

$$\mathsf{E}_{*,a}(f) \leq \varliminf_{n\to\infty} \frac{1}{2}\int \operatorname{lip}_a^2(g_n)\,\mathrm{dm} \leq \varliminf_{n\to\infty} \left(\mathsf{E}_{*,a}(f_n) + \frac{2}{n}\right) = \varliminf_{n\to\infty} \mathsf{E}_{*,a}(f_n).$$

In order to conclude, suppose that E is an $L^2(\mathrm{m})$-lower semicontinuous functional such that $\mathsf{E}(f) \leq \frac{1}{2}\int \operatorname{lip}_a^2(f)\,\mathrm{dm}$ for every $f \in \mathrm{LIP(X)}$. We claim that $\mathsf{E} \leq \mathsf{E}_{*,a}$. Fix $f \in L^2(\mathrm{m})$. Then for any sequence $(f_n)_n \subseteq \mathrm{LIP(X)} \cap L^2(\mathrm{m})$ that converges to f in $L^2(\mathrm{m})$ it holds that

$$\mathsf{E}(f) \leq \varliminf_{n\to\infty} \mathsf{E}(f_n) \leq \varliminf_{n\to\infty} \frac{1}{2}\int \operatorname{lip}_a^2(f_n)\,\mathrm{dm}.$$

By the arbitrariness of $(f_n)_n$, we conclude that $\mathsf{E}(f) \leq \mathsf{E}_{*,a}(f)$, as required. □

Exercise 3.1.3 Let V, W, Z be normed spaces. Let $B : V \times W \to Z$ be a bilinear operator.

i) Suppose V is Banach. Show that B is continuous if and only if both $B(v,\cdot)$ and $B(\cdot,w)$ are continuous for every $v \in V$ and $w \in W$.
ii) Prove that B is continuous if and only if there exists a constant $C > 0$ such that the inequality $\|B(v,w)\|_Z \leq C\,\|v\|_V\,\|w\|_W$ holds for every $(v,w) \in V \times W$.

Solution The proof goes as follows:

i) Sufficiency is obvious. To prove necessity, let us define $T_w \in \mathrm{L}(V,Z)$ as $T_w(v) := B(v,w)$ for all $v \in V$; here $\mathrm{L}(V,Z)$ denotes the space of all linear continuous operators from V to Z. Given any $v \in V$, we have that $B(v,\cdot)$ is linear continuous, so that there exists $C_v > 0$ for which $\|B(v,w)\|_Z \leq C_v\,\|w\|_W$ for all $w \in W$. This grants that

$$\sup_{\|w\|_W\leq 1} \|T_w(v)\|_Z \leq C_v < +\infty \quad \text{for every } v \in V.$$

Then an application of the Banach-Steinhaus theorem yields

$$C := \sup_{\|v\|_V,\|w\|_W\leq 1} \|B(v,w)\|_Z = \sup_{\|w\|_W\leq 1} \|T_w\|_{\mathrm{L}(V,Z)} < +\infty.$$

Therefore $\|B(v,w)\|_Z \leq C\,\|v\|_V\,\|w\|_W$ for all $v \in V$ and $w \in W$, whence B is continuous.

ii) Necessity is trivial. To prove sufficiency, we argue by contradiction: suppose B is continuous and there exists a bounded sequence $\{(v_n,w_n)\}_n \subseteq V \times W$ with $\|B(v_n,w_n)\|_Z \to +\infty$. Now call $\lambda_n := \sqrt{\|B(v_n,w_n)\|_Z}$. Observe that $(v_n/\lambda_n, w_n/\lambda_n) \to 0$ in $V \times W$, because the sequences $(v_n)_n$, $(w_n)_n$ are bounded and $\lambda_n \to \infty$. On the other hand, we clearly have that

$$\left\| B(v_n/\lambda_n, w_n/\lambda_n) \right\|_Z = 1 \text{ for every } n \in \mathbb{N},$$

thus contradicting the continuity of B. □

Exercise 3.2.4 Assume that $\mathfrak{m}$ has no atoms and let $L : \mathscr{M} \to L^\infty(\mathfrak{m})$ be linear, continuous and satisfying $L(fv) = fL(v)$ for every $v \in \mathscr{M}$ and $f \in L^\infty(\mathfrak{m})$. Prove that $L = 0$.

Solution We argue by contradiction: suppose that $L(v) \neq 0$ for some $v \in \mathscr{M}$. Then (possibly taking $-v$ in place of v) we can find a Borel set $A \subseteq \mathrm{X}$ and some $C \geq 1$ such that $\mathfrak{m}(A) > 0$ and $1/C \leq L(v) \leq C$ $\mathfrak{m}$-a.e. on A. Pick $\bar{n} \in \mathbb{N}$ with $\sum_{n \geq \bar{n}} 1/n^4 < \mathfrak{m}(A)$. We claim that:

$$\begin{gathered}\text{There exists a sequence } (A_n)_{n\geq\bar{n}} \text{ of pairwise disjoint}\\ \text{subsets of } A \text{ such that } 0 < \mathfrak{m}(A_n) \leq 1/n^4 \text{ for all } n \geq \bar{n}.\end{gathered} \tag{B.7}$$

To prove it, we use a recursive argument: suppose to have already built $A_{\bar{n}}, \ldots, A_{n-1}$. The set $A' := A \setminus (A_{\bar{n}} \cup \ldots \cup A_{n-1})$ has positive $\mathfrak{m}$-measure by hypothesis on $\bar{n}$. Since $\mathfrak{m}$ is atomless and outer regular, we see that any point of A' is center of some ball whose $\mathfrak{m}$-measure does not exceed $1/n^4$. By the Lindelöf property, countably many of such balls cover the whole A'; call them $(B_i)_{i\in\mathbb{N}}$. Then there exists $i \in \mathbb{N}$ with $\mathfrak{m}(A' \cap B_i) > 0$, otherwise the set A' would be negligible. Hence the set $A_n := A' \cap B_i$ satisfies the required properties. This provides us with a sequence $(A_n)_{n\geq\bar{n}}$ as in the claim (B.7).

Now let us define $w_k := \sum_{n=\bar{n}}^{k} n\,\chi_{A_n} v \in \mathscr{M}$ for every $k \geq \bar{n}$. Notice that for any $k \geq \bar{n}$ and $i, j \geq k$ it holds that

$$\|w_i - w_j\|_{\mathscr{M}} \leq \sum_{n=k}^{\infty} \int_{A_n} n^2 |v|^2 \,\mathrm{d}\mathfrak{m} \leq C \sum_{n=k}^{\infty} n^2\,\mathfrak{m}(A_n) \leq C \sum_{n=k}^{\infty} \frac{1}{n^2}.$$

Since $\sum_{n=k}^{\infty} 1/n^2 \to 0$ as $k \to \infty$, we conclude that the sequence $(w_k)_k$ is Cauchy in $\mathscr{M}$, thus it admits a limit in $\mathscr{M}$. On the other hand, for all $k \geq \bar{n}$ we have that

$$L(w_k) = \sum_{n=\bar{n}}^{k} n\,\chi_{A_n} L(v) \geq \frac{1}{C} \sum_{n=\bar{n}}^{k} n\,\chi_{A_n} \quad \mathfrak{m}\text{-a.e.},$$

thus accordingly $L(w_k)$ cannot converge in $L^\infty(\mathfrak{m})$. This leads to a contradiction, as the operator L is continuous. □

Exercise 3.2.32 Let $T : L^2(\mathfrak{m}) \to L^2(\mathfrak{m})$ be an $L^\infty(\mathfrak{m})$-linear and continuous operator. Prove that there exists a unique $g \in L^\infty(\mathfrak{m})$ such that $T(f) = gf$ for every $f \in L^2(\mathfrak{m})$.

Solution First of all, we claim that:

$$\text{There exists a unique } g \in L^0(\mathfrak{m}) \text{ such that } T(f) = gf \text{ for every } f \in L^2(\mathfrak{m}). \tag{B.8}$$

To prove it, choose a Borel partition $(E_n)_{n\in\mathbb{N}}$ of X into sets of finite positive $\mathfrak{m}$-measure and define the operators $T_n : L^2(\mathfrak{m}) \to L^1(\mathfrak{m})$ as $T_n(f) := \chi_{E_n} T(f)$ for all $f \in L^2(\mathfrak{m})$. It is then clear that each T_n is $L^\infty(\mathfrak{m})$-linear and continuous, thus Riesz Theorem 3.2.14—as already observed in Example 3.2.15—gives us a function $g_n \in L^2(\mathfrak{m})$ such that $T_n(f) = g_n f$ holds for all $f \in L^2(\mathfrak{m})$. In particular, $g_n = 0$ holds $\mathfrak{m}$-a.e. in $\mathrm{X} \setminus E_n$. Therefore it makes sense to define the function $g \in L^0(\mathfrak{m})$ as $g := \sum_{n\in\mathbb{N}} g_n$ and it holds that

$$T(f) = \sum_{n\in\mathbb{N}} T_n(f) = \sum_{n\in\mathbb{N}} g_n f = gf \quad \text{for every } f \in L^2(\mathfrak{m}),$$

which proves the existence part of the claim (B.8). The uniqueness part is trivial.

In order to conclude, it only remains to show that:

$$\text{If } g \in L^0(\mathfrak{m}) \text{ and } gf \in L^2(\mathfrak{m}) \text{ for every } f \in L^2(\mathfrak{m}), \text{ then } g \in L^\infty(\mathfrak{m}). \tag{B.9}$$

We argue by contradiction: suppose g is not essentially bounded. Then we can find a strictly increasing sequence $(k_n)_n \subseteq \mathbb{N}$ and a countable collection $(A_n)_n$ of pairwise disjoint Borel subsets of X such that $k_n \leq g^2 < k_{n+1}$ $\mathfrak{m}$-a.e. on A_n and $0 < \mathfrak{m}(A_n) < +\infty$ for all $n \in \mathbb{N}$. Hence let us define

$$f := \sum_{n\in\mathbb{N}} \frac{1}{\sqrt{n\, k_n\, \mathfrak{m}(A_n)}} \chi_{A_n} \in L^0(\mathfrak{m}).$$

Given that $k_n \geq n$, we see that $\int f^2 \,\mathrm{d}\mathfrak{m} = \sum_n 1/(n\, k_n) \leq \sum_n 1/n^2 < +\infty$, i.e. $f \in L^2(\mathfrak{m})$. On the other hand, the function gf does not belong to $L^2(\mathfrak{m})$, indeed

$$\int (gf)^2 \,\mathrm{d}\mathfrak{m} = \sum_{n\in\mathbb{N}} \int_{A_n} \frac{g^2}{n\, k_n\, \mathfrak{m}(A_n)} \,\mathrm{d}\mathfrak{m} \geq \sum_{n\in\mathbb{N}} \int_{A_n} \frac{1}{n\, \mathfrak{m}(A_n)} \,\mathrm{d}\mathfrak{m} = \sum_{n\in\mathbb{N}} \frac{1}{n} = +\infty.$$

This leads to a contradiction, thus (B.9) and accordingly the statement follow. □

Exercise 4.2.11 Prove that Dual is single-valued and linear if and only if $\mathbb{B}$ is a Hilbert space. In this case, Dual is the Riesz isomorphism.

Solution To prove necessity, suppose $\mathbb{B}$ is Hilbert. We show that Dual is single-valued arguing by contradiction: if not, there exist $v \in \mathbb{B}$ and $L_1, L_2 \in \mathsf{Dual}(v)$ with $L_1 \neq L_2$. By Riesz theorem we know that there exist $v_1, v_2 \in \mathbb{B}$ such that $v_1 \neq v_2$ and $L_i(\cdot) = \langle v_i, \cdot \rangle$ for $i = 1, 2$. Hence $\|v_i\|_{\mathbb{B}} = \|L_i\|_{\mathbb{B}'} = \|v\|_{\mathbb{B}}$ and $\langle v_i, v \rangle = L_i(v) = \|v\|_{\mathbb{B}}^2$ for $i = 1, 2$. This forces $v_1 = v_2 = v$, thus leading to a contradiction. Moreover, this shows that Dual coincides with the Riesz isomorphism, so in particular it is linear.

To prove sufficiency, suppose Dual is single-valued and linear. Fix any two $v_1, v_2 \in \mathbb{B}$ and call $L_i := \mathsf{Dual}(v_i)$ for $i = 1, 2$. By linearity of Dual we know that $\mathsf{Dual}(v_1 \pm v_2) = L_1 \pm L_2$, whence

$$\begin{aligned}(L_1 + L_2)(v_1 + v_2) &= \|v_1 + v_2\|_{\mathbb{B}}^2, \\ (L_1 - L_2)(v_1 - v_2) &= \|v_1 - v_2\|_{\mathbb{B}}^2.\end{aligned} \tag{B.10}$$

By summing the two identities in (B.10) we thus deduce that

$$\|v_1 + v_2\|_{\mathbb{B}}^2 + \|v_1 - v_2\|_{\mathbb{B}}^2 = 2\,L_1(v_1) + 2\,L_2(v_2) = 2\,\|v_1\|_{\mathbb{B}}^2 + 2\,\|v_2\|_{\mathbb{B}}^2,$$

which shows that $\mathbb{B}$ is a Hilbert space. □

Exercise 4.2.13 Prove that the multi-valued map Dual on $(\mathbb{R}^n, \|\cdot\|)$ is single-valued at any point if and only if the norm $\|\cdot\|$ is differentiable.

Solution It is well-known that the subdifferential of $\|\cdot\|$ at $v \in \mathbb{R}^n$ is single-valued if and only if $\|\cdot\|$ is differentiable at v, thus it is enough to show that

$$\mathsf{Dual}(v) = \|v\|\big(\partial^-\|\cdot\|(v)\big) \quad \text{for every } v \in \mathbb{R}^n. \tag{B.11}$$

Let $L \in \mathsf{Dual}(v)$. Hence for any $w \in \mathbb{R}^n$ it holds that

$$\|v\| + \frac{L}{\|v\|}(w - v) = \|v\| - \frac{L(v)}{\|v\|} + \frac{L(w)}{\|v\|} = \frac{L(w)}{\|v\|} \le \frac{\|L\|\,\|w\|}{\|v\|} \le \|w\|,$$

which shows that $L/\|v\| \in \partial^-\|\cdot\|(v)$. This proves that $\mathsf{Dual}(v) \subseteq \|v\|\big(\partial^-\|\cdot\|(v)\big)$.

Conversely, let $L \in \partial^-\|\cdot\|(v)$. This means that $\|v\| + L(w - v) \le \|w\|$ for all $w \in \mathbb{R}^n$, or equivalently $L(w) - \|w\| \le L(v) - \|v\|$ for all $w \in \mathbb{R}^n$. In other words,

$$\|\cdot\|^*(L) := \sup_{w \in \mathbb{R}^n} \big[L(w) - \|w\|\big] \le L(v) - \|v\|. \tag{B.12}$$

(The function $\|\cdot\|^*$ is usually called *Fenchel conjugate* of $\|\cdot\|$.) We can compute $\|\cdot\|^*(L)$:

- If $\|L\| \le 1$ then $L(w) - \|w\| \le 0$ for all $w \in \mathbb{R}^n$, so that $\|\cdot\|^*(L) \le 0$. But $L(0) - \|0\| = 0$, whence we conclude that $\|\cdot\|^*(L) = 0$.
- If $\|L\| > 1$ then $L(w) > 1$ for some $w \in \mathbb{R}^n$ with $\|w\| = 1$. Hence

$$\|\cdot\|^*(L) \le L(tw) - \|tw\| = t\big(L(w) - 1\big) \to +\infty \quad \text{as } t \to +\infty,$$

thus showing that $\|\cdot\|^*(L) = +\infty$.

Therefore we proved that

$$\|\cdot\|^*(L) = \begin{cases} 0 & \text{if } \|L\| \leq 1, \\ +\infty & \text{if } \|L\| > 1. \end{cases}$$

Accordingly we deduce from (B.12) that $\|L\| \leq 1$ and $L(v) \geq \|v\|$, which force the validity of the identities $L(v) = \|v\|$ and $\|L\| = 1$. This implies that $\|v\| L \in \mathsf{Dual}(v)$, whence also the inclusion $\|v\|\big(\partial^-\|\cdot\|(v)\big) \subseteq \mathsf{Dual}(v)$ is proven. This gives (B.11). □

Exercise 4.2.18 Prove that the norm of a finite-dimensional Banach space is differentiable if and only if its dual norm is strictly convex.

Solution Given a Banach space $\mathbb{B}$, we denote by $\mathsf{Dual}_{\mathbb{B}}$ the multi-valued map defined as in (4.19). Let us prove the following two claims: given any Banach space $\mathbb{B}$, it holds that

$$\mathbb{B}' \text{ is strictly convex} \implies \mathsf{Dual}_{\mathbb{B}} \text{ is single-valued,} \tag{B.13a}$$

$$\mathsf{Dual}_{\mathbb{B}'} \text{ is single-valued} \implies \mathbb{B} \text{ is strictly convex.} \tag{B.13b}$$

In order to prove (B.13a), let us argue by contradiction: suppose to have $v \in \mathbb{B}$ with $\|v\|_{\mathbb{B}} = 1$ and $L_1, L_2 \in \mathsf{Dual}_{\mathbb{B}}(v)$ with $L_1 \neq L_2$. For any $t \in (0,1)$ we have that

$$1 = (1-t)\, L_1(v) + t\, L_2(v) = \big((1-t)\, L_1 + t\, L_2\big)(v) \leq \big\|(1-t)\, L_1 + t\, L_2\big\|_{\mathbb{B}'},$$

while on the other hand $\big\|(1-t)\, L_1 + t\, L_2\big\|_{\mathbb{B}'} \leq (1-t)\,\|L_1\|_{\mathbb{B}'} + t\,\|L_2\|_{\mathbb{B}'} = 1$. Therefore the segment in $\mathbb{B}'$ joining L_1 to L_2 is contained in the boundary of the unit ball of $\mathbb{B}'$, contradicting the strict convexity of $\mathbb{B}'$.

Also (B.13b) can be proven by contradiction: suppose $\|2\,v\|_{\mathbb{B}} = \|2\,w\|_{\mathbb{B}} = \|v+w\|_{\mathbb{B}} = 1$ for some $v, w \in \mathbb{B}$ with $v \neq w$. Choose any $L \in \mathsf{Dual}_{\mathbb{B}}(v+w)$ and notice that $\iota(v+w) \in \mathsf{Dual}_{\mathbb{B}'}(L)$, where $\iota : \mathbb{B} \to \mathbb{B}''$ is the canonical embedding of $\mathbb{B}$ into its bidual $\mathbb{B}''$. Now observe that

$$1 = L(v+w) = \frac{1}{2}\, L(2\,v) + \frac{1}{2}\, L(2\,w) \leq \frac{1}{2}\,\|L\|_{\mathbb{B}'}\big(\|2\,v\|_{\mathbb{B}} + \|2\,w\|_{\mathbb{B}}\big) = 1,$$

which forces the equalities $L(2\,v) = L(2\,w) = 1$. This means that $\iota(v), \iota(w) \in \mathsf{Dual}_{\mathbb{B}'}(L)$, thus contradicting the hypothesis.

The statement of the exercise is a direct consequence of (B.13a) and (B.13b), because any finite-dimensional Banach space is necessarily reflexive. □

Exercise 5.1.1 Consider $H := \mathbb{R}$ and $E(x) := |x|$ for every $x \in \mathbb{R}$. Then

$$\partial^- E(x) := \begin{cases} \{1\} & \text{if } x > 0, \\ [-1, 1] & \text{if } x = 0, \\ \{-1\} & \text{if } x < 0. \end{cases}$$

Solution We first treat the case $x > 0$. Notice that $1 \in \partial^- E(x)$ because $y = x+(y-x) \leq |y|$ for all $y \in \mathbb{R}$. Now take any $z \in \partial^- E(x)$, so that $z(y-x) \leq |y|-x$ for all $y \in \mathbb{R}$. By picking any $y \in (0, x)$ (resp. $y > x$) and dividing by $y - x$, we deduce that $z \geq 1$ (resp. $z \leq 1$). Hence one has $\partial^- E(x) = \{1\}$ for every $x > 0$. Similarly, $\partial^- E(x) = \{-1\}$ for every $x < 0$.

Now consider $x = 0$. We have that $\partial^- E(0) = \big\{z \in \mathbb{R} : zy \leq |y| \text{ for all } y \in \mathbb{R}\big\}$. Then some $z \in \mathbb{R}$ belongs to $\partial^- E(0)$ if and only if $zy \leq y$ for all $y > 0$ and $zy \leq -y$ for all $y < 0$. This shows that $\partial^- E(0) = [-1, 1]$. □

Exercise 5.1.6 Let H be a Hilbert space. Given any $x \in H$ and $\tau > 0$, let us define

$$F_{x,\tau}(\cdot) := E(\cdot) + \frac{|\cdot - x|^2}{2\tau}.$$

Then it holds that $\partial^- F_{x,\tau}(y) = \partial^- E(y) + \frac{y-x}{\tau}$ for every $y \in H$.

Solution First of all, it is clear that $D(E) = D(F_{x,\tau})$. Then let us fix $y \in D(E)$. Notice that the operator $|\cdot - x|^2/(2\tau)$ is differentiable at y and its differential is given by $\langle y - x, \cdot\rangle/\tau$, whence

$$\partial^- \frac{|\cdot - x|^2}{2\tau}(y) = \left\{\frac{y-x}{\tau}\right\}. \tag{B.14}$$

Then from the very definition of subdifferential it immediately follows that

$$\partial^- E(y) + \frac{y-x}{\tau} = \partial^- E(y) + \partial^- \frac{|\cdot - x|^2}{2\tau}(y) \subseteq \partial^- F_{x,\tau}(y).$$

To prove the converse inclusion, fix any $v \in \partial^- F_{x,\tau}(y)$. This means that

$$E(y) + \frac{|y-x|^2}{2\tau} - \langle v, hz\rangle \leq E(y - hz) + \frac{|y - x - hz|^2}{2\tau} \quad \text{for every } z \in H \text{ and } h > 0,$$

which can be rewritten as

$$\frac{E(y) - E(y-hz)}{h} \leq \langle v, z\rangle + \frac{|y-x-hz|^2 - |y-x|^2}{2\tau h} \quad \text{for every } z \in H \text{ and } h > 0. \tag{B.15}$$

Since in (B.15) the left hand side is convex with respect to h and the right hand side converges to $\langle v, z\rangle - \langle y - x, z\rangle/\tau$ as $h \searrow 0$, we conclude that

$$E(y) - E(y-z) \leq \lim_{h \searrow 0} \frac{E(y) - E(y-hz)}{h} \leq \left\langle v - \frac{y-x}{\tau}, z\right\rangle \quad \text{for every } z \in H,$$

which shows that $v - \frac{y-x}{\tau} \in \partial^- E(y)$, as required. □

Exercise 5.2.17 Given $p \in [1, \infty]$ and $t > 0$, we (provisionally) denote by h_t^p the heat flow in $L^p(\mathfrak{m})$ at time t. Prove that $\mathsf{h}_t^p f = \mathsf{h}_t^q f$ for all $p, q \in [1, \infty]$ and $f \in L^p(\mathfrak{m}) \cap L^q(\mathfrak{m})$.

Solution First of all, we aim to prove that $\mathsf{h}_t^p = \mathsf{h}_t^q$ on $L^p(\mathfrak{m}) \cap L^q(\mathfrak{m})$ whenever $p, q \in [1, \infty)$. To do so, fix $f \in L^p(\mathfrak{m}) \cap L^q(\mathfrak{m})$ and define $f_n := \chi_{B_n(\bar{x}) \cap \{|f| \leq n\}} f$ for all $n \in \mathbb{N}$, where the point $\bar{x} \in \mathrm{X}$ is arbitrary. Note that $f_n \to f$ both in $L^p(\mathfrak{m})$ and in $L^q(\mathfrak{m})$ by dominated convergence theorem. Each function f_n has bounded support and is essentially bounded, so that $(f_n)_n \subseteq L^2(\mathfrak{m})$. This grants that $\mathsf{h}_t^p f = \lim_n \mathsf{h}_t f_n$ in $L^p(\mathfrak{m})$ and $\mathsf{h}_t^q f = \lim_n \mathsf{h}_t f_n$ in $L^q(\mathfrak{m})$, whence necessarily $\mathsf{h}_t^p f = \mathsf{h}_t^q f$.

Now we prove that $\mathsf{h}_t^p = \mathsf{h}_t^\infty$ on $L^p(\mathfrak{m}) \cap L^\infty(\mathfrak{m})$ for all $p \in [1, \infty)$. We begin with the following claim:

$$\mathsf{h}_t^p f \in L^\infty(\mathfrak{m}) \quad \text{for every } f \in L^p(\mathfrak{m}) \cap L^\infty(\mathfrak{m}). \tag{B.16}$$

To prove it, pick any sequence $(f_n)_n \subseteq L^2(\mathfrak{m}) \cap L^p(\mathfrak{m}) \cap L^\infty(\mathfrak{m})$ that converges to f in $L^p(\mathfrak{m})$ and satisfies $\|f_n\|_{L^\infty(\mathfrak{m})} \leq \|f\|_{L^\infty(\mathfrak{m})}$ for all $n \in \mathbb{N}$. Hence we have that $\mathsf{h}_t f_n \to \mathsf{h}_t^p f$ in $L^p(\mathfrak{m})$, while $|\mathsf{h}_t f_n| \leq \|f\|_{L^\infty(\mathfrak{m})}$ holds $\mathfrak{m}$-a.e. by item iii) of Proposition 5.2.14. This implies that the $\mathfrak{m}$-a.e. estimate $|\mathsf{h}_t^p f| \leq \|f\|_{L^\infty(\mathfrak{m})}$, thus obtaining (B.16). Now let us fix $f \in L^p(\mathfrak{m}) \cap L^\infty(\mathfrak{m})$. To prove that $\mathsf{h}_t^p f = \mathsf{h}_t^\infty f$ is clearly equivalent (by Definition 5.2.16) to the following condition:

$$\int f\, \mathsf{h}_t^1 g \, d\mathfrak{m} = \int \mathsf{h}_t^p f\, g \, d\mathfrak{m} \quad \text{for every } g \in L^1(\mathfrak{m}). \tag{B.17}$$

Call q the conjugate exponent of p. Choose two sequences $(f_i)_i \subseteq L^2(\mathfrak{m}) \cap L^p(\mathfrak{m}) \cap L^\infty(\mathfrak{m})$ and $(g_j)_j \subseteq L^1(\mathfrak{m}) \cap L^2(\mathfrak{m}) \cap L^q(\mathfrak{m})$ such that $f_i \to f$ in $L^p(\mathfrak{m})$ and $g_j \to g$ in $L^1(\mathfrak{m})$. We know from Corollary 5.2.9 that $\int f_i\, \mathsf{h}_t g_j \, d\mathfrak{m} = \int \mathsf{h}_t f_i \, g_j \, d\mathfrak{m}$. Given that $g_j, \mathsf{h}_t g_j \in L^q(\mathfrak{m})$, we can let $i \to \infty$ and obtain $\int f\, \mathsf{h}_t g_j \, d\mathfrak{m} = \int \mathsf{h}_t^p f\, g_j \, d\mathfrak{m}$. Since $f, \mathsf{h}_t^p f \in L^\infty(\mathfrak{m})$ by (B.16), we can let $j \to \infty$ and obtain (B.17). This concludes the proof. □

Exercise 6.4.8 Let H_1, H_2 be Hilbert spaces. Let $\varphi : H_1 \to H_2$ be a linear and continuous operator. Then there exists a unique linear and continuous operator $\Lambda^k \varphi : \Lambda^k H_1 \to \Lambda^k H_2$ such that $\Lambda^k \varphi(v_1 \wedge \ldots \wedge v_k) = \varphi(v_1) \wedge \ldots \wedge \varphi(v_k)$ is satisfied for every $v_1, \ldots, v_k \in H_1$. Prove that $\|\Lambda^k \varphi\|_{\mathsf{op}} \leq \|\varphi\|_{\mathsf{op}}^k$.

Solution First of all, note that there is at most one linear continuous map $T : H_1^{\otimes k} \to \Lambda^k H_2$ such that $T(v_1 \otimes \ldots \otimes v_k) = \varphi(v_1) \wedge \ldots \varphi(v_k)$ for all $v_1, \ldots, v_k \in H_1$. Such map is well-posed, linear and continuous as a consequence of the following estimate:

$$\begin{aligned} \big\|\varphi(v_1) \wedge \ldots \wedge \varphi(v_k)\big\|_{\Lambda^k H_2} &= \prod_{i=1}^k \big\|\varphi(v_i)\big\|_{H_2} \leq \|\varphi\|_{\mathsf{op}}^k \prod_{i=1}^k \|v_i\|_{H_1} \\ &= \|\varphi\|_{\mathsf{op}}^k \|v_1 \otimes \ldots \otimes v_k\|_{H_1^{\otimes k}}. \end{aligned} \tag{B.18}$$

Moreover, if some $v_1, \ldots, v_k \in H_1$ satisfy $v_i = v_j$ for some $i \neq j$, then $T(v_1 \otimes \ldots \otimes v_k) = 0$. This shows that the operator T passes to the quotient, thus yielding a (uniquely determined) linear and continuous map $\Lambda^k \varphi$ as in the claim. Finally, the estimate $\|\Lambda^k \varphi\|_{\mathsf{op}} \leq \|\varphi\|_{\mathsf{op}}^k$ follows from (B.18). □

Exercise 6.4.26 Prove that

$$\mathsf{h}_{\mathsf{H},t}(\mathrm{d}\omega) = \mathrm{d}\mathsf{h}_{\mathsf{H},t}\omega \quad \text{for every } \omega \in W^{1,2}_{\mathrm{d}}(\Lambda^k T^*\mathrm{X}) \text{ and } t \geq 0.$$

Moreover, an analogous property is satisfied by the codifferential δ.

Solution Let us consider the curve $t \mapsto \mathrm{d}\mathsf{h}_{\mathsf{H},t}\omega$. Since d is a closed operator, we see that

$$\frac{\mathrm{d}}{\mathrm{d}t}\mathrm{d}\mathsf{h}_{\mathsf{H},t}\omega = \mathrm{d}\left(\frac{\mathrm{d}}{\mathrm{d}t}\mathsf{h}_{\mathsf{H},t}\omega\right) = -\mathrm{d}\Delta_{\mathsf{H}}\mathsf{h}_{\mathsf{H},t}\omega \quad \text{for a.e. } t > 0. \tag{B.19}$$

On the other hand, given any $\alpha \in \mathrm{TestForm}_k(\mathrm{X})$ it holds that

$$\begin{aligned}
\int \langle \Delta_{\mathsf{H}}\mathrm{d}\mathsf{h}_{\mathsf{H},t}\omega, \alpha\rangle \,\mathrm{d}\mathfrak{m} &= \int \big\langle \mathrm{d}(\mathrm{d}\mathsf{h}_{\mathsf{H},t}\omega), \mathrm{d}\alpha\big\rangle \,\mathrm{d}\mathfrak{m} + \int \big\langle \delta(\mathrm{d}\mathsf{h}_{\mathsf{H},t}\omega), \delta\alpha\big\rangle \,\mathrm{d}\mathfrak{m} \\
&= \int \big\langle \delta(\mathrm{d}\mathsf{h}_{\mathsf{H},t}\omega), \delta\alpha\big\rangle \,\mathrm{d}\mathfrak{m} = \int \big\langle \mathrm{d}\mathsf{h}_{\mathsf{H},t}\omega, \mathrm{d}(\delta\alpha)\big\rangle \,\mathrm{d}\mathfrak{m} \\
&= \int \big\langle \mathrm{d}\mathsf{h}_{\mathsf{H},t}\omega, \mathrm{d}(\delta\alpha)\big\rangle \,\mathrm{d}\mathfrak{m} + \int \big\langle \delta\mathsf{h}_{\mathsf{H},t}\omega, \delta(\delta\alpha)\big\rangle \,\mathrm{d}\mathfrak{m} \\
&= \int \langle \Delta_{\mathsf{H}}\mathsf{h}_{\mathsf{H},t}\omega, \delta\alpha\rangle \,\mathrm{d}\mathfrak{m} = \int \langle \mathrm{d}\Delta_{\mathsf{H}}\mathsf{h}_{\mathsf{H},t}\omega, \alpha\rangle \,\mathrm{d}\mathfrak{m},
\end{aligned}$$

which shows that $\Delta_{\mathsf{H}}\mathrm{d}\mathsf{h}_{\mathsf{H},t}\omega = \mathrm{d}\Delta_{\mathsf{H}}\mathsf{h}_{\mathsf{H},t}\omega$. By recalling (B.19) we thus see that

$$\frac{\mathrm{d}}{\mathrm{d}t}\mathrm{d}\mathsf{h}_{\mathsf{H},t}\omega = -\Delta_{\mathsf{H}}\mathrm{d}\mathsf{h}_{\mathsf{H},t}\omega \quad \text{for a.e. } t > 0.$$

Since the gradient flow is unique, we can conclude that $\mathsf{h}_{\mathsf{H},t}(\mathrm{d}\omega) = \mathrm{d}\mathsf{h}_{\mathsf{H},t}\omega$ for all $t \geq 0$. □

References

1. Ambrosio, L.: Calculus, heat flow and curvature-dimension bounds in metric measure spaces. In: Proceedings of the ICM 2018, 2018
2. Ambrosio, L., Gigli, N., Savaré, G.: Gradient flows in metric spaces and in the space of probability measures. In: Lectures in Mathematics ETH Zürich, 2nd edn. Birkhäuser Verlag, Basel (2008)
3. Ambrosio, L., Gigli, N., Savaré, G.: Density of Lipschitz functions and equivalence of weak gradients in metric measure spaces. Rev. Mat. Iberoam. **29**, 969–996 (2013)
4. Ambrosio, L., Gigli, N., Savaré, G.: Calculus and heat flow in metric measure spaces and applications to spaces with Ricci bounds from below. Invent. Math. **195**, 289–391 (2014)
5. Ambrosio, L., Gigli, N., Savaré, G.: Metric measure spaces with Riemannian Ricci curvature bounded from below. Duke Math. J. **163**, 1405–1490 (2014)
6. Ambrosio, L., Gigli, N., Savaré, G.: Bakry-Émery curvature-dimension condition and Riemannian Ricci curvature bounds. Ann. Probab. **43**, 339–404 (2015). arXiv:1209.5786
7. Ambrosio, L., Mondino, A., Savaré, G.: On the Bakry-Émery condition, the gradient estimates and the Local-to-Global property of $RCD^*(K, N)$ metric measure spaces. J. Geom. Anal. **26**, 1–33 (2014)
8. Ambrosio, L., Mondino, A., Savaré, G.: Nonlinear diffusion equations and curvature conditions in metric measure spaces. Memoirs Am. Math. Soc. (2015, accepted)
9. Bakry, D.: Transformations de Riesz pour les semi-groupes symétriques. II. Étude sous la condition $\Gamma_2 \geq 0$. In: Séminaire de probabilités, XIX, 1983/84, vol. 1123 of Lecture Notes in Math., pp. 145–174. Springer, Berlin (1985)
10. Bakry, D., Émery, M.: Diffusions hypercontractives. In: Séminaire de probabilités, XIX, 1983/84, vol. 1123 of Lecture Notes in Math., pp. 177–206. Springer, Berlin (1985)
11. Bogachev, V.I.: Measure Theory. Vol. I, II. Springer, Berlin (2007)
12. Burago, D., Burago, Y., Ivanov, S.: A Course in Metric Geometry. vol. 33 of Graduate Studies in Mathematics. American Mathematical Society, Providence, RI (2001)
13. Cheeger, J.: Differentiability of Lipschitz functions on metric measure spaces. Geom. Funct. Anal. **9**, 428–517 (1999)
14. Diestel, J., Uhl, J. J., Jr.: Vector Measures. American Mathematical Society, Providence, RI (1977). With a foreword by B. J. Pettis, Mathematical Surveys, No. 15

N. Gigli, E. Pasqualetto, *Lectures on Nonsmooth Differential Geometry*,
SISSA Springer Series 2, https://doi.org/10.1007/978-3-030-38613-9

15. Erbar, M., Kuwada, K., Sturm, K.-T.: On the equivalence of the entropic curvature-dimension condition and Bochner's inequality on metric measure spaces. Invent. Math. **201**, 1–79 (2014). arXiv:1303.4382
16. Fuglede, B.: Extremal length and functional completion. Acta Math. **98**, 171–219 (1957)
17. Gigli, N.: Nonsmooth differential geometry - an approach tailored for spaces with Ricci curvature bounded from below. Mem. Am. Math. Soc. (2014, Accepted). arXiv:1407.0809
18. Gigli, N.: On the differential structure of metric measure spaces and applications. Mem. Am. Math. Soc. **236**, vi+91 (2015). arXiv:1205.6622
19. Gigli, N.: Lecture notes on differential calculus on RCD spaces. Preprint, arXiv:1703.06829, 2017
20. Gigli, N., Rigoni, C.: Recognizing the flat torus among $\mathsf{RCD}^*(0, N)$ spaces via the study of the first cohomology group. Calc. Var. Partial Differ. Equ. **57**(4), (2017)
21. Hajłasz, P.: Sobolev spaces on an arbitrary metric space. Potential Anal. **5**, 403–415 (1996)
22. Koskela, P., MacManus, P.: Quasiconformal mappings and Sobolev spaces. Studia Math. **131**, 1–17 (1998)
23. Levi, B.: Sul principio di Dirichlet. Rend. Circ. Mat. Palermo (1906)
24. Lott, J., Villani, C.: Weak curvature conditions and functional inequalities. J. Funct. Anal. **245**, 311–333 (2007)
25. Sauvageot, J.-L.: Tangent bimodule and locality for dissipative operators on C^*-algebras. In: Quantum Probability and Applications, IV (Rome, 1987), vol. 1396 of Lecture Notes in Math., pp. 322–338. Springer, Berlin (1989)
26. Sauvageot, J.-L.: Quantum Dirichlet forms, differential calculus and semigroups. In: Quantum Probability and Applications, V (Heidelberg, 1988), vol. 1442 of Lecture Notes in Math., pp. 334–346. Springer, Berlin (1990)
27. Savaré, G.: Self-improvement of the Bakry-Émery condition and Wasserstein contraction of the heat flow in $\mathrm{RCD}(K, \infty)$ metric measure spaces. Discrete Contin. Dyn. Syst. **34**, 1641–1661 (2014)
28. Shanmugalingam, N.: Newtonian spaces: an extension of Sobolev spaces to metric measure spaces. Rev. Mat. Iberoamericana **16**, 243–279 (2000)
29. Sturm, K.-T.: On the geometry of metric measure spaces. I. Acta Math. **196**, 65–131 (2006)
30. Sturm, K.-T.: On the geometry of metric measure spaces. II. Acta Math. **196**, 133–177 (2006)
31. Villani, C.: Inégalités isopérimétriques dans les espaces métriques mesurés [d'après F. Cavalletti & A. Mondino]. Séminaire Bourbaki, available at: http://www.bourbaki.ens.fr/TEXTES/1127.pdf
32. Villani, C.: Synthetic theory of ricci curvature bounds. Jpn. J. Math. **11**, 219–263 (2016)
33. Weaver, N.: Lipschitz Algebras. World Scientific Publishing, River Edge, NJ (1999)
34. Weaver, N.: Lipschitz algebras and derivations. II. Exterior differentiation. J. Funct. Anal. **178**, 64–112 (2000)

Notation

$\mathbb{R}^n$	n-dimensional Euclidean space
$\mathbb{Q}$	Set of rational numbers
$\mathbb{N}$	Set of natural numbers
$\mathcal{L}^n$	n-dimensional Lebesgue measure
$\mathcal{L}_1$	1-dimensional Lebesgue measure restricted to the interval $[0, 1]$
$f \vee g$	Maximum between two real-valued functions f and g
$f \wedge g$	Minimum between two real-valued functions f and g
f^+	Positive part of a real-valued function f
f^-	Negative part of a real-valued function f
$\overline{\lim}$	limsup
$\underline{\lim}$	liminf
χ_E	Characteristic function of a set E
V'	Dual of a normed space V
ℓ^∞	Space of bounded sequences in $\mathbb{R}$
$\mathrm{Graph}(T)$	Graph of a map T
(X, d)	Metric space (typically complete and separable)
$\mathscr{P}(\mathrm{X})$	Space of Borel probability measures on X
$C(\mathrm{X})$	Space of real-valued continuous functions on X
$C_b(\mathrm{X})$	Space of bounded continuous functions on X
$B_r(x)$	Open ball of center $x \in \mathrm{X}$ and radius $r > 0$
$\bar{B}_r(x)$	Closed ball of center $x \in \mathrm{X}$ and radius $r > 0$
$\mathrm{cl}_{\mathrm{X}}(E)$	Closure of a set E in X
$\mathrm{dist}(E, F)$	Distance between two sets $E, F \subseteq \mathrm{X}$
δ_x	Dirac delta measure at a point $x \in \mathrm{X}$
$\|\cdot\|_{\mathsf{TV}}$	Total variation norm
$\mathcal{M}(\mathrm{X})$	Space of signed Radon measures on X

N. Gigli, E. Pasqualetto, *Lectures on Nonsmooth Differential Geometry*,
SISSA Springer Series 2, https://doi.org/10.1007/978-3-030-38613-9

μ^+	Positive part of a measure μ
μ^-	Negative part of a measure μ
$\mathrm{spt}(\mu)$	Support of a measure μ
$\frac{\mathrm{d}\mu}{\mathrm{d}\nu}$	Radon-Nikodým derivative of μ with respect to ν
$T_*\mu$	Pushforward of a measure μ under the map T
$\mathrm{spt}(f)$	Support of a real-valued Lipschitz function f on X
LIP(X)	Space of real-valued Lipschitz functions on X
$\mathrm{LIP}_{bs}(\mathrm{X})$	Elements of LIP(X) having bounded support
$\mathrm{Lip}(f)$	(Global) Lipschitz constant of a function $f \in$ LIP(X)
$\mathrm{lip}(f)$	Local Lipschitz constant of a function $f \in$ LIP(X)
$\mathrm{lip}_a(f)$	Asymptotic Lipschitz constant of a function $f \in$ LIP(X)
$\underline{\mathrm{d}}$	Sup distance on the space $C([0,1],\mathrm{X})$ of continuous curves in X
$\lvert\dot{\gamma}\rvert$	Metric speed of an absolutely continuous curve $\gamma : [0,1] \to \mathrm{X}$
KE	Kinetic energy functional on $C([0,1],\mathrm{X})$
Geo(X)	Space of (constant speed) geodesics of X
$(\mathrm{X},\mathsf{d},\mathfrak{m})$	Metric measure space
$\mathfrak{m}'$	Borel probability measure on X having the same null sets as $\mathfrak{m}$
$L^p(\mathfrak{m})$	Space of p-integrable functions on X, with $p \in [1,\infty]$
$L^p_{loc}(\mathfrak{m})$	Space of locally p-integrable functions on X
$L^0(\mathfrak{m})$	Space of Borel functions on X (modulo $\mathfrak{m}$-a.e. equality)
e_t	Evaluation map at time $t \in [0,1]$
$\boldsymbol{\pi}$	Test plan
$\mathrm{Comp}(\boldsymbol{\pi})$	Compression constant of a test plan $\boldsymbol{\pi}$
Restr_t^s	Restriction operator between t and s
$\mathrm{S}^2(\mathrm{X})$	Sobolev class over $(\mathrm{X},\mathsf{d},\mathfrak{m})$
$\mathrm{S}^2_{loc}(\mathrm{X})$	Local Sobolev class over $(\mathrm{X},\mathsf{d},\mathfrak{m})$
$W^{1,2}(\mathrm{X})$	Sobolev space over $(\mathrm{X},\mathsf{d},\mathfrak{m})$
$\lvert Df\rvert$	Minimal weak upper gradient of a Sobolev function f
$\mathsf{Der}_{\boldsymbol{\pi}}(f)$	Derivative of a Sobolev function f in the direction of a test plan $\boldsymbol{\pi}$
$\mathsf{E}_{*,a}$, E_*, E_{Ch}	Cheeger energies
$W^{1,2}_{*,a}(\mathrm{X})$, $W^{1,2}_*(\mathrm{X})$, $W^{1,2}_{\mathrm{Ch}}(\mathrm{X})$	Sobolev spaces associated to the Cheeger energies

$\Gamma(X)$	Space of absolutely continuous curves in X
$\mathrm{Dom}(\gamma)$	Interval of definition of a curve $\gamma \in \Gamma(X)$
$\mathrm{Mod}_2(\Gamma)$	2-modulus of a curve family Γ
$W^{1,2}_{\mathrm{Sh}}(X)$	Sobolev space obtained via the 2-modulus
$\mathscr{M}$	$L^2(\mathfrak{m})$-normed $L^\infty(\mathfrak{m})$-module
$\lvert v\rvert$	Pointwise norm of an element $v \in \mathscr{M}$
$\mathscr{M}\lvert_E$	Restriction of $\mathscr{M}$ to a Borel set $E \subseteq X$
$\mathscr{M}(S)$	Submodule generated by a set $S \subseteq \mathscr{M}$
$\mathscr{M}^0$	$L^0(\mathfrak{m})$-normed $L^0(\mathfrak{m})$-module
$\mathscr{M}^*$	Dual of $\mathscr{M}$ (in the sense of modules)
$\lvert L\rvert_*$	(Dual) pointwise norm of an element $L \in \mathscr{M}^*$
$\mathrm{Int}_{\mathscr{M}}$	Natural map from $\mathscr{M}^*$ to $\mathscr{M}'$ obtained by integration
$I_{\mathscr{M}}$	Canonical embedding $\mathscr{M} \hookrightarrow \mathscr{M}^{**}$ in the (module) bidual
$\mathscr{H}$	Hilbert module
$\langle\cdot,\cdot\rangle$	Pointwise scalar product on a Hilbert module $\mathscr{H}$
$\mathscr{H}_1 \otimes \mathscr{H}_2$	Tensor product of two Hilbert modules $\mathscr{H}_1$ and $\mathscr{H}_2$
$v \otimes w$	Tensor product between $v \in \mathscr{H}_1$ and $w \in \mathscr{H}_2$
$\lvert A\rvert_{\mathsf{HS}}$	Pointwise Hilbert-Schmidt norm of a tensor $A \in \mathscr{H}_1 \otimes \mathscr{H}_2$
t	Transposition operator from $\mathscr{H}_1 \otimes \mathscr{H}_2$ to $\mathscr{H}_2 \otimes \mathscr{H}_1$
$\Lambda^k \mathscr{H}^0$	k-th exterior product of a Hilbert L^0-module $\mathscr{H}^0$
$v \wedge w$	Wedge product between $v, w \in \mathscr{H}^0$
$\mathrm{Comp}(\varphi)$	Compression constant of a map of bounded compression
$\varphi^*\mathscr{M}$	Pullback module of $\mathscr{M}$ under the map φ
$\varphi^*, [\varphi^*]$	Pullback map
$L^2(T^*X)$	Cotangent module associated to a metric measure space $(X, \mathsf{d}, \mathfrak{m})$
$\mathrm{d}f$	Differential (as an element of the cotangent module) of a Sobolev function $f \in S^2(X)$
$L^2(TX)$	Tangent module associated to $(X, \mathsf{d}, \mathfrak{m})$
$\mathsf{Grad}(f)$	Set of gradients of an element $f \in S^2(X)$
∇f	The only element of $\mathsf{Grad}(f)$ when $(X, \mathsf{d}, \mathfrak{m})$ is infinitesimally strictly convex
div	Divergence operator
$\boldsymbol{\pi}'$	Speed of a test plan $\boldsymbol{\pi}$
$\partial^- E$	Subdifferential of an operator E
$\lvert\partial^- E\rvert$	Slope of an operator E

$D(E)$	Domain of an operator E
Δ	Laplacian operator
$(\mathsf{h}_t)_{t\geq 0}$	Heat flow (for functions)
h_φ	'Mollified' heat flow
$\mathrm{Test}^\infty(\mathrm{X})$	Space of test functions on an $\mathsf{RCD}(K,\infty)$ space $(\mathrm{X},\mathsf{d},\mathfrak{m})$
$\mathrm{Test}^\infty_+(\mathrm{X})$	Space of non-negative test functions on X
$L^2\big((T^*)^{\otimes 2}\mathrm{X}\big)$	The tensor product $L^2(T^*\mathrm{X})\otimes L^2(T^*\mathrm{X})$
$L^2(T^{\otimes 2}\mathrm{X})$	The tensor product $L^2(T\mathrm{X})\otimes L^2(T\mathrm{X})$
$W^{2,2}(\mathrm{X}),\ H^{2,2}(\mathrm{X})$	Second-order Sobolev spaces over X
$\mathrm{H}f$	Hessian of a function $f\in W^{2,2}(\mathrm{X})$
$\mathbf{\Delta}$	Measure-valued Laplacian operator
Γ_2	Bakry-Émery curvature operator
$\mathrm{ess\,int}(E)$	Essential interior of a Borel set $E\subseteq \mathrm{X}$
$W^{1,2}_C(T\mathrm{X}),\ H^{1,2}_C(T\mathrm{X})$	Spaces of Sobolev vector fields on X
∇X	Covariant derivative of a Sobolev vector field $X\in W^{1,2}_C(T\mathrm{X})$
$\nabla_Z X$	Covariant derivative of X in direction Z
$\sharp$	Riesz (musical) isomorphism from $L^2\big((T^*)^{\otimes 2}\mathrm{X}\big)$ to $L^2(T^{\otimes 2}\mathrm{X})$
$[X,Y]$	Lie brackets between X and Y
$\mathrm{TestV}(\mathrm{X})$	Space of test vector fields on X
$\Lambda^k L^0(T^*\mathrm{X})$	k-th exterior power of the cotangent L^0-module $L^0(T^*\mathrm{X})$
$\mathrm{d}\omega$	Exterior differential of a k-form ω
$W^{1,2}_{\mathrm{d}}(\Lambda^k T^*\mathrm{X}),\ H^{1,2}_{\mathrm{d}}(\Lambda^k T^*\mathrm{X})$	Spaces of k-forms admitting an exterior differential
$\mathrm{TestForm}_k(\mathrm{X})$	Space of test k-forms on X
$\mathsf{H}^k_{\mathrm{dR}}(\mathrm{X})$	k-th de Rham-cohomology group of X
Δ_{H}	Hodge Laplacian operator
$W^{1,2}_{\mathrm{H}}(\Lambda^k T^*\mathrm{X}),\ H^{1,2}_{\mathrm{H}}(\Lambda^k T^*\mathrm{X})$	Spaces of k-forms admitting a Hodge Laplacian
Δ_{C}	Connection Laplacian operator
$\delta\omega$	Codifferential of a k-form ω
$\mathrm{Harm}_k(\mathrm{X})$	Space of harmonic k-forms on X
$(\mathsf{h}_{\mathrm{H},t})_{t\geq 0}$	Heat flow (for k-forms)
$(\mathsf{h}_{\mathrm{C},t})_{t\geq 0}$	Heat flow (for vector fields)
$\mathbf{Ric}$	Ricci curvature operator

Index

N. Gigli, E. Pasqualetto, *Lectures on Nonsmooth Differential Geometry*,
SISSA Springer Series 2, https://doi.org/10.1007/978-3-030-38613-9

Printed in the United States
By Bookmasters